P9-DDM-959

THE BLACK HOLE WAR

ALSO BY LEONARD SUSSKIND

The Cosmic Landscape:
String Theory and the Illusion of Intelligent Design

THE BLACK HOLE WAR

MY BATTLE WITH STEPHEN HAWKING TO MAKE THE WORLD SAFE FOR QUANTUM MECHANICS

Leonard Susskind

LB 1837
LITTLE, BROWN AND COMPANY
New York Boston London

Little, Brown and Company
Hachette Book Group USA
237 Park Avenue, New York, NY 10017
Visit our Web site at www.HachetteBookGroupUSA.com

First Edition: July 2008

Little, Brown and Company is a division of Hachette Book Group USA, Inc.
The Little, Brown name and logo are trademarks of Hachette Book Group USA, Inc.

Please address physics questions to *Susskind's Blog: Physics for Everyone,* http://susskindsblogphysicsforeveryone.blogspot.com/.

Library of Congress Cataloging-in-Publication Data

Susskind, Leonard.
The black hole war : my battle with Stephen Hawking to make the world safe for quantum mechanics / Leonard Susskind. — 1st ed.
p. cm.
Includes index.
ISBN 978-0-316-01640-7
1. Quantum theory. 2. General relativity (Physics). 3. Black holes (Astronomy). 4. Space and time. 5. Hawking, S. W. (Stephen W.). I. Title.
QC174.12.S896 2008
530.12 — dc22 2007048355

10 9 8 7 6 5 4 3 2 1

RRD-IN

Printed in the United States of America

What is it that breathes fire into the equations and makes a universe for them to describe?

— Stephen Hawking

CONTENTS

THE BLACK HOLE WAR

INTRODUCTION

There was so much to grok, so little to grok from.

—ROBERT A. HEINLEIN,
STRANGER IN A STRANGE LAND

Somewhere on the East African savanna, an aging lion spies her intended dinner. She prefers older, slower victims, but the young, healthy antelope is her only choice. The wary eyes of the prey are placed on the side of his head, ideally suited for scouring the landscape in search of dangerous predators. The predator's eyes look straight ahead, perfect for locking onto her victim and gauging the distance.

This time the antelope's wide-angle scanners miss the predator, and he wanders within striking range. The powerful rear legs of the lion thrust her forward toward the panicked victim. The timeless race begins anew.

Though burdened by age, the big cat is the superior sprinter. At first the gap closes, but the lion's powerful fast-twitch muscles gradually give way to oxygen deprivation. Soon the antelope's natural endurance wins out, and at some point the relative velocity of the cat and her prey switches sign; the closing gap begins to open. The moment she senses this reversal of fortune, Her Royal Highness is defeated. She slinks back into the underbrush.

. . .

Fifty thousand years ago, a tired hunter spots a cave opening blocked by a boulder: a safe place to rest if he can move the heavy obstruction. Unlike his apish forebears, the hunter stands upright. In that straight-up posture, he pushes mightily against the boulder, but nothing happens. To get a better angle, the hunter places his feet at a greater distance from the rock. When his body is almost horizontal, the applied force has a much larger component in the right direction. The boulder moves.

Distance? Velocity? Change of sign? Angle? Force? Component? What incredibly sophisticated calculations took place in the untutored brain of the hunter, let alone the cat? These are technical concepts that one ordinarily first meets in college physics textbooks. Where did the cat learn to gauge not only the velocity of its prey but also, more important, the relative velocity? Did the hunter take a physics course to learn the concept of force? And trigonometry to reckon the sines and cosines for computing components?

The truth, of course, is that all complex life-forms have built-in, instinctive physics concepts that have been hardwired into their nervous systems by evolution.[1] Without this preprogrammed physics software, survival would be impossible. Mutation and natural selection have made us all physicists, even animals. In humans the large size of the brain has allowed these instincts to evolve into concepts that we carry at the conscious level.

1. No one really knows how much is hardwired and how much is learned in early life, but the distinction is not important here. The point is that by the time our nervous systems are mature, experience, of either the personal or the evolutionary kind, has given us a lot of instinctual knowledge of how the physical world behaves. Whether hardwired or learned at a very young age, the knowledge is very difficult to unlearn.

Rewiring Ourselves

In fact, we're all *classical*[2] physicists. We feel force, velocity, and acceleration at a gut level. In the science fiction novel *Stranger in a Strange Land* (1961), Robert Heinlein invented a word to express this kind of deeply intuitive, almost visceral understanding of a phenomenon: grok.[3] I grok force, velocity, and acceleration. I grok three-dimensional space. I grok time and the number 5 . The trajectories of a stone or a spear are grokable. But my built-in, standard-issue groker breaks down when I try to apply it to ten-dimensional space-time, or to the number $10^{1,000}$, or even worse to the world of electrons and the Heisenberg Uncertainty Principle.

At the turn of the twentieth century, a wholesale breakdown of intuition occurred; physics suddenly found itself flummoxed by totally unfamiliar phenomena. My paternal grandfather was already ten years old when Albert Michelson and Edward Morley discovered that the Earth's orbital motion through the hypothetical ether could not be detected.[4] The electron was unknown until he was in his twenties; he was thirty the year Albert Einstein published the Special Theory of Relativity, and he was already well into middle age when Heisenberg discovered the Uncertainty Principle. There is no way that evolutionary pressure could have created an instinctive comprehension of these radically different worlds. But something in our neural networks, at least in some of us, has been primed for a fantastic rewiring that allows us not only to ask about these obscure phenomena but also to create mathematical abstractions — deeply unintuitive new concepts — to handle and explain them.

Speed created the first need to rewire — speed so fast that it

2. The word *classical* refers to physics that does not require the considerations of Quantum Mechanics.

3. *Grok* means to understand thoroughly and intuitively.

4. The famous experiment by Michelson and Morley was the first to show that the velocity of light does not depend on the motion of the Earth. It led to the paradoxes that Einstein eventually resolved in the Special Theory of Relativity.

almost rivaled the velocity of an evanescent beam of light. No animal had ever moved faster than 100 miles per hour before the twentieth century, and even today light travels so fast that for all but scientific purposes, it doesn't travel at all: it just appears instantaneously when the lights are switched on. Early humans had no need for hardwired circuits, attuned to ultrahigh speeds such as the speed of light.

Rewiring for speed happened suddenly. Einstein was no mutant; he had struggled in obscurity for a decade to replace his old Newtonian wiring. But to physicists of the time, it must have seemed that a new kind of human had spontaneously appeared among them — someone who could see the world not in terms of three-dimensional space, but in terms of four-dimensional *space-time*.

Einstein struggled for another decade — this time in plain view of physicists — to unify what he had called Special Relativity with Newton's theory of gravity. What emerged — the General Theory of Relativity — profoundly changed all the traditional ideas about geometry. Space-time became flexible, curved, or warped. It responded to the existence of matter almost like a sheet of rubber under stress. Previously, space-time had been passive, its geometric properties fixed. In the General Theory, space-time became an active player: it could be deformed by massive objects such as planets and stars, but it could not be visualized — not without a lot of additional mathematics anyway.

In 1900, five years before Einstein appeared on the scene, another much weirder paradigm shift was set in motion with the discovery that light is composed of particles called photons, or sometimes light quanta. The photon[5] theory of light was only a hint of the revolution to come; the mental gymnastics would be far more abstract than anything yet seen. Quantum Mechanics was more than a new law of nature. It involved changing the rules of classical

5. The term *photon* was not used until 1926, when the chemist Gilbert Lewis coined it.

logic, the ordinary rules of thought that every sane person uses to make deductions. It seemed crazy. But crazy or not, physicists were able to rewire themselves with a new logic called quantum logic. In chapter 4, I will explain everything you need to know about Quantum Mechanics. Be prepared to be mystified by it. Everyone is.

Relativity and Quantum Mechanics have been reluctant companions from the beginning. As soon as they were brought together in a shotgun wedding, violence broke out — the mathematics unleashing furious infinities for every question a physicist could ask. It took half a century for Quantum Mechanics and Special Relativity to be reconciled, but eventually the mathematical inconsistencies were eliminated. By the early 1950s, Richard Feynman, Julian Schwinger, Sin-Itiro Tomanaga, and Freeman Dyson[6] had laid the groundwork for a synthesis of *Special* Relativity and Quantum Mechanics called Quantum Field Theory. But the *General* Theory of Relativity (Einstein's synthesis of Special Relativity and Newton's theory of gravity) and Quantum Mechanics remained irreconcilable, though not from lack of trying. Feynman, Steven Weinberg, Bryce DeWitt, and John Wheeler all attempted to "quantize" Einstein's gravity equations, but all that came out was mathematical rubbish. Perhaps that was not surprising. Quantum Mechanics ruled the world of very light objects. Gravity, by contrast, seemed important for only very heavy lumps of matter. It seemed safe to assume that nothing was light enough for Quantum Mechanics to be important and also heavy enough for gravity to be important. As a result, many physicists throughout the second half of the twentieth century considered the pursuit of such a unifying theory to be worthless, fit only for crackpots and philosophers.

But others thought this was a myopic view. For them the idea of two incompatible — even contradictory — theories of nature was intellectually intolerable. They believed that gravity almost surely

6. In 1965 Feynman, Schwinger, and Tomanaga received the Nobel Prize for their work. But the modern way of thinking about Quantum Field Theory owes as much to Dyson as to the others.

played a role in determining the properties of the smallest building blocks of matter. The problem was that physicists had not probed deeply enough. Indeed, they were correct: down in the basement of the world, where distances are far too small to be directly observed, nature's smallest objects exert powerful gravitational forces on one another.

Today it is widely believed that gravity and Quantum Mechanics will play equally important roles in determining the laws of elementary particles. But the size of nature's basic building blocks is so inconceivably small that no one should be surprised if a radical rewiring will be needed to understand them. The new wiring, whatever it is, will be called *quantum gravity,* but even without knowing its detailed form, we can safely say that the new paradigm will involve very unfamiliar concepts of space and time. The objective reality of points of space and instants of time is on its way out, going the way of simultaneity,[7] determinism,[8] and the dodo. Quantum gravity describes a much more subjective reality than we ever imagined. As we will see in chapter 18, it is a reality that in many ways is like the ghostly three-dimensional illusion cast by a hologram.

Theoretical physicists are struggling to gain a foothold in a strange land. As in the past, thought experiments have brought to light paradoxes and conflicts between fundamental principles. This book is about an intellectual battle over a single thought experiment. In 1976 Stephen Hawking imagined throwing a bit of information — a book, a computer, even an elementary particle — into a black hole. Black holes, Hawking believed, were the ultimate traps, and the bit of information would be irretrievably lost to the outside world. This apparently innocent observation was hardly as innocent as it sounds; it threatened to undermine and topple the entire

7. One of the first things to go with the 1905 relativity revolution was the idea that two events can objectively be simultaneous.

8. Determinism is the principle that the future is completely determined by the past. According to Quantum Mechanics, the laws of physics are statistical and nothing can be predicted with certainty.

edifice of modern physics. Something was terribly out of whack; the most basic law of nature — the conservation of information — was seriously at risk. To those who paid attention, either Hawking was wrong or the three-hundred-year-old center of physics wasn't holding.

At first very few people paid attention. For almost two decades, the controversy took place largely below the radar. The great Dutch physicist Gerard 't Hooft and I were an army of two on one side of the intellectual divide. Stephen Hawking and a small army of relativists were on the opposite side. It was not until the early 1990s that most theoretical physicists — especially string theorists — woke up to the threat that Hawking had posed, and then they mostly got it wrong. Wrong for a while anyway.

The Black Hole War was a genuine scientific controversy — nothing like the pseudodebates over intelligent design, or the existence of global warming. Those phony arguments, cooked up by political manipulators to confuse a naive public, don't reflect any real scientific differences of opinion. By contrast, the split over black holes was very real. Eminent theoretical physicists could not agree on which principles of physics to trust and which to give up. Should they follow Hawking, with his conservative views of space-time, or 't Hooft and myself, with our conservative views of Quantum Mechanics? Every point of view seemed to lead only to paradox and contradiction. Either space-time — the stage on which the laws of nature play out — could not be what we thought it was, or the venerable principles of entropy and information were wrong. Millions of years of cognitive evolution, and a couple of hundred years of physics experience, once again had fooled us, and we found ourselves in need of new mental wiring.

The Black Hole War is a celebration of the human mind and its remarkable ability to discover the laws of nature. It is an explanation of a world far more remote to our senses than Quantum Mechanics and relativity. Quantum gravity deals with objects a hundred billion billion times smaller than a proton. We have never directly experienced such small things, and we probably never will, but human ingenuity has allowed us to deduce their existence, and surprisingly, the portals into that world are objects of huge mass and size: black holes.

The Black Hole War is also a chronicle of a discovery. The Holographic Principle is one of the most unintuitive abstractions in all of physics. It was the culmination of more than two decades of intellectual warfare over the fate of information that falls into a black hole. It was not a war between angry enemies; indeed the main participants are all friends. But it was a fierce intellectual struggle of ideas between people who deeply respected each other but also profoundly disagreed.

There is a widespread opinion that needs to be dispelled. The public image of physicists, especially theoretical physicists, is often one of nerdy, narrow people whose interests are alien, nonhuman, and boring. Nothing could be further from the truth. The great physicists I have known, and there have been many of them, are extremely charismatic people with powerful passions and fascinating minds. The diversity of personalities and ways of thinking is endlessly interesting to me. Writing about physics for a general audience without including the human element seems to me to leave out something interesting. In writing this book, I have tried to capture some of the emotional side of the story as well as the scientific side.

A Note About Big Numbers and Small Numbers

Throughout this book, you will find lots of very big and very small numbers. The human brain was not constructed to visualize numbers much bigger than 100 or much smaller than 1/100, but we can train ourselves to do better. For example, being very used to dealing with numbers, I can more or less picture a million, but the difference between a trillion and a quadrillion is beyond my powers of visualization. Many of the numbers in this book are far beyond trillions and quadrillions. How do we keep track of them? The answer involves one of the greatest rewiring feats of all time: the invention of *exponents* and *scientific notation.*

Let's begin with a fairly big number. The population of the Earth is about 6 billion. One billion is 10 multiplied by itself nine times. It can also be expressed as 1 followed by nine 0s.

$$\text{One billion} = 10 \times 10 \times 10 \times 10 \times 10 \times 10 \times 10 \times 10 \times 10 = 1{,}000{,}000{,}000$$

A shorthand notation for 10 multiplied by itself nine times is 10^9, or *ten to the ninth power*. Thus, the Earth's population is roughly given by this equation:

$$6 \text{ billion} = 6 \times 10^9$$

In this case, 9 is called the exponent.

Here is a much bigger number: the total number of protons and neutrons in the Earth.

$$\text{Number of protons and neutrons in Earth (approximately)} = 5 \times 10^{51}$$

That's obviously a lot bigger than the number of people on Earth. How much bigger? Ten to the fifty-first power has 51 factors of ten, but 1 billion has only 9. So 10^{51} has 42 more factors of ten than 10^9. That makes the number of nuclear particles in the Earth about 10^{42} times bigger than the number of people. (Notice that I've ignored the multipliers 5 and 6 in the previous equations. Five and 6 are not very different from each other, so if you just want a rough "order of magnitude estimate," you can ignore them.)

Let's take two really big numbers. The total number of electrons in the portion of the universe that we can see with the most powerful telescopes is about 10^{80}. The total number of photons[9] is about 10^{90}. Now, 10^{90} may not sound so much bigger than 10^{80}, but that's deceptive: 10^{90} is 10^{10} times bigger, and 10,000,000,000 is a very big number. In fact, 10^{80} and 10^{81} look almost the same, but the second number is ten times bigger than the first. So a modest change in the exponent can mean an enormous change in the number it represents.

Now let's consider very small numbers. The size of an atom is about one ten-billionth of a meter (a meter is about a yard). In decimal notation,

$$\text{Size of atom} = .0000000001 \text{ meters}$$

Note that the 1 appears in the tenth decimal place. Scientific notation for one ten-billionth involves a negative exponent, namely -10.

$$.0000000001 = 10^{-10}$$

Numbers with negative exponents are small, and numbers with positive exponents are large.

Let's do one more small number. Elementary particles, such as the electron, are very light compared to ordinary objects. A kilogram is the mass of a liter (roughly a quart) of water. The mass of an electron is vastly smaller. In fact, the mass of a single electron is about 9×10^{-31} kilograms.

Finally, multiplying and dividing is very easy in scientific notation. All you have to do is add or subtract the exponents. Here are some examples.

9. Don't confuse photons with protons. *Photons* are particles of light. *Protons*, together with neutrons, make up the atomic nucleus.

$$10^{51} = 10^{42} \times 10^{9}$$
$$10^{81} \div 10^{80} = 10$$
$$10^{-31} \times 10^{9} = 10^{-22}$$

Exponents aren't the only shorthand people use to describe immensely large numbers. Some of these numbers have their own names. For example, a *googol* is 10^{100} (1 followed by one hundred 0s), and a *googolplex* is 10^{googol} (1 followed by a googol 0s), a tremendously bigger number.

With these basics out of the way, let's turn to the somewhat less abstract world — in this case, San Francisco three years into President Ronald Reagan's first term — the cold war at fever pitch and a new war about to begin.

PART I

The Gathering Storm

History will be kind to me, for I intend to write it.

—WINSTON CHURCHILL[1]

1. The titles of the first and fourth parts of this book are taken from the first and fifth volumes of Churchill's history of World War II.

1

THE FIRST SHOT

San Francisco, 1983

The dark clouds of war had been gathering for more than eighty years by the time the initial skirmish took place in the attic of Jack Rosenberg's San Francisco mansion. Jack, also known as Werner Erhard, was a guru, a supersalesman, and a bit of a con man. Prior to the early 1970s, he had been just plain Jack Rosenberg, encyclopedia salesman. Then one day, while crossing the Golden Gate Bridge, he had an epiphany. He would save the world and, while he was at it, make a huge fortune. All he needed was a classier name and a new pitch. His new name would be Werner (for Werner Heisenberg) Erhard (for the German statesman Ludwig Erhard); the new pitch would be Erhard Seminars Training, aka EST. And he did succeed, if not in saving the world, at least in making his fortune. Thousands of shy, insecure people paid several hundred dollars each to be harangued, harassed, and (according to legend) told that they couldn't go to the toilet during the sixteen-hour motivational seminars run by Werner or one of his many disciples. It was a lot cheaper and faster than psychotherapy, and in a way it was effective. Shy and uncertain going in, the attendees appeared confident, strong, and friendly — just like Werner — coming out. Never mind that they sometimes seemed like manic, hand-shaking robots. They felt better. "The training" was even the subject of a very funny movie called *Semi-Tough* with Burt Reynolds.

EST groupies surrounded Werner. *Slaves* would definitely be too strong a term; let's call them volunteers. There were EST-trained chefs to cook his food, chauffeurs to drive him around town, and all manner of house servants to staff his mansion. But ironically, Werner himself was a groupie — a physics groupie.

I liked Werner. He was smart, interesting, and fun. And he was fascinated by physics. He wanted to be part of it, so he spent wads of money bringing groups of elite theoretical physicists to his mansion. Sometimes just a few of his special physics buddies — Sidney Coleman, David Finkelstein, Dick Feynman, and I — would meet in his home for spectacular dinners catered by celebrity chefs. But more to the point, Werner liked to host small, elite conferences. With a well-equipped seminar room in the attic, a staff of volunteers to cater to our every whim, and San Francisco as the venue, the mini-conferences were lots of fun. Some physicists were suspicious of Werner. They thought he would use the physics connection in some devious way to promote himself, but he never did. As far as I can tell, he just liked hearing about the latest ideas from the characters who were hatching them.

I think there were three or four EST conferences altogether, but only one left an indelible imprint on me, and on my physics research. The year was 1983. The guests included, among other notables, Murray Gell-Mann, Sheldon Glashow, Frank Wilczek, Savas Dimopoulos, and Dave Finkelstein. But for this story, the most important participants were the three main combatants in the Black Hole War: Gerard 't Hooft, Stephen Hawking, and myself.

Although I had met Gerard only a few times before 1983, he had made a big impression on me. Everyone knew that he was brilliant, but I sensed much more than that. He seemed to have a steel core, an intellectual toughness that exceeded that of anyone else I knew, with the possible exception of Dick Feynman. Both of them were showmen. Dick was an American showman — brash, irreverent, and full of macho one-upmanship. Once, among a group of young physicists at Cal Tech, he described a joke that the graduate students had played on him. There was a sandwich place in Pasadena where they served "celebrity" sandwiches. You could get a Humphrey Bogart, a Marilyn Monroe, and so on. The students had taken him to lunch there — I think for his birthday — and one after another ordered the Feynman sandwich. They had conspired with the manager beforehand, and the guy behind the counter didn't bat an eye.

After he finished the story, I said, "Gee, Dick, I wonder what the difference would be between a Feynman sandwich and a Susskind sandwich."

"Oh, they'd be about the same," he replied, "except the Susskind sandwich would have more ham."

"Yeah," I responded, "but a lot less baloney." That was probably the only time I beat him at that game.

Gerard is a Dutchman. The Dutch are the tallest people in Europe, but Gerard is short and solidly built, with a mustache and the look of a burgher. Like Feynman, 't Hooft has a strong competitive streak, but I am sure that I never got the better of him. Unlike Feynman, he is a product of old Europe — the last great European physicist, inheritor of the mantle of Einstein and Bohr. Although he is six years younger than I am, I was in awe of him in 1983, and rightfully so. In 1999 he was awarded the Nobel Prize for his work leading to the Standard Model of elementary particles.

But it wasn't Gerard whom I most remember from Werner's attic. It was Stephen Hawking, whom I first met there. It's where Stephen dropped the bomb that set the Black Hole War in motion.

Stephen is also a showman. He is a physically tiny man — I doubt that he weighs a hundred pounds — but his small body contains a prodigious intellect and an equally outsized ego. At that time, Stephen was in a more or less ordinary powered wheelchair, and he could still talk using his own voice, though he was very hard to understand unless you spent a lot of time with him. He traveled with an entourage that included a nurse and a young colleague who would listen to him very carefully and then repeat what he said.

In 1983 his translator was Martin Rocek, now a well-known physicist and one of the pioneers in an important subject called Supergravity. At the time of the EST conference, however, Martin was quite young and not so well known. Nevertheless, from previous meetings I knew that he was a very capable theoretical physicist. At some point in our conversation, Stephen (through Martin) said something that I thought was wrong. I turned to Martin and asked him for clarification of the physics. He looked at me like a deer caught in the headlights. Later he told me what had happened. It seems that translating Stephen's speech required such intense concentration that he was usually unable to keep track of the conversation. He barely knew what we were talking about.

Stephen is an unusual sight. I am not talking about his wheelchair or the obvious physical limitations of his body. Despite the immobility of his facial muscles, his faint smile is unique, simultaneously angelic and devilish, projecting a sense of secret amusement. During the EST conference, I found talking to Stephen very difficult. It took a long time for him to answer, and his answers were usually very brief. These short, sometimes single-word answers, his smile, and his almost disembodied intellect were unnerving. It was like communicating with the Oracle at Delphi. When someone submitted a question to Stephen, the initial response was absolute silence, and the eventual output was often incomprehensible. But the knowing smile said, "*You* may not understand what I'm saying, but *I* do, and I am right."

The world sees the diminutive Stephen as a mighty man, a hero of extraordinary courage and fortitude. Those who know him see other sides: Stephen the Playful and Stephen the Bold. One evening during the EST conference, a few of us were out walking on one of San Francisco's famous brake-busting hills. Stephen was with us, driving his powered chair. When we reached the steepest section, he turned on the devilish smile. Without a moment's hesitation, he took off down the hill at maximum velocity, the rest of us startled. We chased him, fearing the worst. When we got to the bottom, we found him sitting and smiling. He wanted to know whether there was a steeper hill to try. Stephen Hawking: the Evel Knievel of physics.

Indeed, Hawking is very much a daredevil of a physicist. But perhaps his boldest move ever was the bomb he dropped in Werner's attic.

I can't remember how his lecture worked at EST. Today a physics seminar given by Stephen has him sitting quietly in his chair while a disembodied computer voice lectures from a previous recording. That computerized voice has become Stephen's trademark; as flat as it is, it is full of personality and humor. But back then, maybe he talked and Martin translated. However it happened, the bomb fell with full force on Gerard and me.

Stephen claimed that "information is lost in black hole evaporation," and, worse, he seemed to prove it. If that was true, Gerard and I realized, the foundations of our subject were destroyed. How did the rest of the people in Werner's attic receive the news? Like the coyote in the roadrunner cartoon who overruns the edge of the cliff: the ground had disappeared beneath their feet, but they didn't know it yet.

It is said of cosmologists that they are often in error but never in doubt. If so, Stephen is only half a cosmologist: never in doubt but hardly ever wrong. In this case, he was. But Stephen's "mistake" was one of the most seminal in the history of physics and could ultimately lead to a profound paradigm shift about the nature of space, time, and matter.

Stephen's lecture was the last that day. For about an hour afterward, Gerard stood glaring at the diagram on Werner's blackboard. Everyone else had left. I can still see the intense frown on Gerard's face and the amused smile on Stephen's. Almost nothing was said. It was an electric moment.

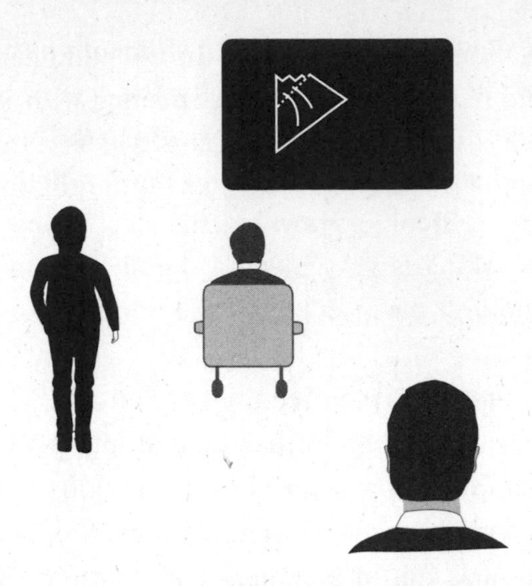

On the blackboard was a *Penrose diagram,* a type of diagram representing a black hole. The horizon (the edge of the black hole) was drawn as a dashed line, and the singularity at the center of the black hole was an ominous-looking jagged line. Lines pointing inward through the horizon represented bits of information falling past the horizon into the singularity. There were no lines coming back out. According to Stephen, those bits were irretrievably lost. To make matters worse, Stephen had proved that black holes eventually evaporate and disappear, leaving no trace of what has fallen in.

Stephen's theory went even further. He postulated that the vacuum – empty space – was full of "virtual" black holes that flashed into and out of existence so rapidly that we didn't notice them. The effect of these virtual black holes, he claimed, was to erase information, even if there was no "real" black hole in the vicinity.

In chapter 7, you will learn exactly what "information" means and also what it means to lose it. For now, just take it from me: this was an unmitigated disaster. 'T Hooft and I knew it, but the response from everyone else who heard about it that day was "Ho hum, information is lost in black holes." Stephen himself was sanguine. For me the toughest part of dealing with Stephen has always been the irritation I feel at his complacency. Information loss was something that just could not be right, but Stephen couldn't see it.

The conference broke up, and we all went home. For Stephen and Gerard, that meant back to Cambridge University and the University of Utrecht, respectively; for me a forty-minute drive south on Route 101 back to Palo Alto and Stanford University. It was hard to concentrate on the traffic. It was a cold day in January, and every time I stopped or slowed down, I would draw the figure from Werner's blackboard on my frosty windshield.

Back at Stanford, I told my friend Tom Banks about Stephen's claim. Tom and I thought about it intensely. To try to learn some more, I even invited one of Stephen's former students to come up from Southern California. We were very suspicious of Stephen's claim, but for a while we weren't sure why. What's so bad about losing a bit of information inside a black hole? Then it dawned on us. Losing information is the same as generating entropy. And generating entropy means generating heat. The virtual black holes that Stephen had so blithely postulated would create heat in empty space. Together with another colleague, Michael Peskin, we made an estimate based on Stephen's theory. We found that if Stephen was right, empty space would heat up to a thousand billion billion billion degrees in a tiny fraction of a second. Although I knew that Stephen was wrong, I couldn't find the hole in his reasoning. Perhaps that was what irritated me the most.

The ensuing Black Hole War was more than an argument between physicists. It was also a war of ideas, or perhaps a war between fundamental principles. The principles of Quantum Mechanics and those of General Relativity always seemed to be fighting each other, and it was not clear that the two could coexist. Hawking is a general relativist who had put his trust in Einstein's Equivalence

Principle. 'T Hooft and I are quantum physicists who felt certain that the laws of Quantum Mechanics could not be violated without destroying the foundations of physics. In the next three chapters, I will set the stage for the Black Hole War by explaining the basics of black holes, General Relativity, and Quantum Mechanics.

2

THE DARK STAR

There are more things in heaven and earth, Horatio,
Than are dreamt of in your philosophy.

—WILLIAM SHAKESPEARE, *HAMLET*

The earliest glimpse of anything like a black hole came in the late eighteenth century, when the great French physicist Pierre-Simon de Laplace and the English cleric John Michell had the same remarkable thought. All physicists in those days were intensely interested in astronomy. Everything that was known about astronomical bodies was known by the light they emitted or, in the case of the Moon and planets, the light they reflected. In Michell and Laplace's time, Isaac Newton, though dead for half a century, was by far the most powerful influence in physics. Newton believed that light was composed of tiny particles — corpuscles he called them — and if so, why wouldn't light be affected by gravity? Laplace and Michell wondered whether there could be stars so massive and dense that light could not escape their gravitational pull. Wouldn't such stars, if they existed, be completely dark and therefore invisible?

Can a projectile[1] — a stone, a bullet, or even an elementary

1. *The American Heritage Dictionary of the English Language* (4th ed.) defines a projectile as "a fired, thrown, or otherwise propelled object, such as a bullet, having no capacity for self-propulsion." Could a projectile be a single particle of light? According to Michell and Laplace, the answer was yes.

particle — ever escape the gravitational pull of a mass such as the Earth? In one sense yes, and in another sense no. The gravitational field of a mass never ends; it goes on forever, getting weaker and weaker with increasing distance. Thus, a projectile can never completely escape the Earth's gravity. But if a projectile is thrown upward with a large enough velocity, it will continue its outward motion forever, the diminishing gravity being too weak to turn it around and pull it back down to the surface. That is the sense in which a projectile can escape the Earth's gravity.

The strongest human has no chance of throwing a rock into outer space. A professional baseball pitcher might achieve a vertical throw of seventy-five yards, about one-quarter the height of the Empire State Building. Ignoring air resistance, a pistol can fire a bullet to a height of about three miles. But there is a certain velocity — naturally called the *escape velocity* — which is just barely enough to launch an object onto an eternal outbound trajectory. Starting with anything less than the escape velocity, a projectile will fall back to the Earth. Starting with a greater velocity, the projectile will escape to infinity. The escape velocity from the surface of the Earth is a mighty 25,000 miles per hour.[2]

For the moment, let's refer to any massive astronomical body as a *star,* whether it's a planet, an asteroid, or a true star. The Earth is just a small star, the Moon an even smaller star, and so on. According to Newton's law, the gravitational influence of a star is proportional to its mass, so it's entirely natural that the escape velocity should also depend on the star's mass. But mass is only half the story. The other half has to do with the star's radius. Imagine that as you stand on the surface of the Earth, some force begins to squeeze the Earth to a smaller size, but without it losing any of its mass. If you were standing on the surface of the Earth, the compression would move you closer to each and every atom of the Earth. As you

2. The escape velocity is an idealization that ignores effects such as air resistance, which would require the object to have a far greater velocity.

moved closer to the mass, the effect of gravity would become more powerful. Your own weight — a function of gravity — would increase, and, as you might expect, it would become harder to escape the Earth's pull. This illustrates one fundamental rule of physics: shrinking a star (without losing any mass) increases the escape velocity.

Now imagine exactly the opposite situation. For some reason, the Earth expands, so that you move away from the mass. Gravity at the surface would become weaker and therefore easier to escape. The question that Michell and Laplace asked was whether a star could have such large mass and small size that the escape velocity would exceed the speed of light.

When Michell and Laplace first had this prophetic thought, the speed of light (denoted by the letter *c*) had been known for more than one hundred years. The Danish astronomer Ole Rømer had determined c in 1676 and had found that light travels with the stupendous velocity of 186,000 miles (or seven times around the Earth) per second.[3]

$$c = 186{,}000 \text{ miles per second}$$

With that enormous speed, it would take an extremely large or extremely concentrated mass to trap light, but there was no obvious reason why it couldn't happen. Michell's paper to the Royal Society was the first reference to the objects that John Wheeler would later call *black holes*.

It may surprise you to know that as forces go, gravity is extremely weak. A weight lifter or high jumper may feel differently, but a simple experiment shows just how feeble gravity is. Begin with a light weight: a small ball of Styrofoam works well. By one means or another, electrically charge the weight with some static electricity. (Rubbing it against your sweater should work.) Now hang it from

3. In metric units, the speed of light is about 3×10^8 meters per second.

the ceiling by a thread. When it stops swinging, the thread will hang vertically. Next, bring a second, similarly charged object near the hanging weight. The electrostatic force will push the suspended weight and make the thread hang at an angle.

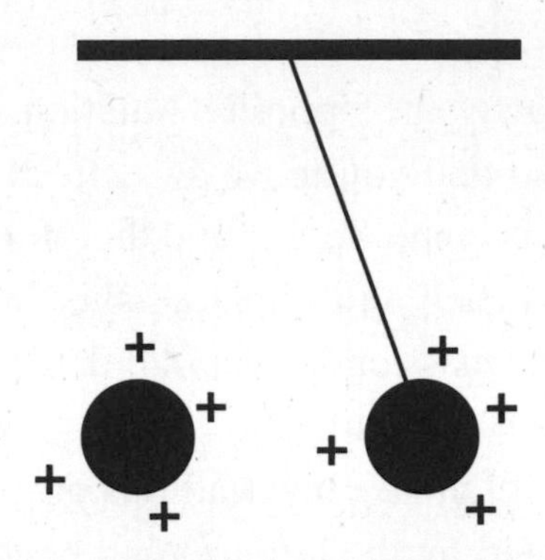

The same thing can be done with a magnet if the hanging weight is made of iron.

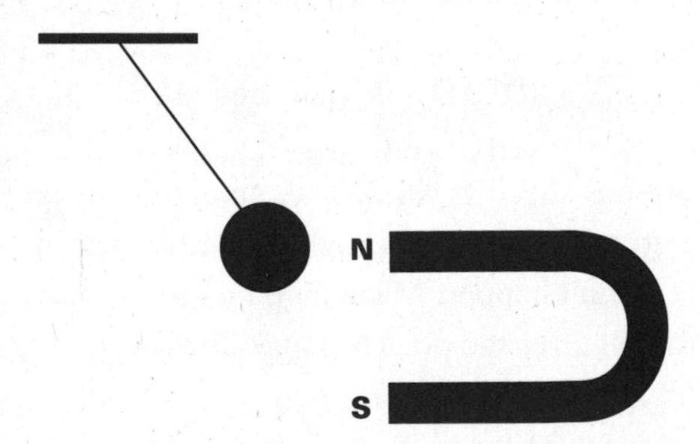

Now get rid of the electric charge and the magnet, and attempt to deflect the small weight by bringing a very heavy mass up close. The gravitational pull of that heavy mass will pull on the hanging weight, but the effect will be far too small to detect. Gravity is extremely feeble by comparison with electric and magnetic forces.

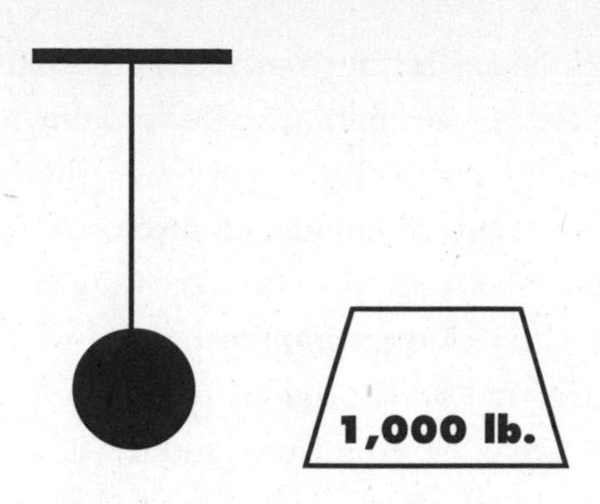

But if gravity is so weak, why can't we jump to the Moon? The answer is that the huge mass of the Earth, 6×10^{24} kilograms, easily compensates the weakness of gravity. But even with that mass, the escape velocity from the Earth's surface is less than one ten-thousandth of the speed of light. The dark star of Michell's and Laplace's imaginations would have to be tremendously massive and tremendously compressed if the escape velocity were to be greater than c.

Just to give you a feel for the magnitudes involved, let's look at the escape velocities from a few astronomical objects. Escaping from the Earth's surface requires an initial velocity of about 8 miles (about 11 kilometers) per second, which, as I said, is about 25,000 miles per hour. By terrestrial standards, that's very fast, but compared to the speed of light, it's a slow crawl.

You would have a much better chance of escaping from an asteroid than from the Earth. An asteroid with a radius of one mile has an escape velocity of about 6 feet (2 meters) per second: an easy jump. By contrast, the Sun is much bigger than the Earth, in both radius and mass.[4] These two things work against each other. The larger mass makes it more difficult to escape from the Sun's surface, while the larger radius makes it easier. The mass wins, however, and

4. The mass of the Sun is about 2×10^{30} kilograms. That's about half a million times the mass of the Earth. The Sun's radius is about 700,000 kilometers, or about one hundred times the Earth's radius.

the escape velocity from the Sun's surface is about fifty times greater than from the Earth's surface. That's still very much slower than the speed of light.

The Sun is not fated to remain the same size forever, however. Eventually, when a star runs out of fuel, the outward pressure caused by its internal heat fails. Like a giant vise, gravity begins to crush the star to a small fraction of its original size. About five billion years from now, the Sun will be exhausted, and it will collapse to what is known as a *white dwarf,* with a radius about the same as the Earth's. Escaping from its surface will require a velocity of 4,000 miles per second — fast, but still only 2 percent of the speed of light.

If the Sun were just a bit heavier — about one and a half times its actual value — the additional mass would crush it right past the white dwarf stage. The electrons in the star would be squeezed into the protons to form an incredibly dense ball of neutrons. A neutron star is so dense that just a single teaspoon of the stuff would weigh more than ten trillion pounds. But a neutron star is not yet a dark star; the escape velocity from its surface would be close to the speed of light (about 80 percent of c), but not quite there.

If the collapsing star were even heavier — say, about five times the Sun's mass — even the dense neutron ball would no longer be able to withstand the inward pull of gravity. In an ultimate implosion, it would be crushed to a *singularity* — a point of almost infinite density and destructive power. The escape velocity from that tiny core would be far beyond the speed of light. Thus, a dark star — or as we would say today, a black hole — would be born.

Einstein so disliked the idea of black holes that he dismissed their possibility, claiming that they could never form. But whether Einstein liked them or not, black holes are real. Astronomers today routinely study them, not only in the form of single collapsed stars but also in the centers of galaxies, where millions and even billions of stars have coalesced into black giants.

The Sun is not heavy enough to compress itself into a black hole, but if we could help it along by squeezing it in a cosmic vise to a radius of just two miles, it would become a black hole. You might think that it would spring back to a five-mile radius if the pressure of the vise were relaxed, but by then it would be too late; the material of the Sun would have gone into a kind of free fall. The surface would have quickly passed the one-mile point, the one-foot point, and the one-inch point. There would be no stopping it until it formed a singularity, and that awful implosion would be irreversible.

Imagine that we found ourselves near a black hole, but at some point well away from the singularity. Would light starting from that point escape the black hole? The answer depends on both the mass of the black hole and exactly where the light began its journey. An imaginary sphere called the *horizon* divides the universe in two. Light that starts inside the horizon will inevitably get pulled back into the black hole, but light that starts outside the horizon can escape the black hole's gravity. If the Sun were ever to become a black hole, the radius of the horizon would be about two miles.

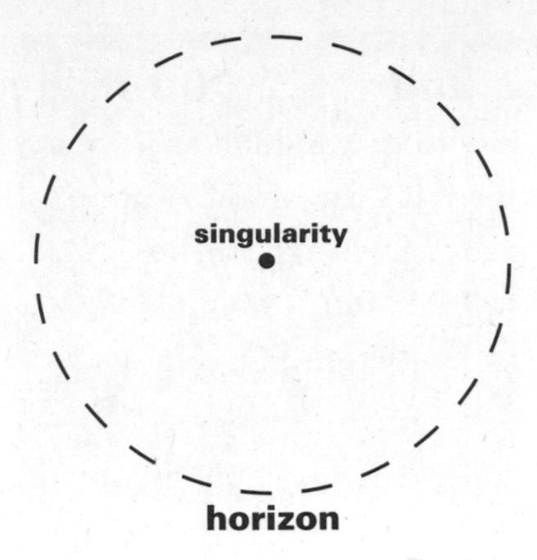

The radius of the horizon is called the *Schwarzschild radius.* It was named for the astronomer Karl Schwarzschild, who was the first to study the mathematics of black holes. The Schwarzschild radius depends on the mass of the black hole; in fact, it is directly proportional to the mass. For example, if the Sun's mass were replaced by a thousand solar masses, a light ray starting two or three miles away would have no chance of escaping, because the radius of the horizon would increase a thousandfold, to two thousand miles.

The proportionality between mass and the Schwarzschild radius is the first fact that a physicist learns about black holes. The Earth is approximately a million times less massive than the Sun, so its Schwarzschild radius is a million times smaller than the Sun's. It would have to be squeezed to about the size of a cranberry to make a dark star. By contrast, lurking at the center of our galaxy is a supersized black hole with a Schwarzschild radius of about a hundred million miles — about the size of the Earth's orbit around the Sun. And in other pockets of the universe, there are even bigger monsters than that.

No place is as nasty as the singularity of a black hole. Nothing can survive its infinitely powerful forces. Einstein was so appalled by the idea of a singularity that he rebelled against it. But there was no way out; if enough mass piles up, nothing can withstand the overwhelming pull toward the center.

Tides and the 2,000-Mile Man

What causes the seas to rise and fall as if they were breathing two big breaths every day? It's the Moon, of course, but how does it do it, and why twice a day? I will explain, but first let me tell you about the fall of the 2,000-Mile Man.

Imagine the 2,000-Mile Man — a giant who measures 2,000 miles from the tip of his head to the bottoms of his feet — as he falls, feet-first, from outer space toward the Earth. Far out in outer space, gravity is weak — so weak that he feels nothing. But as he gets closer to the Earth, strange sensations arise in his long body — sensations not of falling but of being stretched.

The problem is not the giant's overall acceleration toward the Earth. The cause of his discomfort is that gravity is not uniform throughout space. Far from the Earth, it is almost entirely absent. But as he draws closer, the pull of gravity increases. For the 2,000-Mile Man, this presents difficulties even while he is in free fall. The poor man is so tall that the pull on his feet is much stronger than the pull on his head. The net effect is an uncomfortable feeling that his head and feet are being pulled in opposite directions.

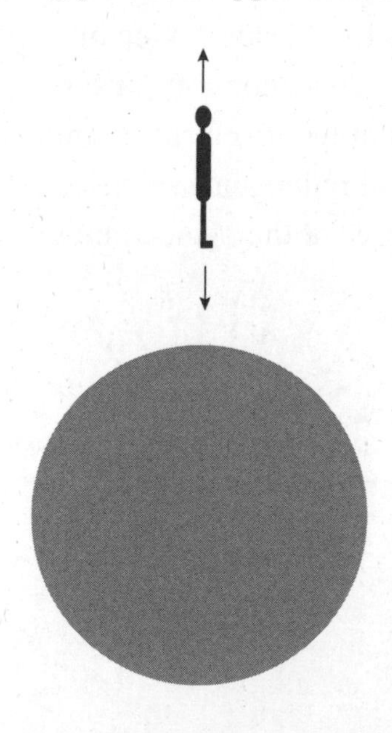

Perhaps he can avoid being stretched by falling in a horizontal position, legs and head at the same altitude. Yet when the giant tries it, he finds a new discomfort; the stretching sensation is replaced by an equal feeling of compression. He feels as if his head is being pressed toward his feet.

To understand why this is so, let's temporarily imagine that the Earth is flat. Here is what it would look like. The vertical lines, together with the arrows, indicate the direction of the gravitational

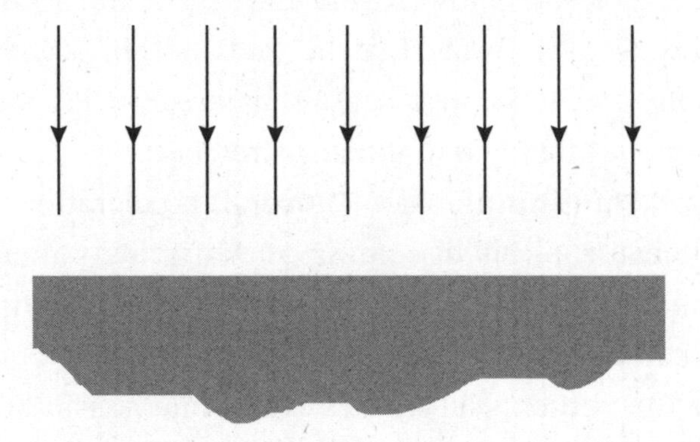

force — not surprisingly, straight down. But more than that, the strength of the gravitational pull is entirely uniform. The 2,000-Mile Man would have no trouble in this environment, whether he fell vertically or horizontally — not until he hit the ground anyway.

But the Earth is not flat. Both the strength and the direction of gravity vary. Instead of pulling in a single direction, gravity pulls directly toward the center of the planet, like this:

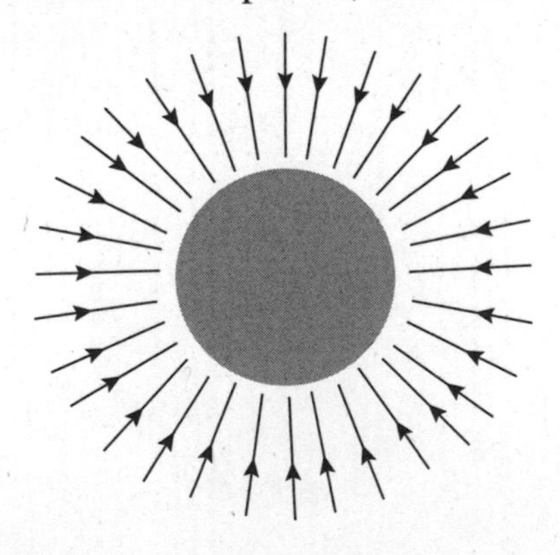

This creates a new problem for the giant if he falls horizontally. The force on his head and feet will not be the same because gravity, as it pulls toward the center of the Earth, will push his head toward his feet, leading to the strange sensation of being compressed.

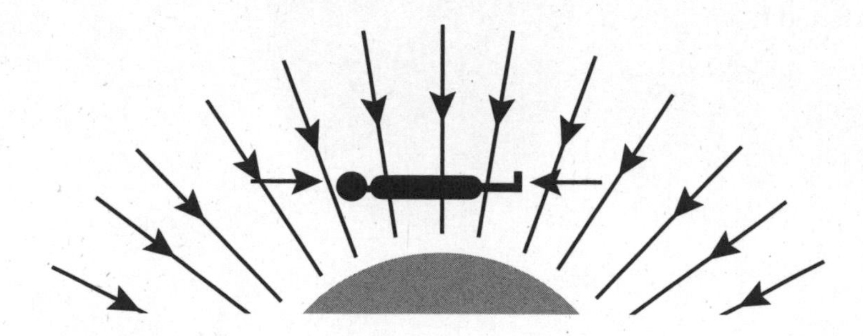

Let's return to the question of the ocean tides. The cause of the twice-daily rising and falling of the seas is exactly the same as the cause of the 2,000-Mile Man's discomfort: the non-uniformity of gravity. But in this case, it's the Moon's gravity, not the Earth's. The Moon's pull on the oceans is strongest on the side of the Earth facing the Moon and weakest on the far side. You might expect the Moon to create a single oceanic bulge on the closer side, but that's wrong. For the same reason that the tall man's head is pulled away from his feet, the water on both sides of the Earth – near and far – bulges away from it. One way to think about this is that on the near side, the Moon pulls the water away from the Earth, but on the far side, it pulls the Earth away from the water. The result is two bulges on opposite sides of the Earth, one facing toward the Moon and the other facing away. As the Earth turns one revolution under the bulges, each point experiences two high tides.

The distorting forces caused by variations in the strength and direction of gravity are called *tidal forces,* whether they are due to the Moon, Earth, Sun, or any other astronomical mass. Can humans of normal size feel tidal forces — for example, when jumping from a diving board? No, we cannot, but only because we are so small that the Earth's gravitational field hardly varies across the length of our bodies.

Descent into Hell

I entered on the deep and savage way.

—DANTE, *THE DIVINE COMEDY*

Tidal forces would not be so benign if you fell toward a solar-mass black hole. All that mass compacted into the tiny volume of the black hole not only makes gravity very strong near the horizon, but it also makes it very non-uniform. Well before you arrived at the

Schwarzschild radius, when you were more than 100,000 miles from the black hole, tidal forces would become quite uncomfortable. Like the 2,000-Mile Man, you would be too big for the rapidly varying gravitational field of the black hole. By the time you approached the horizon, you would be deformed, almost like toothpaste squeezed from the tube.

There are two cures for tidal forces at a black hole horizon: either make yourself smaller or make the black hole bigger. A bacterium wouldn't notice the tidal forces at the horizon of a solar-mass black hole, but neither would you notice the tidal forces at the horizon of a million-solar-mass black hole. That may seem counterintuitive, since the gravitational influence of the more massive black hole would be stronger. But that thinking neglects an important fact: the horizon of the larger black hole would be so large that it would almost appear flat. Near the horizon, the gravitational field would be very strong but practically uniform.

If you know a little about Newtonian gravity, you can work out the tidal forces at the horizon of a dark star. What you will find is that the bigger and more massive the dark star, the weaker the tidal forces at the horizon. For that reason, crossing the horizon of a very big black hole would be uneventful. But ultimately, there is no escape from tidal forces, even for the biggest black hole. The large size would just delay the inevitable. Eventually, the unavoidable descent toward the singularity would be as awful as any torture that Dante imagined or that Torquemada inflicted during the Spanish Inquisition. (The rack springs to mind.) Even the smallest bacterium would be pulled apart along the vertical axis and at the same time squished horizontally. Small molecules would survive longer than bacteria, and atoms even a bit longer. But sooner or later, the singularity would win, even over a single proton. I don't know if Dante was right in claiming that no sinner can escape the torments of hell, but I am quite certain that nothing can escape the awesome tidal forces at the singularity of a black hole.

Despite the alien and brutal properties of the singularity, that is not where the deepest mysteries of black holes lie. We know what happens to any object unlucky enough to get pulled to the singularity, and it's not pretty. But pleasant or not, the singularity isn't nearly as paradoxical as the horizon. Almost nothing in modern physics has created greater confusion than the question, What happens to matter as it falls through the horizon? Whatever your answer, it is probably wrong.

Michell and Laplace lived long before Einstein was born and couldn't have guessed the two discoveries he would make in 1905. The first was the Special Theory of Relativity, which rests on the principle that *nothing* — neither light nor anything else — can ever exceed the speed of light. Michell and Laplace understood that light could not escape from a dark star, but they didn't realize that nothing else could either.

Einstein's second discovery in 1905 was that light really *is* made of particles. Shortly after Michell and Laplace speculated about dark stars, Newton's particle theory of light came into disfavor. Evidence mounted that light was made of waves, similar to sound waves or waves on the surface of the sea. By 1865 James Clerk Maxwell had figured out that light consists of undulating *electric and magnetic fields,* propagating through space with the speed of light, and the particle theory of light was as dead as the proverbial doornail. It seems that no one had yet thought that electromagnetic waves might also be pulled by gravity, and thus dark stars were forgotten.

Forgotten, that is, until 1917, when the astronomer Karl Schwarzschild solved the equations of Einstein's brand-new General Theory of Relativity and rediscovered the dark star.[5]

5. Black holes come in a variety of kinds. In particular, they can rotate about an axis if the original star was rotating (all stars do to some extent), and they can be electrically charged. Dropping electrons into a black hole would charge it. Only the nonrotating, uncharged variety of black holes are called Schwarzschild black holes.

The Equivalence Principle

Like most of Einstein's work, the General Theory of Relativity was difficult and subtle, but it grew out of extremely simple observations. They were, in fact, so elementary that anyone could have made them, but no one did.

It was Einstein's style to draw very far-reaching conclusions from the simplest of thought experiments. (Personally, I have always admired this way of thinking above all others.) In the case of the General Theory, the thought experiment involved an observer in an elevator. Textbooks often update the elevator to a rocket ship, but in Einstein's day elevators were the exciting new technology. He first imagined the elevator to be floating freely in outer space, far from any gravitating object. Everyone in the elevator would experience complete weightlessness,

and projectiles would move along perfectly straight trajectories at uniform velocity. Light rays would do exactly the same, but of course at the speed of light.

Einstein next imagined what would happen if the elevator were accelerated upward, perhaps by means of a cable attached to some distant anchor or by means of rockets bolted to the underside. The passengers would be pushed to the floor, and the trajectories of projectiles would curve downward, in parabolic orbits. All things would be exactly the same as if they were under the influence of gravity. Everyone since Galileo knew this, but it remained for Einstein to make this simple fact into a powerful new physical principle. The Equivalence Principle asserts that there is absolutely no difference between the effects of gravity and the effects of acceleration. No experiment done inside the elevator could distinguish whether the elevator was standing still in a gravitational field or being accelerated in outer space.

In itself, this was not surprising, but the consequences were momentous. At the time that Einstein formulated the Equivalence Principle, very little was known about how gravity affected other phenomena, such as the flow of electricity, the behavior of magnets, or the propagation of light. Einstein's method was to start by first working out how acceleration influenced these phenomena. That usually didn't involve any new or unknown physics. All he had to do was to imagine how known phenomena would be seen from an accelerating elevator. The Equivalence Principle would then tell him the effects of gravity.

The first example involved the behavior of light in a gravitational field. Imagine a beam of light moving horizontally, from left to right, across an elevator. If the elevator were moving freely, far from any gravitating mass, the light would move in a perfectly straight horizontal line.

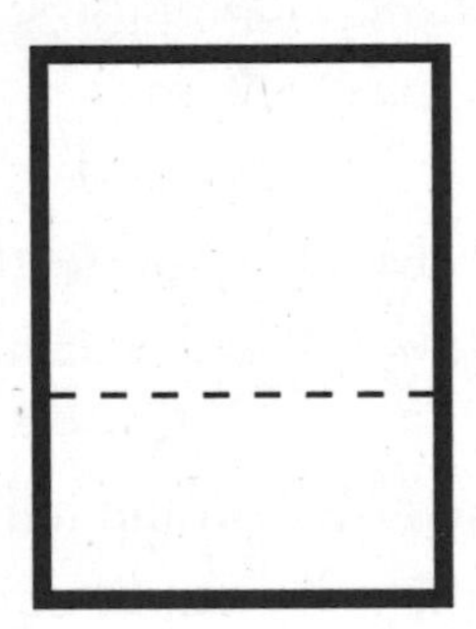

But now let the elevator accelerate upward. The light starts at the left side of the elevator moving horizontally, but because of the elevator's acceleration, by the time it arrives at the other side, it appears to have a downward component of motion. From one point of view, the elevator has accelerated upward, but to a passenger, the light appears to accelerate downward.

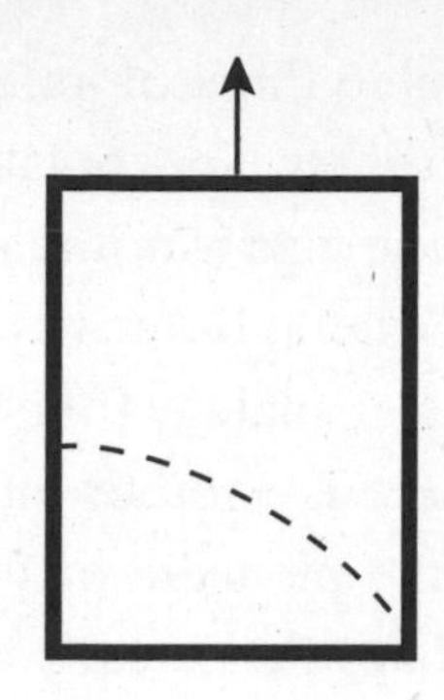

In fact, the path of a light ray curves in the same way as the trajectory of a very fast particle. This effect has nothing to do with whether light is made of waves or particles; it is just the effect of upward acceleration. But, argued Einstein, if acceleration makes the trajectory of a light ray bend, so must gravity. Indeed, you might say that gravity pulls light and makes it fall. This was exactly what Michell and Laplace had guessed.

There is another side of the coin: if acceleration can simulate the effects of gravity, it can also cancel them. Imagine that the same elevator is no longer infinitely far away in outer space but is instead at the top of a skyscraper. If it is standing still, the passengers experience the full effect of gravity, including the bending of light rays as those rays cross the elevator. But then the elevator cable snaps, and the elevator begins to accelerate toward the ground. During the brief interval of free fall, gravity inside the elevator appears to be completely canceled out.[6] The passengers float in the cabin with no sense of up or down. Particles and light beams travel in perfect straight lines. That is the other side of the Equivalence Principle.

Drain Holes, Dumb Holes, and Black Holes

Anyone who tries to describe modern physics without the use of mathematical formulas knows how useful analogies can be. For

6. I am assuming that the elevator is small enough that tidal forces are negligible.

example, it is very helpful to think of an atom as a miniature solar system, and the use of ordinary Newtonian mechanics to describe a dark star can help someone who is not ready to plunge into the advanced mathematics of General Relativity. But analogies have their limitations, and the dark star analogy for a black hole is flawed when we push it too hard. There is another analogy that does better. I learned it from one of the pioneers of black hole Quantum Mechanics, Bill Unruh. Perhaps I especially like it because of my first career, which was as a plumber.

Imagine a shallow, infinite lake. It's only a few feet deep, but it goes on forever in the horizontal directions. A species of blind pollywogs live their entire lives in the lake, without any knowledge of light, but they are very good at using sound to locate objects and to communicate. There is one unbreakable rule: nothing is allowed to move through the water faster than the speed of sound. For most purposes, the speed limit is unimportant, since the pollywogs move much slower than sound.

There is a danger in this lake. Many pollywogs have discovered it too late to save themselves, and none has ever returned to tell the tale. At the center of the lake is a drain hole. The water empties out through the drain into a cave below, where it cascades onto deadly sharp rocks.

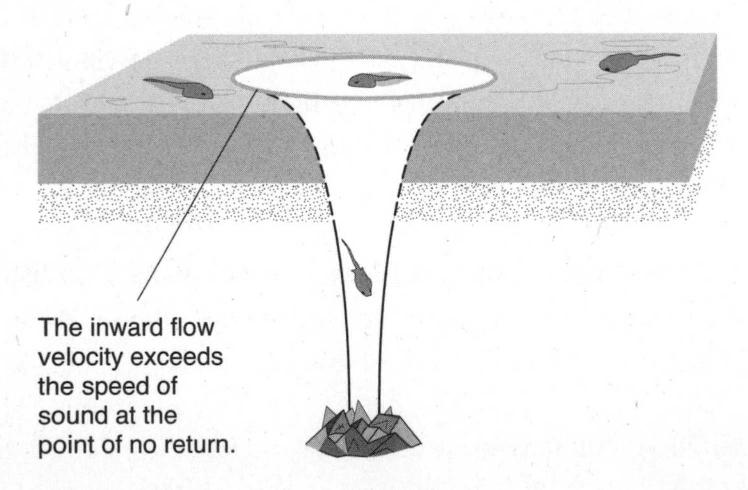

If you look down on the lake from above, you can see the water moving toward the drain. Far from the drain, the velocity is undetectably slow, but closer in, the water picks up speed. Let's assume that the drain draws off water so fast that at some distance, its velocity becomes equal to the speed of sound. Even closer to the drain, the flow becomes supersonic. We now have a very dangerous drain.

The pollywogs floating in the water, experiencing only their own liquid environment, never know how fast they are moving; everything in their vicinity is swept along at the same speed. The big danger is that they may get sucked into the drain and then be destroyed on the sharp rocks. In fact, once one of them has crossed the radius where the inward velocity exceeds the speed of sound, he is doomed. Having crossed the point of no return, he can't outswim the current, nor can he send a warning to anyone in the safe region (no audible signal moves through the water faster than sound). Unruh calls the drain hole and its point of no return a *dumb hole* — dumb in the sense of silent — because no sound can escape from it.

One of the most interesting things about the point of no return is that an unwary observer, floating past it, would initially notice nothing out of the ordinary. There is no signpost or siren to warn him, no obstruction to stop him, nothing to advise him of the imminent danger. One moment everything seems fine, and the next moment everything still seems fine. Passing the point of no return is a non-event.

A freely drifting pollywog, let's call her Alice, floats toward the drain singing to her friend Bob, who is far away. Like her fellow sightless pollywogs, Alice has a very limited repertoire. The only note she can sing is middle C, with a frequency of 262 cycles per second — or in technical jargon, 262 hertz (Hz).[7] While Alice is still far from the drain, her motion is almost imperceptible. Bob listens to the sound of Alice's voice and hears middle C. But as Alice picks up

7. The hertz, named after the nineteenth-century German physicist Heinrich Hertz, is a unit of frequency. One hertz is the same as one cycle per second.

speed, the sound deepens, at least to Bob's ears; C gives way to B, then to A. The cause is the familiar *Doppler shift,* which can be heard when a speeding train passes while blowing its whistle. As the train approaches, the whistle sounds higher-pitched to you than it does to the trainman on board. Then, as the whistle passes and recedes into the distance, the sound deepens. Each successive oscillation has a little farther to travel than the previous one, and it arrives at your ears slightly delayed. The time between successive sound oscillations is drawn out, and you hear a lower frequency. Moreover, if the train picks up speed as it races away, the perceived frequency gets lower and lower.

The same thing happens to Alice's musical note as she drifts toward the point of no return. At first Bob hears the note at 262 Hz. Later it shifts to 200 Hz, then 100 Hz, 50 Hz, and so on. A sound produced very close to the point of no return takes an extremely long time to escape; the motion of the water almost cancels the outward motion of the sound, slowing it nearly to a halt. Soon the sound becomes so low-pitched that without special equipment, Bob can no longer hear it.

Bob may have special equipment that allows him to focus sound waves and form images of Alice as she approaches the point of no return. But the successive sound waves take longer and longer to reach Bob, thus making everything about Alice appear to slow down. Her voice deepens, but not only that; the waving of her arms slows almost to a halt. The very last wave that Bob can detect seems to take an eternity. In fact, it seems to Bob that it takes forever for Alice to reach the point of no return.

Meanwhile, Alice doesn't notice anything strange. She happily drifts past the point of no return without any sense of slowing down or speeding up. Only later, as she is swept to the deadly rocks, does she realize the danger. Here we see one of the key features of a black hole: different observers have paradoxically different perceptions of the same events. To Bob, at least judging by the sound that he hears, it takes an eternity for Alice to reach the point of no return, but to Alice it may take no more than the blink of an eye.

By now you may have guessed that the point of no return is an analog for the horizon of a black hole. Substitute light for sound (recall that nothing can exceed the speed of light), and you have a fairly accurate picture of the properties of a Schwarzschild black hole. As in the case of the drain hole, anything that passes the horizon is incapable of escaping, or even of standing still. In the black hole, the danger is not sharp rocks but the singularity at the center. All matter inside the horizon will be dragged to the singularity, where it will be squeezed to infinite pressure and density.

When we are armed with our dumb hole analogy, many paradoxical things about black holes become clear. For example, consider Bob, no longer a pollywog but now an astronaut on a space station orbiting a huge black hole at a safe distance. Meanwhile, Alice is falling toward the horizon, not singing — there is no air in outer space to carry her voice — but signaling instead with a blue flashlight. As she falls, Bob sees the light shift in frequency from blue to red to infrared to microwave and finally to low-frequency radio waves. Alice herself seems to get more and more lethargic, slowing almost to a standstill. Bob never sees her fall through the horizon; to him it takes an infinite time for Alice to reach the point of no return. But in Alice's frame of reference, she falls right past the horizon and begins to feel funny only when she approaches the singularity.

The horizon of a Schwarzschild black hole is at the Schwarzschild radius. Alice may be doomed when she crosses the horizon, but just like the pollywogs, she still has some time before being destroyed at the singularity. How much time? That depends on the size, or mass, of the black hole. The larger the mass, the larger the Schwarzschild radius and the more time Alice has. For a black hole with the mass of the Sun, Alice would have only about ten microseconds. For a black hole at the center of a galaxy, which may be the size of a billion solar masses, Alice would have a billion microseconds, or roughly half an hour. One could imagine even bigger black holes, in which Alice could live out her whole life, and maybe even several generations of Alice's progeny could live and die, before the singularity destroyed them.

Of course, according to Bob's observations, Alice will never even get to the horizon. Who is right? Does she or doesn't she get to the horizon? What really happens? *Is* there a *really?* Physics is, after all, an observational and experimental science, so one would have to credit Bob's observations with having their own validity, although they apparently conflict with Alice's description of events. (We will come back to Alice and Bob in later chapters, after we have discussed the amazing quantum properties of black holes discovered by Jacob Bekenstein and Stephen Hawking.)

The drain analogy is a good one for many purposes, but like all analogies, it has its limits. For example, when an object falls through the horizon, its mass gets added to that of the black hole. The increase in mass implies that the horizon grows. No doubt we could model this in the drain analogy by hooking up a pump to the drainpipe to control the flow. Every time something falls into the drain, the pump would be turned up a bit, speeding up the flow and pushing the point of no return farther out. But the model quickly loses its simplicity.[8]

Another property of black holes is that they themselves are movable objects. If you place a black hole in the gravitational field of another mass, it will be accelerated, just like any other mass. It can even fall into a bigger black hole. If we tried to represent all these features of real black holes, the drain analogy would be more complicated than the mathematics it was supposed to avoid. But despite its limitations, the drain is a very useful picture that allows us to un-

8. Professor George Ellis has reminded me of a subtlety when the flow is variable. In that case, the point of no return does not exactly coincide with the point where the water velocity reaches the speed of sound. In the case of black holes, the analogous subtlety is the difference between an apparent horizon and a true horizon.

derstand basic features of black holes without mastering the equations of General Relativity.

A Few Formulas for Those Who Like Them

I've written this book for the less mathematically inclined reader, but for those who enjoy a bit of math, here are a few formulas and their meaning. If you don't like them, just go on to the next chapter. There won't be a test.

According to Newton's law of gravity, every object in the universe attracts every other object, with a gravitational force *proportional to the product of their masses and inversely proportional to the square of the distance between them.*

$$F = \frac{mMG}{D^2}$$

This equation is one of the most famous in physics, almost as famous as $E = mc^2$ (Einstein's famous equation relating energy, E, to mass, m, and the speed of light, c). On the left side is the force, F, between two masses, such as the Moon and the Earth, or the Earth and the Sun. On the right side, the bigger mass is M and the smaller mass is m. For example, the Earth's mass is 6×10^{24} kilograms, and the Moon's is 7×10^{22} kilograms. The distance between the masses is denoted by D. From the Earth to the Moon, the distance is about 4×10^8 meters.

The last symbol in the equation, G, is a numerical constant called *Newton's constant.* Newton's constant is not something that can be deduced from pure mathematics. To find its value, the gravitational force between two known masses at some known distance must be measured. Once you've done that, you can calculate the force between any two masses at any distance. Ironically, Newton never knew the value of his own constant. Because gravity is so feeble, G was too small to measure until the end of the eighteenth century. At that time, an English physicist named Henry Cavendish devised a clever way to measure extremely small forces. Cavendish found that the force between a pair of one-kilogram masses separated by one meter is approximately 6.7×10^{-11} newtons. (The newton is the unit of force in the metric system. It's equal to about one-fifth of a pound.) Thus, the value of Newton's constant, in metric units, is

$$G = 6.7 \times 10^{-11}$$

Newton had one lucky break in working out the consequences of his theory: the special mathematical properties of the inverse square law. When you weigh yourself, some of the gravitational force pulling you toward the Earth is due to mass just beneath your feet, some is due to mass deep within the Earth, and some comes from the antipodal point eight thousand miles away. But by a miracle of mathematics, you can pretend that the entire mass is concentrated at a single point, right at the geometric center of the planet.

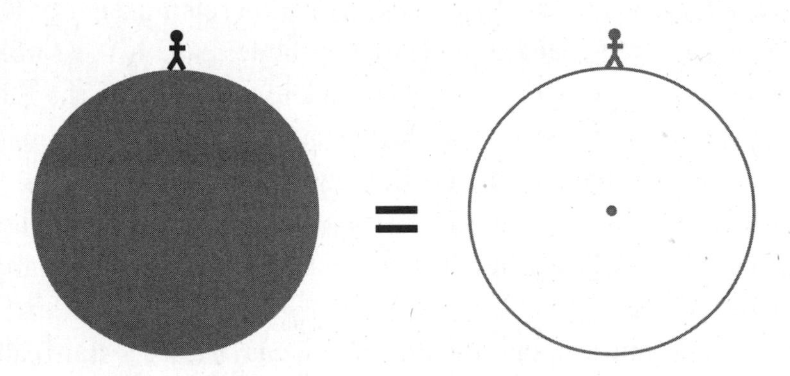

The gravity of a ball of mass is exactly the same as if all the mass were concentrated at a point in the center.

This convenient fact allowed Newton to calculate the escape velocity from a large object by replacing the large mass with a tiny mass point. Here is the result.

$$\text{Escape velocity} = \sqrt{2MG/R}$$

The formula clearly shows that the bigger the mass and the smaller the radius, R, the larger the escape velocity.

It's now an easy exercise to compute the Schwarzschild radius, R_s. All you have to do is plug in the speed of light for the escape velocity and then solve the equation for the radius.

$$R_s = \frac{2MG}{c^2}$$

Note the important fact that the Schwarzschild radius is proportional to the mass.

That's all there is to dark stars — at least at the level that Laplace and Michell were able to understand them.

3

NOT YOUR GRANDFATHER'S GEOMETRY

In the old days, before mathematicians such as Gauss, Bolyai, Lobachevski, and Riemann[1] messed around with it, geometry meant Euclidean geometry — the same geometry that we all learned in high school. First came plane geometry, the geometry of a perfectly flat, two-dimensional surface. The basic concepts were points, straight lines, and angles. We learned that three points define a triangle, unless they are on a common line; parallel lines never intersect; and the sum of the angles of any triangle is 180 degrees.

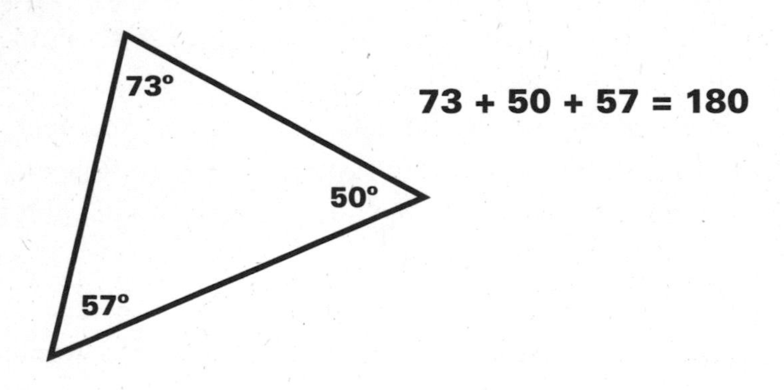

1. Carl Friedrich Gauss (1777–1855), János Bolyai (1802–1860), Nikolai Lobachevski (1792–1856), and Georg Friedrich Bernhard Riemann (1826–1866).

Later, if you had the same course that I did, you stretched your powers of visualization to three dimensions. Some things remained the same as in two dimensions, but other things had to change, or there wouldn't be any difference between two and three dimensions. For example, there are straight lines in three dimensions that never meet but that are not parallel; they are called skew lines.

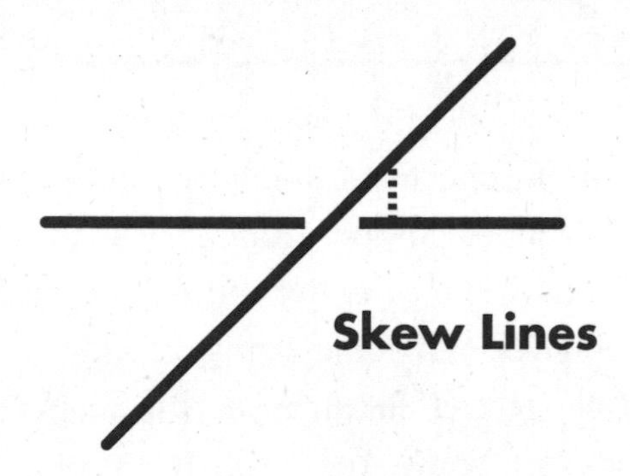

Whether it is two dimensions or three dimensions, the rules of geometry remain the ones that Euclid laid down sometime around 300 B.C. However, other kinds of geometry — geometries with different axioms — are possible even in two dimensions.

The word *geometry* literally means "measurement of the Earth." It's ironic that had Euclid actually taken the trouble to measure triangles on the Earth's surface, he would have discovered that Euclidean geometry doesn't work. The reason is that the Earth's surface is a sphere,[2] not a plane. Spherical geometry certainly has points and angles, but it's not so obvious that it has anything that we should call straight lines. Let's see if we can make any sense out of the words "straight line on a sphere."

A familiar way to describe a straight line in Euclidean geometry is that it is the shortest route between two points. If I wanted to construct a straight line on a football field, I would put two stakes in the ground and then stretch a string between them as tightly as pos-

2. I am, of course, referring to an idealized, perfectly round Earth.

sible. Pulling the string taut would ensure that the line was as short as possible.

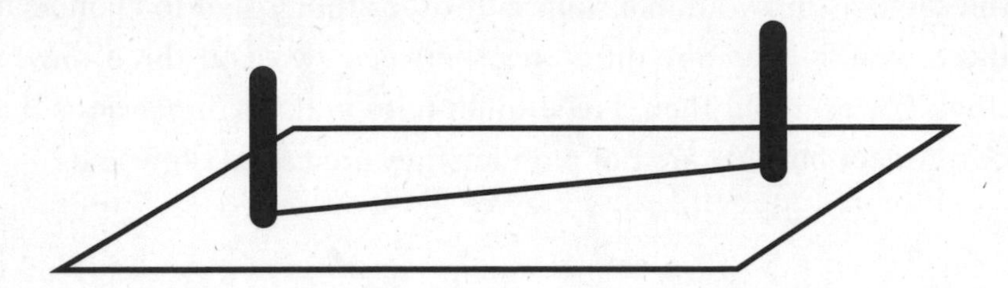

That concept of the shortest route between two points can easily be extended to a sphere. Suppose our goal is to find the shortest possible air route between Moscow and Rio de Janeiro. We need a globe, two thumbtacks, and some string. Placing the tacks at Moscow and Rio, we can stretch the string across the globe's surface and determine the shortest route. These shortest routes are called *great circles,* such as the equator and the meridians. Does it make sense to call them the straight lines of spherical geometry? It doesn't really matter what we call them. The important thing is the logical relationship between points, angles, and lines.

Being the shortest route between two points, such lines are in some sense the straightest possible on a sphere. The correct mathematical name for such routes is *geodesics*. Whereas geodesics in a flat plane are ordinary straight lines, geodesics on a sphere are great circles.

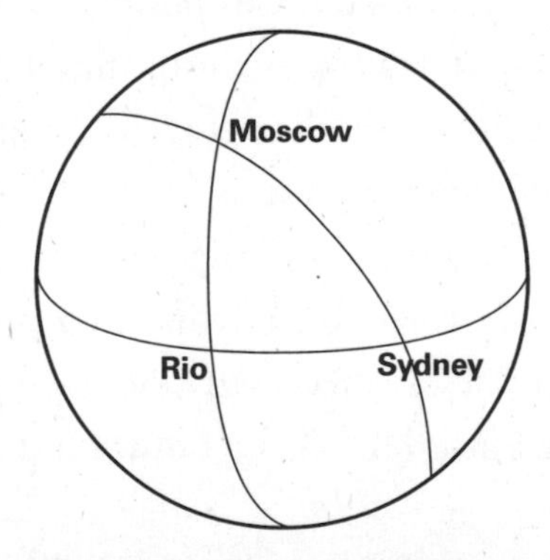

Great Circles on a Sphere

Having these spherical replacements for straight lines, we can go about constructing triangles. Pick three points on the sphere — say, Moscow, Rio, and Sydney. Next, draw the three geodesics connecting the points pair-wise: the Moscow-Rio geodesic, the Rio-Sydney geodesic, and finally the Sydney-Moscow geodesic. The result is a *spherical triangle.*

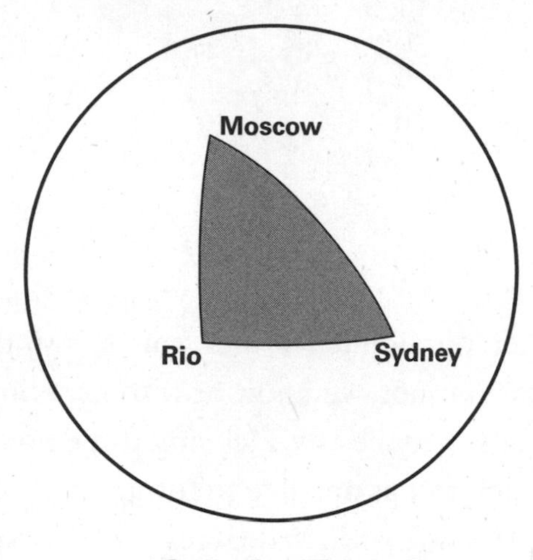

Spherical Triangle

In plane geometry, if we add the angles of any triangle, we get exactly 180 degrees. But looking carefully at the spherical triangle, we can see that the sides bow out and make the angles somewhat larger than they would be on a plane. The result is that the sum of the angles of a spherical triangle is always larger than 180 degrees. A surface whose triangles have this property is said to be *positively curved.*

Can there be surfaces with the opposite property — namely that the sum of the angles of a triangle is less than 180 degrees? An example of this kind of surface is a saddle. Saddle-shaped surfaces are *negatively curved;* instead of bowing out, the geodesics forming a triangle on a negatively curved surface are pinched in.

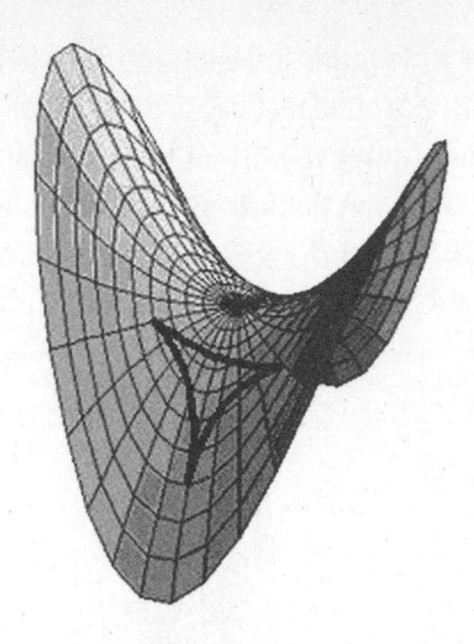

Thus, whether our limited brains can envision curved three-dimensional space or not, we know how to experimentally test for curvature. Triangles are the key. Pick any three points in space and stretch strings as tightly as possible to form a three-dimensional triangle. If the angles add up to 180 degrees for every such triangle, the space is flat. If not, it's curved.

Geometries far more complicated than spheres or saddles can exist — geometries with irregular hills and valleys having regions with both positive and negative curvature. But the rule for constructing geodesics is always simple. Imagine yourself crawling on such a surface and following your nose forward, never turning your head. Don't look around; don't worry about where you came from or where you are going; just myopically crawl straight ahead. Your path will be a geodesic.

Imagine a man in a mechanized wheelchair attempting to navigate a desert of sand dunes. With only limited water, he must get out of the desert quickly. The rounded hills, saddle-shaped passes, and deep valleys define a land of positive and negative curvature, and it's not entirely obvious how best to steer the chair. The driver reasons that the high hills and deep valleys will slow him down, so at first he steers around them. The steering mechanism is simple — if he slows one of the wheels relative to the other, the chair will turn in that direction.

But after a few hours, the driver begins to suspect that he is passing the same shapes that he passed earlier. Steering the chair has led him in a dangerous random walk. He now realizes that the best strategy is to go absolutely straight ahead, turning neither to the left nor to the right. "Just follow your nose," he says to himself. But how to make sure that he is not wavering?

The answer soon becomes obvious. The wheelchair has a mechanism that locks the two wheels together so that they turn like a rigid dumbbell. Locking the wheels in this manner, the man takes off, making a beeline for the edge of the desert.

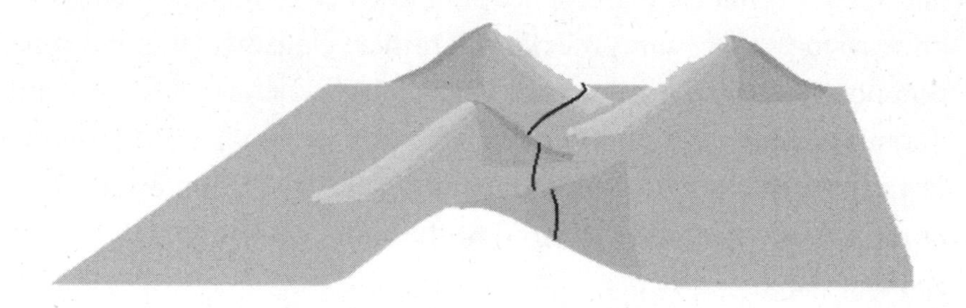

At every point along the trajectory, the traveler seems to be going in a straight line, but looked at as a whole, his path is a complicated, winding curve. Nevertheless, it is as straight and short as possible.

It wasn't until the nineteenth century that mathematicians began to study new kinds of geometry with alternative axioms. A few, such as Georg Friedrich Bernhard Riemann, entertained the idea that "real" geometry — the geometry of real space — might not be exactly Euclidean. But Einstein was the first to take the idea seriously. In the General Theory of Relativity, the geometry of space (or, more correctly, space-time) became a question for experimenters, not philosophers or even mathematicians. Mathematicians can tell you what kinds of geometries are possible, but only measurement can determine the "true" geometry of space.

In crafting the General Theory of Relativity, Einstein built on the mathematical work of Riemann, who had envisioned geometries beyond spheres and saddle surfaces: spaces with lumps and bumps; some places being positively curved, others negatively curved; geodesics winding over and between these features in curving, irregular paths. Riemann thought only of three-dimensional space, but Einstein and his contemporary Hermann Minkowski introduced something new: time as the *fourth dimension*. (Try to visualize that. If you can, you have a very unusual brain.)

The Special Theory of Relativity

Even before Einstein began to think about curved space, Minkowski had the idea that time and space should be combined to form four-dimensional *space-time,* declaring rather elegantly if somewhat pompously, "Henceforth space by itself and time by itself are doomed to fade away into mere shadows, and only a kind of union of the two will preserve an independent reality."[3] Minkowski's flat or uncurved version of space-time became known as *Minkowski space*.

In a 1908 lecture to the 80th Assembly of German Natural Scientists and Physicians, Minkowski represented time as the vertical axis, making do with a single horizontal axis to represent all three dimensions of space. The audience had to use a bit of imagination.

3. Minkowski was the first to realize that a new four-dimensional geometry was the proper framework for Einstein's Special Theory of Relativity. The quote is from "Space and Time," an address delivered at the 80th Assembly of German Natural Scientists and Physicians, on September 21, 1908.

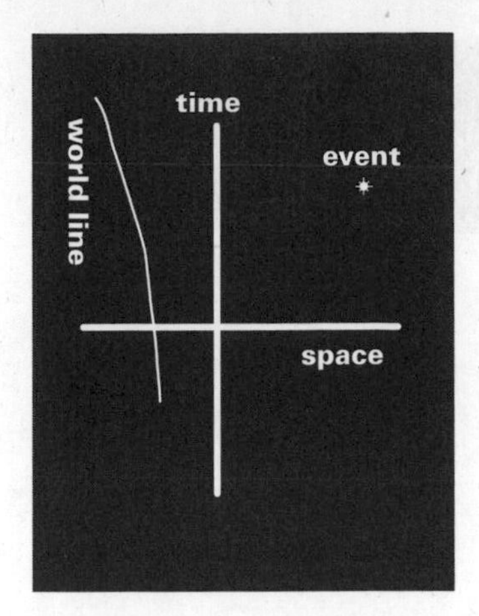

Minkowski called the points of space-time *events*. The common usage of the word *event* implies not only a time and a place but also something happening there. For example: "An event of momentous importance took place at 5:29:45 a.m., July 16, 1945, at Trinity, New Mexico, when the first atomic weapon was tested." Minkowski intended a little less by his use of the word *event.* He meant only a specified time and place, regardless of whether anything actually happened there. What he really meant was *a place and time where an event might or might not happen,* but that was a bit of a mouthful, so he just called it an event.

Lines or curves through space-time play a special role in Minkowski's work. A dot in space represents the position of a particle. But to draw the motion of a particle through space-time, one draws a line or curve sweeping out a trajectory called a *world line*. Some sort of movement is inevitable. Even if the particle stands perfectly still, it nevertheless travels through time. The trajectory of such a stationary particle would be a vertical straight line. The trajectory of a particle moving to the right would be a world line tilting to the right.

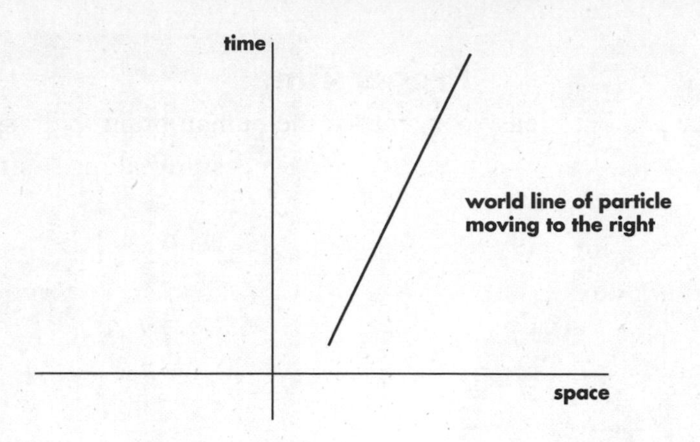

Similarly, tilting the world line to the left would describe a particle moving to the left. The greater the tilt away from the vertical, the faster the particle is moving. Minkowski represented the motion of light rays — the fastest of all objects — as lines drawn at a 45-degree tilt. Since no particle can move faster than light, the trajectory of a real object cannot tilt more than 45 degrees from the vertical.

Minkowski called the world line of a particle moving slower than light *time-like,* because it is close to the vertical. He called the trajectories of light rays inclined at 45 degrees *light-like.*

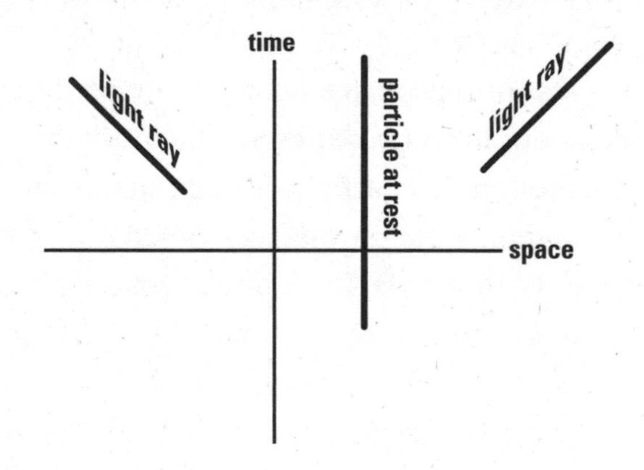

Proper Time

Distance is a fairly easy concept for the human brain to grasp. It's especially easy when distance is being measured along a straight line. To measure it, you need only an ordinary ruler. Measuring distance along a curve is a little harder, but not much. Just replace the ruler with a flexible tape measure. Distances in space-time, however, are more subtle, and it is not immediately clear how to measure them. In fact, no such concept existed until Minkowski invented it.

Minkowski was particularly interested in defining a concept of distance along a world line. For example, take the world line of a particle at rest. Since the trajectory covers no spatial distance, a ruler or tape measure must not be the right tool to measure it. But as Minkowski realized, even a perfectly stationary object moves through time. The right way to measure its world line is not with a ruler but with a clock. He called the new measure of world line distance *proper time*.

Imagine that every object carries a small clock with it, wherever it goes, just as a person might carry a pocket watch. The proper time between two events along a world line is the amount of elapsed time between the two events, as measured by the clock that moved along that world line. The ticks of the clock are analogous to the inch marks along a tape measure, but instead of measuring ordinary distance, they measure Minkowski's proper time.

Here's a concrete example. Mr. Tortoise and Mr. Hare decide to have a race across Central Park. Officials are stationed at each end with carefully synchronized watches so that they can time the winner. The racers start at exactly 12:00 p.m., and halfway across the park, Hare is so far ahead that he decides to take a nap before continuing. But he oversleeps and wakes up just in time to see Tortoise approaching the finish line. Desperate not to lose the race, Hare takes off like a flash and just manages to catch Tortoise as they simultaneously cross the line.

Tortoise pulls out his highly reliable pocket watch and proudly shows the waiting crowd that the proper time, along the segment of his world line from start to finish, is 2 hours and 56 minutes. But why the new term *proper time?* Why didn't Tortoise just say that his time from start to finish was 2 hours and 56 minutes? Isn't time just time?

Newton certainly thought so. He believed that God's master clock defined a universal flow of time and that all clocks could be synchronized to it. Picture Newton's universal time by imagining that space is filled with small clocks that have all been synchronized. The clocks are all good, honest clocks that run at exactly the same rate, so that once they are synchronized, they stay synchronized. Wherever Tortoise or Hare happens to be, he can check the time by looking at the clock in his immediate neighborhood. Or he can look at his own pocket watch. For Newton it was axiomatic that no matter where you went, at any speed, along a straight line or a curved trajectory, your pocket watch — assuming that it was also a good, honest clock — would agree with the local clock in your neighborhood. Newtonian time has an absolute reality; there is nothing relative about it.

But in 1905 Einstein made hash of Newton's absolute time. According to the Special Theory of Relativity, the rates at which clocks tick depend on how they move, even if they are perfect replicas of one another. In ordinary situations, the effect is imperceptible, but when clocks move with velocities near the speed of light, it becomes very noticeable. According to Einstein, every clock moving along its own world line ticks at its own rate. Thus, Minkowski was led to define the new concept of proper time.

Just to illustrate the point, when Hare pulls out his watch (also a good, honest clock), the proper time of his world line shows 1 hour and 36 minutes.[4] Although they began and ended at the same space-time points, the world lines of Tortoise and Hare have quite different proper times.

4. This is an extreme exaggeration, which would have required Hare to move with close to the speed of light.

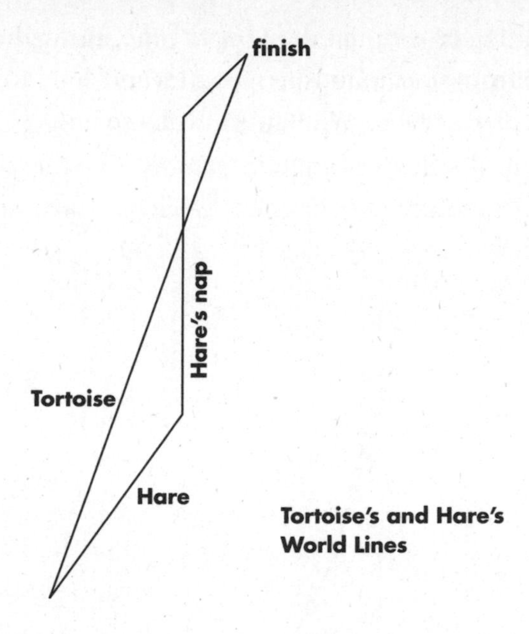

Before discussing proper time any further, it's instructive to think a little more about ordinary distance measured along a curve by a tape measure. Take any two points in space and draw a curve between them. How far apart, along the curve, are the points? The answer obviously depends on the curve. Here are two curves, connecting the same two points (a and b) with quite different lengths. Along the upper curve, the distance between a and b is five inches; along the lower curve, it is eight inches.

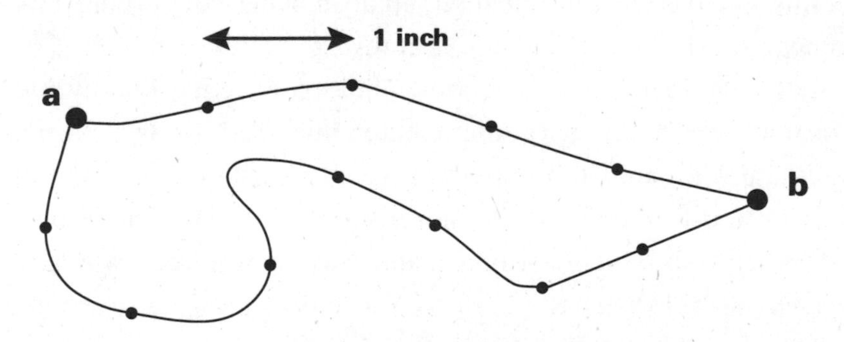

There is, of course, absolutely nothing surprising in the fact that different curves between a and b have different lengths.

Now come back to the problem of measuring world lines in space-time. Here is a picture of a typical world line. Notice that the world line curves. That means that the velocity along the trajectory is not uniform. In this example, a rapidly moving particle slows down. The dots indicate the ticks of the clock. Each interval represents one second.

Notice that seconds appear to tick more slowly when the angle is closer to the horizontal. That's not a mistake; it represents Einstein's famous discovery of *time dilation:* rapidly moving clocks run slowly compared to slow clocks or clocks at rest.

Let's consider two curved world lines connecting two events. Einstein, ever the thought experimenter, imagined two twins — I'll call them Alice and Bob — born at the same instant. The event of their birth is labeled a. At the moment of birth, the twins become separated; Bob remains at home, and Alice is whisked away at a tremendous velocity. After a period, Einstein has Alice turn around and head back home. Eventually, Bob and Alice meet again at b.

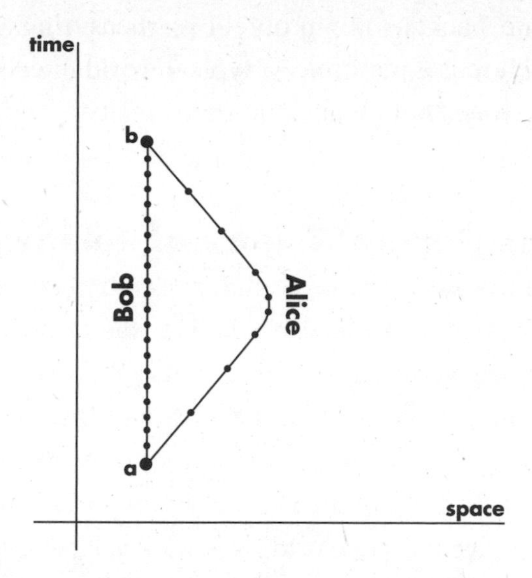

At birth Einstein gave the twins identical pocket watches, which were perfectly synchronized. When Bob and Alice finally meet at b, they compare their watches and discover something that would have surprised Newton. First of all, Bob has a long gray beard, but Alice is the image of youth. According to their pocket watches, the proper time along Alice's world line is a good deal less than that along Bob's. Just as the ordinary distance between two points depends on the curve that connects them, the proper time between two events depends on the world line that connects them.

Did Alice notice that her clock ran slow during her journey? Not at all. Her watch was not the only thing that ran slow; so did her heartbeat, her brain function, and her entire metabolism. During the trip, Alice had nothing to compare her clock to, but when she finally met Bob for the second time, she discovered that she was noticeably younger than he was. This "twin paradox" has puzzled physics students for more than one hundred years.

There is one curiosity that you may have discovered by yourself. Bob travels through space-time on a straight line, while Alice travels on a curved trajectory. Yet the proper time along Alice's trajectory is smaller than the proper time along Bob's. That's an example of a counterintuitive fact about the geometry of Minkowski space: a straight world line has the *longest* proper time between two events. Put that in your rewiring kit.

The General Theory of Relativity

Like Riemann, Einstein believed that geometry (not just space, but space-time) was curved and variable. He was referring not just to space but to the geometry of space-time. Following Minkowski, Einstein let one axis stand for time and the other for all three dimensions of space, but instead of picturing space-time as a flat plane, he imagined it as a warped surface, bent with bulges and bumps. Particles still moved along world lines, and clocks ticked off proper time, but the geometry of space-time was much more irregular.

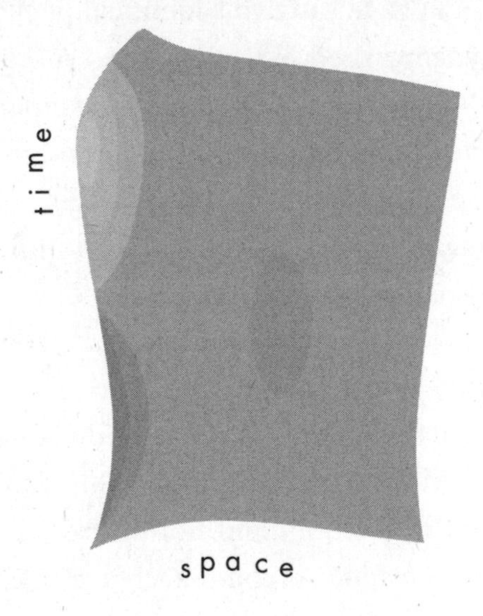

Einstein's Laws

Surprisingly, in many ways the laws of physics are simpler in curved space-time than they are in Newtonian physics. Take, for example, the motion of particles. Newton's laws begin with the principle of inertia:

In the absence of forces, every object will remain in a state of uniform motion.

This simple-sounding rule, with its phrase "uniform motion," hides two separate ideas. First, uniform motion means motion along a straight line in space. But Newton meant something stronger: uniform motion also implies constant, unchanging velocity – that is, no acceleration.[5]

But what about gravitational forces? Newton added a second law – a law of non-uniform motion – that says force equals mass times acceleration, or to put it differently:

The acceleration of an object is the force exerted on that object divided by its mass.

A third rule applies when the force is due to gravity:

The gravitational force on any object is proportional to its mass.

Minkowski simplified Newton's notion of uniform motion with a clever insight that summarized both conditions:

In the absence of forces, every object moves through space-time along a straight world line.

Straightness of a world line implies not only straightness in space but also constant velocity.

Minkowski's straight world line hypothesis was a beautiful synthesis of the two aspects of uniform motion, but it applied only in the complete absence of forces. Einstein took Minkowski's idea to another level when he applied it to curved space-time.

Einstein's new law of motion was stunningly simple. At every point along its world line, a particle does the simplest possible thing: it goes straight ahead (in space-time). If space-time is flat, Einstein's

5. The term *acceleration* refers to any change in velocity, including the slowing down that we ordinarily call deceleration. To a physicist, deceleration is merely negative acceleration.

law is the same as Minkowski's, but if space-time is curved — in regions where massive bodies deform and warp space-time — the new law instructs particles to move along space-time geodesics.

As Minkowski had explained, a curved world line indicates that a force is acting on an object. According to Einstein's new law, particles in curved space-time move as straight as they can, but geodesics inevitably curve and bend to match the local space-time terrain. Einstein's mathematical equations demonstrate that a geodesic in curved space-time behaves exactly like the curved world line of a particle moving through a gravitational field. Gravitational force is thus nothing but the curving of geodesics in curved space-time.

In one almost laughably simple law, Einstein combined Newton's laws of motion with Minkowski's world line hypothesis and explained how gravity acts on all objects. What Newton had taken as an unexplained fact of nature — gravitational forces — Einstein explained as the effect of non-Euclidean space-time geometry.

The principle that particles move along geodesics provided a powerful new way to think about gravity, but it said nothing about the *cause* of curvature. To complete his theory, Einstein had to explain what controls the warped bulges and other irregularities of space-time. In the old Newtonian theory, the source of the gravitational field was mass: the presence of a mass such as the Sun creates a gravitational field around it, which in turn influences the motion of the planets. It was therefore natural for Einstein to conjecture that the presence of mass — or, equivalently, of energy — causes space-time to become warped or curved. John Wheeler, one of the great pioneers and teachers of modern relativity theory, summed it up in one concise slogan: "Space tells bodies how to move and bodies tell space how to curve." (He meant space-time.)

Einstein's new idea means that space-time is not passive; it has properties such as curvature, which respond to the presence of masses. It is almost as though space-time is an elastic material or even a fluid that can be affected by the objects moving through it.

The connection between massive objects, gravity, curvature, and the motion of particles is sometimes described by an analogy that I have mixed feelings about. The idea is to think of space as a horizontal rubber sheet, something like a trampoline. When there are no masses to deform it, the sheet remains flat. But place a heavy mass such as a bowling ball on the sheet, and the weight of the bowling ball will deform it. Now add a much smaller mass — a marble will do nicely — and watch as the marble falls toward the heavier bowling ball. The marble can also be given some tangential velocity, so that it orbits the larger mass, much as the Earth orbits the Sun. The depression in the surface keeps the smaller mass from flying off, just as the Sun's gravity tethers the Earth.

There are some misleading things about this analogy. First of all, the curvature of the rubber sheet is spatial curvature, not space-time curvature. It fails to explain the peculiar effects that masses have on nearby clocks (we will see those effects later in the chapter). Even worse, the model uses gravity to explain gravity. It's the pull of the real Earth on the bowling ball that causes the depression in the rubber surface. In any technical sense, the rubber sheet model is all wrong.

Nevertheless, the analogy does capture some of the spirit of General Relativity. Space-time is deformable, and heavy masses do deform it. The motion of small objects is affected by the curvature created by heavy objects. And the depressed rubber sheet looks quite a lot like the mathematical embedding diagrams that I will soon explain. Use the analogy if it helps, but keep in mind that it is only an analogy.

Black Holes

Take an apple and slice it through the center. The apple is three-dimensional, but the newly exposed cross section is two-dimensional. If you stack up all the two-dimensional cross sections obtained by thinly slicing the apple, you can reconstruct the apple. One might say that each thin slice is *embedded* in the higher-dimensional stack of slices.

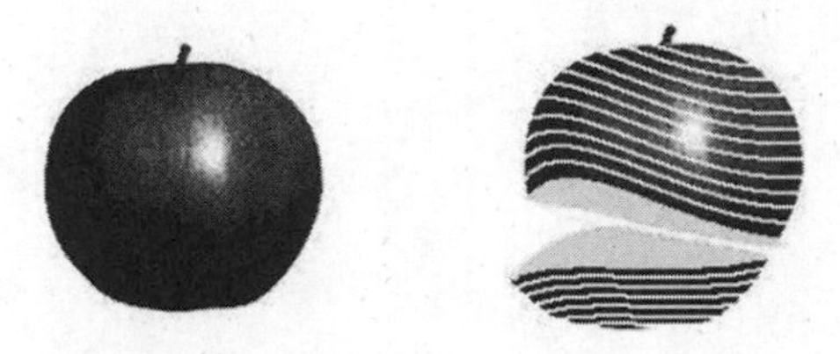

Space-time is four-dimensional, but by slicing it, we can expose three-dimensional slices of space. It can be visualized as a stack of thin slices, each one representing three-dimensional space at a single instant of time. Visualizing three dimensions is a lot easier than visualizing four. The pictures of the slices are called *embedding diagrams,* and they help give an intuitive picture of curved geometry.

Let's take the case of the geometry created by the mass of the Sun. Forget time for the moment and concentrate on visualizing curved space in the Sun's vicinity. The embedding diagram looks like a mild depression in a rubber sheet, centered at the Sun, more or less similar to the trampoline with the bowling ball resting on it.

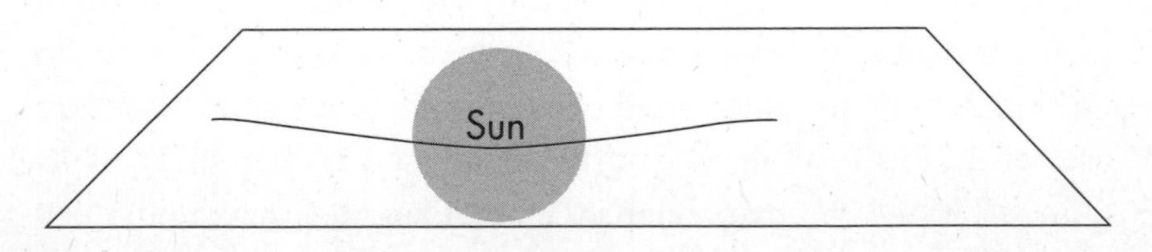

The distortion near the Sun would become more pronounced if the same mass were concentrated in a smaller volume.

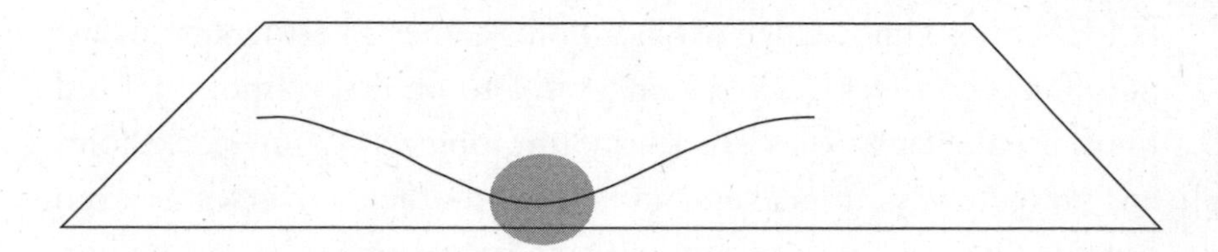

The geometry near a white dwarf or a neutron star is even more curved, though still smooth.

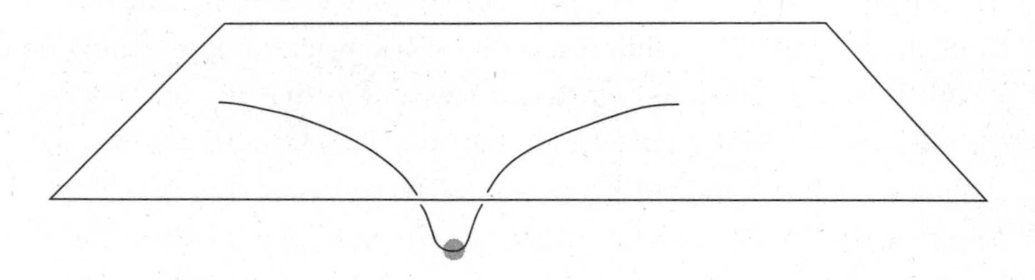

As we learned earlier, if the collapsing star grows small enough to be contained within its Schwarzschild radius (two miles for the Sun), then just like the pollywogs trapped in the drain, the particles of the Sun will be irresistibly drawn in, collapsing until they form a singularity — a point of infinite curvature.[6]

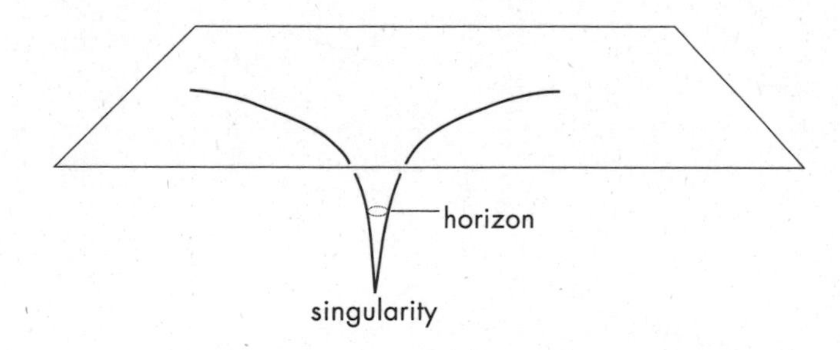

6. Note to the experts: The embedding diagram that follows is not in constant Schwarzschild time. It is obtained by using Kruskal coordinates and choosing the surface T = 1.

What Black Holes Are Not

I'm expecting that this section will generate some irate mail from readers whose knowledge of black holes comes solely from the Disney film *The Black Hole.* I don't want to be a spoilsport – Lord knows that black holes are fascinating objects – but black holes are not gateways to heaven and hell or to other universes, or even tunnels that lead back to our own universe. Since all is fair in love, war, and science fiction, I don't really mind that the moviemakers took a trip to la-la land. But understanding black holes requires more than careful study of B movies.

The premise of *The Black Hole* actually originated with the work of Einstein and his collaborator Nathan Rosen, later popularized by John Wheeler. Einstein and Rosen speculated that the interior of a black hole might connect to a very distant place, through what Wheeler would later call a *wormhole.* The idea was that two black holes, perhaps billions of light-years apart, could be joined at their horizons, forming a fantastic shortcut across the universe. Instead of the embedding diagram of the black hole ending at a sharp singularity, once it passed the horizon, it would open out into a large new region of space-time.

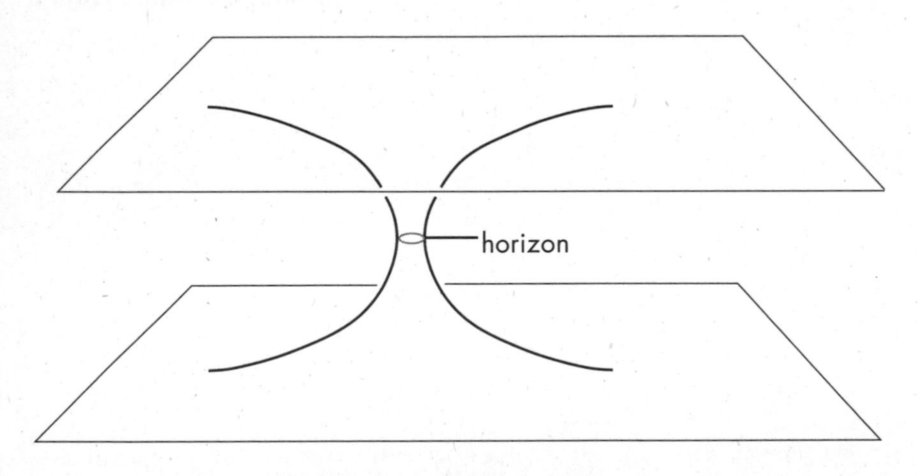

Einstein-Rosen Bridge

Entering one end and leaving the other would be like going into a tunnel in New York and emerging, after no more than a couple of miles, in Beijing, or even on Mars. Wheeler's wormhole was based on genuine mathematical solutions of the General Theory of Relativity.

That's the origin of the urban myth that black holes are tunnels to other worlds. But there are two things wrong with this fantasy. The first is that Wheeler's wormhole can stay open for only a short period of time, and then it pinches off. The wormhole opens and closes so quickly that it is impossible for anything to pass through, including light. It is as though the short tunnel to Beijing would collapse before anyone could traverse it. Some physicists have speculated that Quantum Mechanics might somehow stabilize the wormhole, but there is no evidence for this.

Even more to the point, Einstein and Rosen were studying an "eternal black hole" — one that exists not only into the infinite future but also into the infinite past. But even the universe is not infinitely old. Real black holes almost certainly have their origins in the collapse of stars (or other massive objects) that took place long after the Big Bang. When Einstein's equations are applied to the formation of black holes, the black holes simply do not have wormholes attached to them. The embedding diagram looks like the one on the bottom of page 69.

Now that I've spoiled your day, I suggest that you rent the movie and have some fun.

How to Build a Time Machine

The future ain't what it used to be.

—YOGI BERRA

How about time machines, another stock science fiction gimmick and the subject of numerous books, television shows, and movies? Personally, I'd love to have one. I'm very curious about what the future will be like. Will people survive a million years from now? Will they colonize space? Will sex survive as the preferred method of procreation? I'd love to know, and so, I suspect, would you.

Be careful what you wish for. There would be some downsides to traveling to the future. All your friends and family would be long dead. Your clothes would look ridiculous. Your language would be useless. In short, you would be a freak. A one-way trip to the future sounds depressing, if not tragic.

No problem. Just climb into your time machine and set the dial back to the present. But what if your time machine's transmission had no reverse gear? What if you could only go forward? Would you do it anyway? You might think that's an idle question; everyone knows that time machines are science fiction. But that's actually not true.

One-way time machines to the future are quite possible, at least in principle. In the Woody Allen movie *Sleeper,* the protagonist is transported two hundred years into the future by a technique that is almost feasible today. He simply has himself frozen into a state of suspended animation, something that has already been done with dogs and pigs for a few hours. When he wakes from his frozen state, he is in the future.

Of course, that technique is not really a time machine. It can slow a person's metabolism, but it won't slow the motion of atoms and other physical processes. But we can do better. Remember the twins, Bob and Alice, who were separated at birth? When Alice returns from her trip into space, she finds that the rest of the world has aged far more than she has. Thus, a round-trip on a very fast spaceship is an example of time travel.

A large black hole would be another very handy time machine. Here's how it would work. First of all, you would need an orbiting space station and a long cable to lower yourself down to the vicinity of the horizon. You wouldn't want to get too close, and you certainly wouldn't want to fall through the horizon, so the cable must be very strong. A winch on the space station would lower you down and, after a specified time, reel you back up.

Let's say you wanted to go one thousand years into the future, and you were willing to spend one year suspended on the cable without too much discomfort from the gravitational acceleration. It could be done, but you would need to find a black hole with a horizon about as big as our galaxy. If you didn't mind being uncomfortable, you could do it with the much smaller black hole at the center of our galaxy. The downside would be that you would feel as if you weighed ten billion pounds during that year near the horizon. After a year on the cable, you would be reeled in to find that a millennium had passed. In principle, at least, black holes really are time machines to the future.

But what about getting back? For that you would need a time machine to the past. Alas, going backward in time is probably impossible. Physicists sometimes speculate about time travel to the past by passing through quantum wormholes, but going back in time always leads to logical contradictions. My guess is that you would be stuck in the future, and there would be nothing you could do about it.

Gravitational Slowing of Clocks

What is it about black holes that makes them time machines? The answer is the strong distortion of space-time geometry that they produce. This distortion affects the flow of proper time along world lines in different ways, depending on where the world lines are located. Far from a black hole, its influence is very weak, and the flow of proper time is almost unaffected by its presence. But a clock suspended on a cable just above the horizon would be slowed down a great deal by the distortion of space-time. In fact, all clocks, including your own heartbeat, your metabolism, and even the internal motion of atoms, would be slowed down. You wouldn't notice this in the least, but when you returned to the space station and compared your watch with clocks that remained on board, you would notice the discrepancy. More time would have elapsed on the space station than on your watch.

Indeed, it would not even be necessary to return to the space station to see the effects of the black hole on time. If you, suspended near the horizon, and I, in the space station, had telescopes, we could watch each other. I would see you and your clock in slow motion, while you would see me speeded up like an old Keystone Kops movie. This relative slowing of time near a heavy mass is called *gravitational red shift.* Discovered by Einstein as a consequence of his General Theory of Relativity, it doesn't happen in Newton's theory of gravity, in which clocks all tick at exactly the same rate.

The following space-time picture illustrates gravitational red shift near the horizon of a black hole. The object on the left is the black hole. Remember, the picture represents space-time, with the vertical axis being time. The gray surface is the horizon, and the vertical lines at various distances from the horizon represent a group of stationary identical clocks. The tick marks represent the flow of proper time along the world lines. The units are not important; they could represent seconds, nanoseconds, or years. The closer the clock is to the black hole horizon, the slower it seems to tick. Right at the horizon, time comes to a complete standstill for clocks that remain outside the black hole.

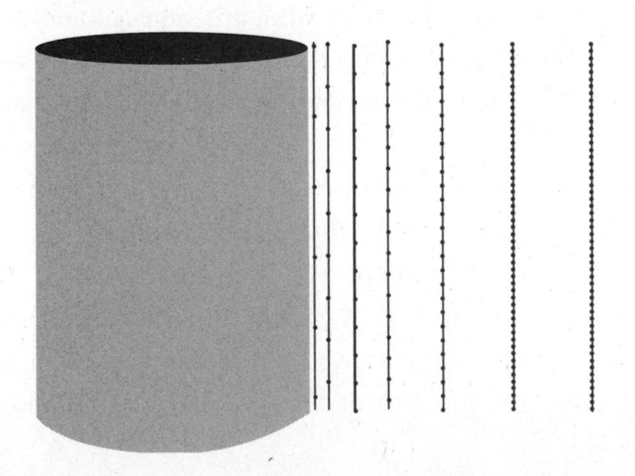

The gravitational slowing of clocks happens in less exotic circumstances than near black hole horizons. A mild version of it happens on the surface of the Sun. Atoms are miniature clocks — the electrons whizzing around the nucleus like the hands of a clock. When observed from the Earth, atoms on the Sun appear to run a bit slow.

The loss of simultaneity, the twin paradox, curved space-time, black holes, time machines — so many far-out, stranger-than-fiction ideas, and these are the reliable ones, the noncontroversial concepts that physicists all agree on. It took some painful rewiring — differential geometry, tensor calculus, space-time metrics, differential forms — to understand the new physics of space-time. Yet even the difficult transition to the Alice-in-Wonderland quantum realm was nothing compared to the conceptual difficulties that baffle us now as we try to reconcile General Relativity with Quantum Mechanics. At times in the past, it looked as if Quantum Mechanics could not coexist with Einstein's theory of gravity and would have to be abandoned. But perhaps one could say that the Black Hole War was "The War That Made the World Safe for Quantum Mechanics."

In the next chapter, I will attempt the impossibly quixotic task of rewiring you for Quantum Mechanics — more or less without equations. The real tools for groking the quantum universe are abstract mathematics: infinite dimensional Hilbert spaces, projection operators, unitary matrices, and a lot of other advanced principles that take a few years to learn. But let's see how we do in just a few pages.

4

"EINSTEIN, DON'T TELL GOD WHAT TO DO"

Putting down her cup of tea, she asked in a timid voice, "Is light made of waves, or is it made of particles?"

There was a table set out under a tree in front of the house, and the March Hare and the Hatter were having tea at it: a Dormouse was sitting between them, fast asleep, and the other two were using it as a cushion, resting their elbows on it, and talking over its head. "Very uncomfortable for the Dormouse," thought Alice; "only, as it's asleep, I suppose it doesn't mind."[1]

Ever since her last science class, Alice had been deeply puzzled by something, and she hoped one of her new acquain-

1. Lewis Carroll, *Alice's Adventures in Wonderland,* illustrations by John Tenniel (London: Macmillan and Company, 1865).

tances might straighten out the confusion. Putting down her cup of tea, she asked in a timid voice, "Is light made of waves, or is it made of particles?" "Yes, exactly so," replied the Mad Hatter. Somewhat irritated, Alice asked in a more forceful voice, "What kind of answer is that? I will repeat my question: Is light particles or is it waves?" "That's right," said the Mad Hatter.

Welcome to the fun house — the mad, insane, topsy-turvy world of Quantum Mechanics, where uncertainty rules and nothing makes sense to the sensible.

Answering Alice — Sort Of

Newton believed that a ray of light was a stream of tiny particles, more or less like little bullets shot from a rapid-fire machine gun. Although the theory was almost completely wrong, he invented remarkably clever explanations for many of the properties of light. By 1865, the Scottish mathematician and physicist James Clerk Maxwell had thoroughly discredited Newton's bullet theory. Maxwell argued that light consists of waves — *electromagnetic waves*. Maxwell's constructions were overwhelmingly confirmed and soon became the accepted theory.

Maxwell pointed out that when electric charge moves — for example, when electrons vibrate in a wire — the moving charge creates wavelike disturbances, in much the same way that wiggling a finger in a pool of water creates waves on the surface.

Light waves are composed of electric and magnetic fields — the same fields that surround electrically charged particles, electric currents in wires, and ordinary magnets. When those charges and currents vibrate, they shake off waves, which spread out through empty space with the speed of light. Indeed, if you project a light beam through two tiny slits, you can see a distinct *interference* pattern formed by the overlapping waves.

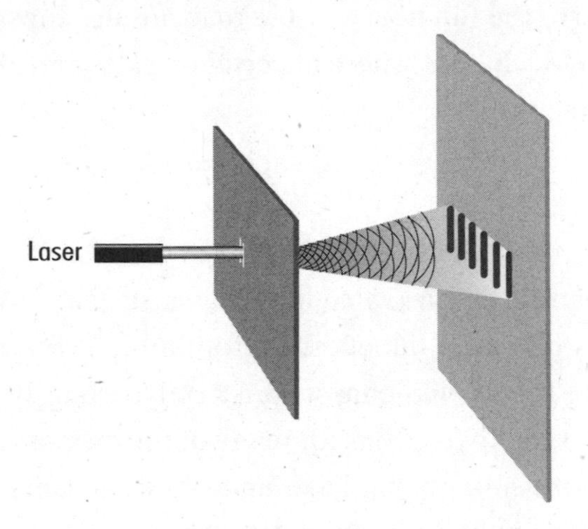

Maxwell's theory even explained how light could come in different colors. Waves are characterized by their wavelength — the distance from one crest to the next. Here are two waves, the first of longer wavelength than the second.

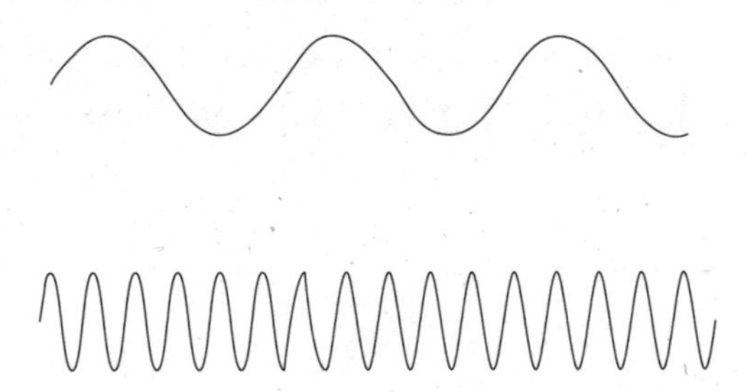

Imagine the two waves moving right past your nose with the speed of light. As they pass, the waves oscillate from maximum to minimum and back again: the shorter the wavelength, the more rapid the oscillation. The number of full cycles (from maximum to minimum to maximum) per second is called the *frequency,* and it is obviously larger for the shorter wavelength.

When light enters the eye, different frequencies affect the rods and cones in the retina in distinctive ways. A signal is transmitted to the brain that says red, orange, yellow, green, blue, or violet, depending on the frequency (or wavelength). The red end of the spectrum consists of longer waves (lower frequency) than the blue or violet end: the wavelength of red light is about 700 nanometers,[2] while the wavelength of violet light is only half that. Because light moves so fast, the frequency of oscillation is enormous. Blue light oscillates a quadrillion (10^{15}) times per second: red light oscillates about half as fast. In physics jargon, the frequency of blue light is 10^{15} Hz.

Can the wavelength of light be longer than 700 or shorter than 400 nanometers? Yes, but then it's not called light; the eye is insensitive to such wavelengths. Ultraviolet rays and X-rays are shorter than violet waves, and the shortest of all rays are called gamma rays. On the longer wavelength side, we have infrared rays, microwaves, and radio waves. The entire spectrum, from gamma rays to radio waves, is known as *electromagnetic radiation*.

So, Alice, the answer to your question is that light is definitely composed of waves.

But wait, not so fast. Between 1900 and 1905, a very disturbing surprise upset the foundations of physics and left the subject in a state of complete confusion for more than twenty years. (Some would say it is still in confusion.) Building on the work of Max Planck, Einstein completely "subverted the dominant paradigm." We don't have time or space for the history of how he got there, but by 1905 Einstein was convinced that light was composed of particles that he called *quanta*. Later they would be known as *photons*. To shorten a fascinating story to its bare bones, light, when it is ex-

2. A nanometer is one-billionth of a meter, or 10^{-9} meters.

tremely dim, behaves like particles, arriving one at a time as if they were intermittent bullets. Go back to the experiment in which light passes through two slits, eventually arriving at a screen. Imagine dimming the light source to a bare trickle. A wave theorist would expect a very dim, wavelike pattern, just barely visible, or perhaps not visible at all. But visible or not, the expected pattern would be wavelike.

That's not what Einstein predicted, and as usual he was right. Instead of continuous illumination, his theory predicted sudden point-flashes of light. The first flash would appear randomly at some unpredictable point on the screen.

Another flash would appear randomly somewhere else, then another. If the flashes were photographed and superimposed, a pattern would begin to emerge out of the random flashes — a wavelike pattern.

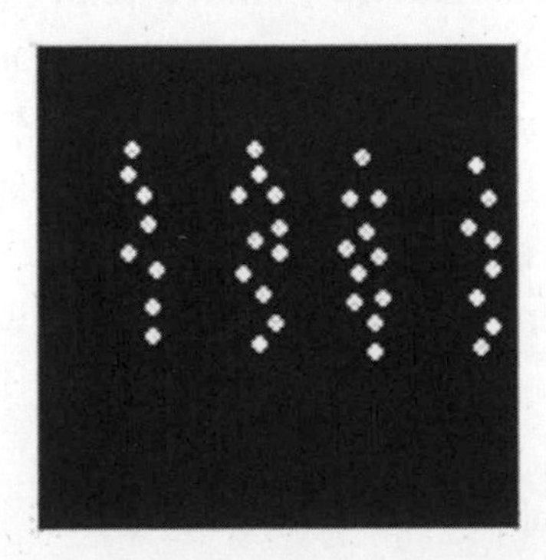

So is light a particle, or is it a wave? The answer depends on the experiment and the question that you ask. If the experiment involves light so dim that the photons trickle through one at a time, light appears to be unpredictable, randomly arriving photons. But if there are enough photons so that they form a pattern, light behaves like waves. The great physicist Niels Bohr described this confusing situation by saying that the wave and particle theories of light are *complementary*.

Einstein argued that photons must have energy. There was certainly evidence of this. Sunlight — photons emitted by the Sun — warms the Earth. Solar panels convert the Sun's photons to electricity. The electricity can be used to run a motor and raise a heavy weight. If light has energy, so too must the photons that make it up.

It's clear that a single photon has only a very small amount of energy, but exactly how much? How many photons does it take to boil a cup of tea or run a 100-watt motor for an hour? The answer depends on the wavelength of the radiation. Longer-wavelength photons are less energetic than shorter-wavelength ones. Thus, more longer-wavelength photons are needed to do a given job. A very famous formula — not as famous as $E = mc^2$, but still very famous — gives the energy of a single photon in terms of its frequency.[3]

$$E = hf$$

The left side of the equation, E, represents the energy of the photon, measured in a unit called the joule. On the right side, f is the frequency. For blue light, the frequency is 10^{15} Hz. That leaves h, the famous Planck's constant that Max Planck introduced in 1900. Planck's constant is a very small number, but it is one of the most important constants of nature, controlling all quantum phenomena.

3. This formula was introduced by Max Planck in 1900. However, it was Einstein who understood that light is made of particle-like quanta and that the formula applies to the energy of a single photon.

It ranks alongside the speed of light, c, and Newton's gravitational constant, G.

$$h = 6.62 \times 10^{-34}$$

Because Planck's constant is so small, the energy of a single photon is minute. To calculate the energy of a photon of blue light, multiply Planck's constant by the frequency, 10^{15} Hz, and you get 6.62×10^{-19} joules. That doesn't sound like much energy, and it's not. It would take about 10^{39} photons of blue light to boil your tea. You would need about twice that number of photons of red light. By contrast, with the highest-energy gamma rays that have ever been detected, boiling the same cup of tea would take a mere 10^{18} photons.

Out of all these formulas and numbers, I want you to remember only one thing: the shorter the wavelength of a light ray, the higher the energy of an individual photon. High energy means short wavelength; low energy means long wavelength. Say it over a few times and write it down. Now say it again: high energy means short wavelength; low energy means long wavelength.

Predicting the Future?

Einstein pompously declared, "God does not play dice."[4] Niels Bohr's response was sharp: "Einstein," Bohr scolded, "don't tell God what to do." Both physicists were pretty close to being atheists; it would seem unlikely that either of them contemplated a deity sitting on a cloud trying to roll a seven. But both Bohr and Einstein were struggling with something totally new in physics — something that Einstein simply could not accept: the unpredictability that the strange new rules of Quantum Mechanics implied. Einstein's intellect rebelled at the idea of a random, uncontrollable element in the

4. Letter to Max Born, December 12, 1926.

laws of nature. The idea that the arrival of a photon was truly an unpredictable event went deeply against his grain. Bohr, by contrast, may or may not have liked the idea, but he accepted it. He also understood that future physicists would have to rewire themselves for Quantum Mechanics, and part of that rewiring would include the unpredictability that Einstein dreaded.

It wasn't that Bohr was better at visualizing quantum phenomena or that he was more comfortable with it. "Anyone who is not shocked by quantum theory has not understood it," he once declared. Many years later, Richard Feynman opined, "I think it's safe to say that no one understands Quantum Mechanics." He added, "The more you see how strangely Nature behaves, the harder it is to make a model that explains how even the simplest phenomena actually work. So theoretical physics has given up on that." I don't think that Feynman really meant that physicists should give up on explaining quantum phenomena; after all, he was constantly explaining them. What he meant was that one can't explain quantum phenomena in terms that the human mind can visualize with the standard-issue wiring. Feynman, no less than anyone else, had to resort to abstract mathematics. Obviously, reading one chapter in a book with no equations cannot rewire you, but with some patience, I think, you can grasp the main points.

The first thing that physicists had to free themselves of — the thing that Einstein held so dear — was the notion that the laws of nature are deterministic. Determinism means that the future can be predicted if enough is known about the present. Newtonian mechanics, as well as everything that followed, was all about predicting the future. Pierre de Laplace — the same Laplace who imagined dark stars — firmly believed that the future could be predicted. Here is what he wrote.

> We may regard the present state of the universe as the effect of its past and the cause of its future. An intellect which at a certain moment would know all forces that set nature in

> motion, and all positions of all items of which nature is composed, if this intellect were also vast enough to submit these data to analysis, it would embrace in a single formula the movements of the greatest bodies of the universe and those of the tiniest atom; for such an intellect nothing would be uncertain and the future just like the past would be present before its eyes.

Laplace was simply laying out the implications of Newton's laws of motion. Indeed, the Newton-Laplace view of nature is the purest form of *determinism*. To predict the future, all you needed to know was the position and the velocity of every particle in the universe at some initial instant of time. Oh, yes, and one more thing: you would need to know the forces acting on every particle. Notice that it is not enough to know the position at an instant. Knowing the location of a particle tells you nothing about where it's going. But if you also know the velocity[5] — both its magnitude and its direction — you can tell where it will be next. Physicists speak of *initial conditions,* meaning everything you need to know at one instant in order to predict the future motion of a system.

To understand what determinism means, let's imagine the simplest possible world — a world so simple that it has only two states of being. A coin is a good model, the two states being heads and tails. We also need to specify a law that dictates how things change from one instant to the next. Here are two possibilities for such a law.

- This first example is very tedious. The rule is — nothing happens. If the coin shows heads at one instant, it shows heads at the next instant (say, a nanosecond later). Likewise, if it shows tails at one instant, it shows tails at the

5. The term *velocity* means not only how fast an object is moving but also the direction of motion. Thus, 60 miles per hour is not complete information about velocity; 60 miles per hour in the direction NNW is.

next instant. The law can be condensed to a simple pair of "formulas":

$$H \rightarrow H \quad T \rightarrow T$$

The history of the world is either H H H H H . . . or T T T T T . . . endlessly repeated.

- If the first rule is boring, the next one is only slightly less so: whatever the state at one instant, a nanosecond later it flips to the opposite state. Symbolically, it can be expressed this way:

$$H \rightarrow T \quad T \rightarrow H$$

History would take the form H T H T H T H T . . . or T H T H T H T H. . . .

Both rules are deterministic, meaning that the future is completely determined by the starting point. In either case, if you know the initial condition, you can predict with certainty what will happen after any length of time.

Deterministic laws are not the only possibility. Random laws also are possible. The simplest random law would be that whatever the initial state happens to be, at the next instant it is randomly heads or tails. A possible history, beginning with tails, would be T T T H H H T T H H T H H T T. . . . But T T H T H H T H H H T T . . . also would be possible. In fact, any sequence would be possible. You can think of either a world without a law or a world in which the law is random updating of the initial condition.

The law needn't be purely deterministic or purely random. These are extremes. A mostly deterministic law, with just a touch of randomness, is possible. The law might say that with probability nine-tenths the state stays unchanged, and with probability one-tenth it flips. A typical history would look like this:

H H H H H H H T T T T T T T T T T T T H H H H H H H H H H H H H T T T T T T . . .

In this case, a gambler could make a pretty good guess about the immediate future: the next state will most likely be the same as the present state. He might even be a bit bolder and guess that the next two states will be the same as the present. His chances of being right would be good, provided he didn't push it too long. If he tried to guess too far into the future, his odds of being right would not be much better than even. This unpredictability is exactly what Einstein was railing against when he said God doesn't play dice.

You may be a bit puzzled by one point: a sequence of real coin tosses is a lot more like the totally random law than either of the deterministic laws. Randomness seems like a very common feature of the natural world. Who needs Quantum Mechanics to make the world unpredictable? But the reason that an ordinary coin toss is unpredictable — even in the absence of Quantum Mechanics — is just plain sloppiness. Keeping track of every relevant detail is usually too difficult. A coin is not really an isolated world. The details of the muscles that move the hand that flips the coin; the air currents in the room; the thermal vibrations of the molecules in both the coin and the air — all are relevant to the outcome, and in most cases all this information is far too much to deal with. Remember, Laplace spoke of knowing "*all* forces that set nature in motion, and *all* positions of *all* items of which nature is composed." Just the tiniest mistake in the position of a single molecule could ruin the ability to predict the future. But this ordinary sort of randomness was not what bothered Einstein. By God playing dice, Einstein meant that the deepest laws of nature have an unavoidable element of randomness that can never be overcome, even if every detail that can be known, is known.

Information Never Dies

One compelling reason for not allowing randomness is that in most cases it would violate the *conservation of energy* (see chapter 7). This law states that although energy comes in many forms and it can change from one form to another, the total amount never changes. The conservation of energy is one of the most accurately confirmed facts of nature, and there is not much room to tamper with it. Random kicks would change the energy of an object by suddenly speeding it up or slowing it down.

There is another very subtle law of physics that may be even more fundamental than energy conservation. It's sometimes called reversibility, but let's just call it *information conservation*. Information conservation implies that if you know the present with perfect precision, you can predict the future for all time. But that's only half of it. It also says that if you know the present, you can be absolutely sure of the past. It goes in both directions.

In the heads-tails world of a single coin, a purely deterministic law would ensure that information is perfectly conserved. For example, if the law is

$$H \rightarrow T \quad T \rightarrow H$$

both the past and the future can be perfectly predicted. But even the smallest amount of randomness would ruin this perfect predictability.

Let me give another example, this time with a fictitious three-sided coin (a die is a six-sided coin). Call the three sides heads, tails, and feet or H, T, and F. Here is a perfectly deterministic law.

$$H \rightarrow T \quad T \rightarrow F \quad F \rightarrow H$$

To visualize the law, it helps to draw a diagram.

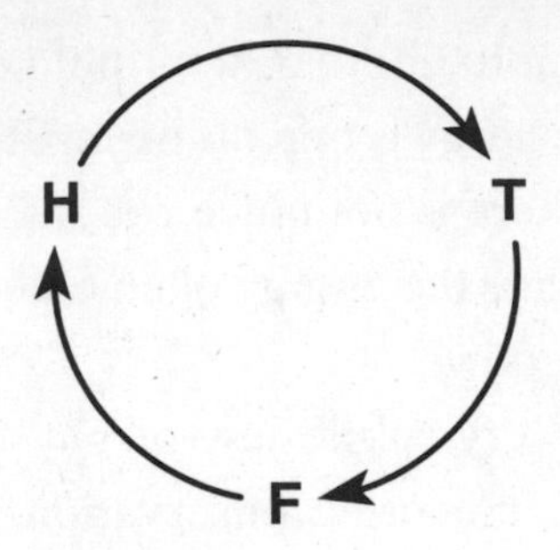

With this law, the history of the world, starting with H, would look like this:

HTFHTFHTFHTFHTFHTFHTFHTFHTF...

Is there a way to experimentally test information conservation? In fact, there are many ways, some feasible and some not. If you were able to control the law and change it at will, there would be a very simple way to test it. In the case of the three-sided coin, here is how it would work. Start the coin in one of its three states and let it run for some definite length of time. Assume that every nanosecond, the state flips from H to T to F, cycling among the three possibilities. At the end of the time interval, change the law. The new law is just the old law run in reverse — counterclockwise instead of clockwise.

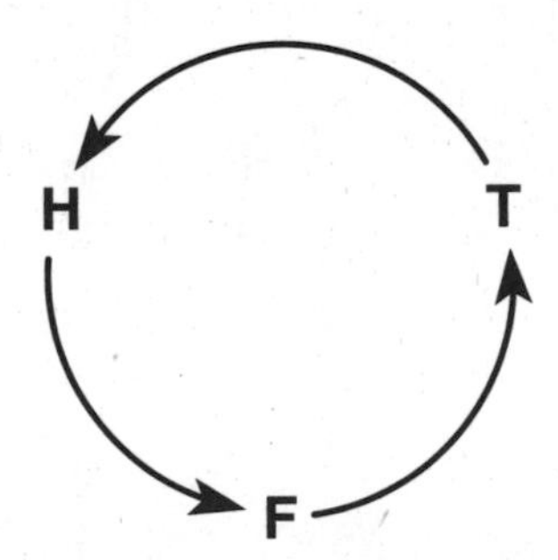

Now run the system in reverse for exactly the same length of time that you ran it forward. The original history will undo itself, and the coin will return to the starting point. No matter how long you wait, the deterministic law will retain perfect memory and always return to the initial conditions. To check information conservation, you would not even need to know the precise law, as long as you knew how to reverse it. The experiment will always work as long as the law is deterministic. But it will fail if there is any randomness — unless the randomness is of some very subtle kind.

Now let's come back to Einstein, Bohr, God (read: the laws of physics), and Quantum Mechanics. Another of Einstein's more famous quotes is "The Lord God is subtle, but malicious he is not." I don't know what made him think that the laws of physics are not malicious. Personally, I occasionally find the law of gravity quite malicious, especially as I get older. But Einstein was right about subtlety. The laws of Quantum Mechanics are very subtle — so subtle that they allow randomness to coexist with both energy conservation and information conservation.

Consider a particle: any particle will do, but a photon is a good choice. The photon is produced by a light source — a laser, for example — and is directed toward an opaque sheet of metal with a tiny hole in it. Behind the hole is a phosphorescent screen that flashes when a photon hits it.

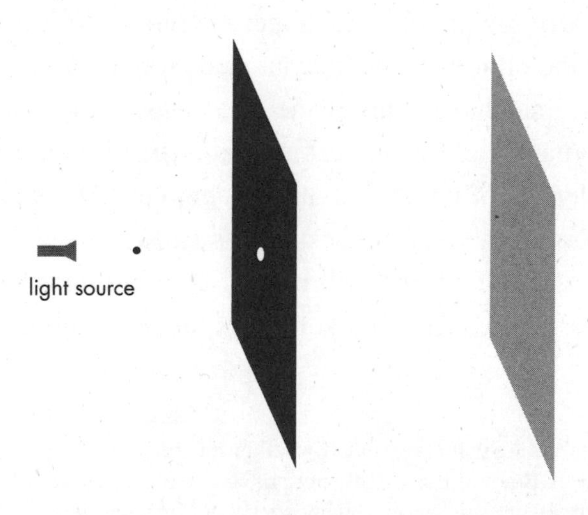

After some time, the photon may pass through the small hole, or it may miss the hole and bounce off the obstruction. If it goes through the hole, it will hit the screen, but not necessarily directly opposite the hole. Instead of going in a straight line, the photon may receive a random impulse as it passes through the hole. Thus, the final position of the flash is unpredictable.

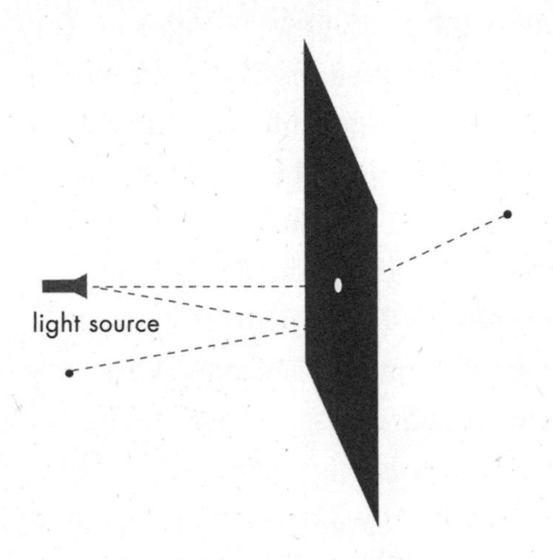

Now remove the phosphorescent screen and redo the experiment. After a short time, the photon will either hit the metal sheet and bounce off, or it will pass through the hole, having gotten a random kick. Without anything to detect the photon, it is impossible to say where the photon is and in what direction it's moving.

But imagine that we intervene and reverse the law of photon motion.[6] What should we expect if we run the photon in reverse for the same length of time? The obvious expectation is that the randomness (randomness run backward is still randomness) will ruin any hope that the photon will return to its original location. The randomness of the second half of the experiment should compound

6. The experts among you may wonder whether it is really possible to intervene and reverse a law. In practice it is usually not possible, but for certain simple systems, it is not difficult. In any case, as a thought experiment or a mathematical exercise, it is entirely doable.

the randomness of the first half and make the photon's motion even more unpredictable.

Yet the answer is much more subtle. Before I explain it, let's briefly go back to the experiment involving the three-sided coin. In that case, we also ran a law in one direction and then reversed it. There was one detail of the experiment that I left out: whether or not anyone looked at the coin just before we reversed the law. What difference would it make if someone had? It would make no difference, as long as looking at the coin doesn't flip it to a new state. That doesn't seem like a very stringent condition; I have yet to see a coin jump into the air and flip just because someone looked at it. But in the delicate world of Quantum Mechanics, it's not possible to look at something without disturbing it.

Take the photon. When we run the photon in reverse, does it reappear at its original location, or does the randomness of Quantum Mechanics ruin the conservation of information? The answer is weird: it all depends on whether or not we look at the photon when we intervene. By "look at the photon" I mean check where it is located or in what direction it is moving. If we do look, the final result (after running backward) will be random, and the conservation of information will fail. But if we ignore the location of the photon — do absolutely nothing to determine its position or direction of motion — and just reverse the law, the photon will magically reappear at the original location after the prescribed period of time. In other words, Quantum Mechanics, despite its unpredictability, nevertheless respects the conservation of information. Whether the Lord is malicious or not, he is certainly subtle.

Running the laws of physics backward is perfectly feasible — mathematically speaking. But really doing it? I very much doubt that anyone will ever be able to reverse any but the simplest systems. However, whether we can carry it out in practice or not, the mathematical reversibility of Quantum Mechanics (physicists call it *unitarity*) is critical to its consistency. Without it, quantum logic would not hold together.

Then why did Hawking think that information was destroyed when quantum theory was combined with gravity? To boil the arguments down to a slogan,

Information that falls into a black hole is lost information.

To put it another way, the law can never be reversed because nothing can return from behind the horizon of a black hole.

If Hawking had been right, the laws of nature would have had an increased element of randomness, and the entire foundation of physics would have collapsed. But we'll return to that later.

The Uncertainty Principle

Laplace believed that he could predict the future if only he knew enough about the present. Unfortunately for all the would-be fortune-tellers in the world, it's not possible to know both the position and the velocity of an object at the same time. When I say that it's not possible, I don't mean that it's very hard or that present technology isn't up to the task. No technology that obeys the laws of physics can *ever* be up to the task, no more than improved technology can allow faster-than-light travel. Any experiment designed to simultaneously measure both the position and the velocity of a particle will come up against the Heisenberg Uncertainty Principle.

The Uncertainty Principle was the great divide that separated physics into the prequantum *classical* era and the postmodern era of quantum "weirdness." Classical physics consists of everything that came before Quantum Mechanics, including Newton's theory of motion, Maxwell's theory of light, and Einstein's theories of relativity. Classical physics is deterministic; quantum physics is full of uncertainty.

The Uncertainty Principle is a strange and audacious claim that was made in 1927 by the twenty-six-year-old Werner Heisenberg shortly after he and Erwin Schrödinger discovered the mathematics of Quantum Mechanics. Even in an era of many unfamiliar ideas, it stood out as especially bizarre. Heisenberg made no claim that there was a limitation on how accurately one could measure the position of an object. The coordinates that locate a particle in space can be determined to any desired degree of precision. He also put no limitation on how accurately the velocity of an object could be measured. What he claimed was that no experiment, however complex or ingenious, could ever be designed to measure both position and velocity simultaneously. It is as though Einstein's God had made sure that no one could ever know enough to predict the future.

The Uncertainty Principle is all about fuzziness, but paradoxically, there is nothing fuzzy about it. Uncertainty is a precise concept involving probability measures, integral calculus, and other fancy mathematics. But to paraphrase a well-known expression, a picture is worth a thousand equations. Let's begin with the idea of a probability distribution. Suppose that a very large number of particles — let's say a trillion — are studied by measuring their positions along the horizontal axis, also called the x-axis. The first particle is found at $x = 1.3257$, the second at $x = .9134$, and so on. We could make a long list of the location of every particle. Unfortunately, the list would fill about ten million books as big as this one, and for most purposes, we would not be terribly interested in that list. It would be much more enlightening to have a statistical graph showing the fraction of particles found at each value of x. That graph might look like this:

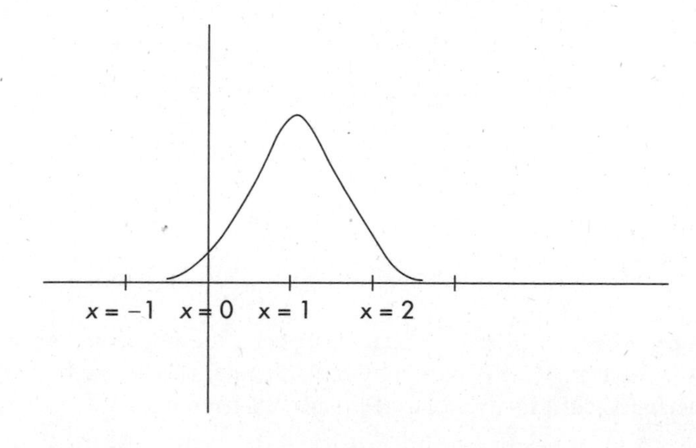

A glance at the graph tells us that most of the particles were found near $x = 1$. For some purposes, that might be enough. But just by eyeballing the graph, we can be a good deal more precise. About 90 percent of the particles were between $x = 0$ and $x = 2$. If we had to gamble as to where a particular particle was found, the best guess would be at $x = 1$, but the uncertainty — a mathematical measure of how wide the curve is — would be about 2 units.[7] The Greek letter delta (Δ) is a standard mathematical symbol for uncertainty. In this example Δx would represent the uncertainty in the x coordinate of the particles.

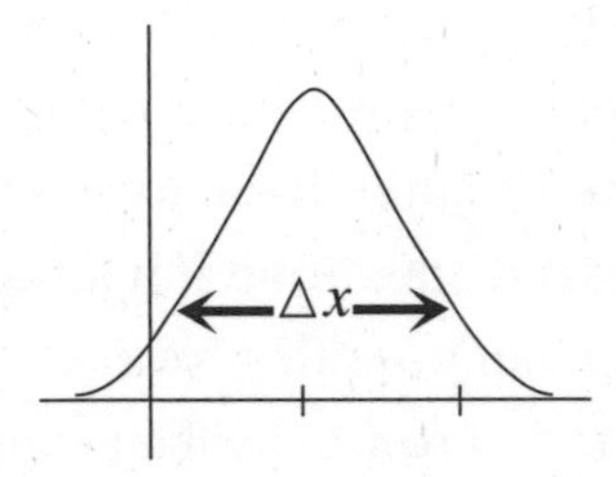

Let's do another thought experiment. Instead of measuring the locations of the particles, we measure their velocities, counting them positive if a particle moves to the right and negative if it moves to the left. This time, the horizontal axis represents the velocity, v.

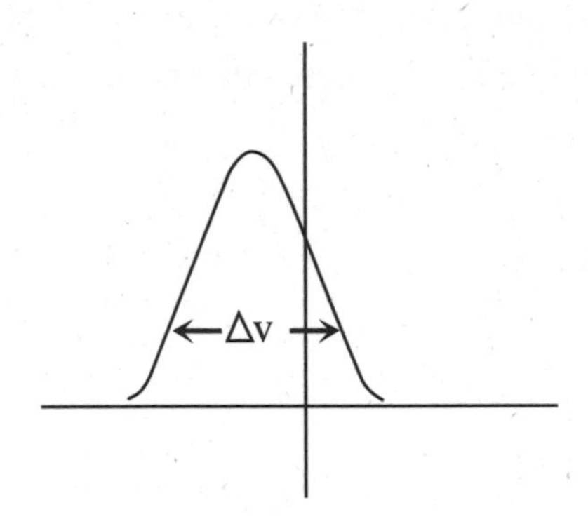

7. Of course, the bell-shaped curve extends beyond the tips of the arrows in the graph below, so there is some possibility of finding particles in the outlier regions. The mathematical uncertainty tells us the range of *likely* values.

From the graph you can see that most of the particles move to the left, and you can also get a good idea of the uncertainty in the velocity, Δv.

Roughly, what the Uncertainty Principle says is this: any attempt to shrink the uncertainty of the position will inevitably expand the uncertainty of the velocity. For example, we might purposefully select only those particles in a narrow range of x — say, between $x = .9$ and $x = 1.1$ — throwing out all the rest. For this more precisely chosen subset of particles, the uncertainty would be only .2, ten times smaller than the original Δx. We might hope in this way to beat the Uncertainty Principle, but it doesn't work.

It turns out that if we take that same subset of particles and measure their velocities, we find the velocities much more scattered than in the original sample. You may wonder why this is so, but I'm afraid it is just one of those incomprehensible quantum facts that has no classical explanation — one of those things about which Feynman said, "So theoretical physics has given up on [explaining] that."

Though incomprehensible, it is an experimental fact that whatever we do to shrink Δx inevitably results in an increase in Δv. Likewise, anything that shrinks Δv, results in an increase in Δx. The more we try to pin down the location of a particle, the more uncertain we make its velocity, and vice versa.

That's the rough idea, but Heisenberg was able to make his Uncertainty Principle more precisely quantitative. The Uncertainty Principle asserts that the product of Δv, Δx, and the mass of the particle, m, is always greater than (>) Planck's constant, h.

$$m \Delta v \Delta x > h$$

Let's see how it works. Suppose that we are very careful to prepare the particles so that Δx is extremely small. That forces Δv to be large enough so that the product is bigger than h. The smaller we make Δx, the larger Δv has to be.

How is it that we don't notice the Uncertainty Principle in daily life? While driving, have you ever experienced an increased fuzziness in your location when you looked carefully at the speedometer? Or seen the speedometer go crazy when you were checking the map to see where you were? Of course not. But why not? After all, the Uncertainty Principle doesn't play favorites; it applies to everything, including you and your car, as well as to electrons. The answer involves the mass that appears in the formula and the smallness of Planck's constant. For an electron, the very small mass tends to cancel the smallness of h, and thus the combined uncertainties Δv, and Δx must be fairly large. But the mass of a car is huge compared to Planck's constant. For that reason, Δv and Δx can both be unmeasurably small without violating the Uncertainty Principle. You can now appreciate why nature didn't prepare our brains for quantum uncertainty. There was no need; in ordinary life, we never encounter objects light enough for the Uncertainty Principle to matter.

So that's the Heisenberg Uncertainty Principle: an ultimate Catch-22 guaranteeing that no one can ever know enough to predict the future. We'll revisit the Uncertainty Principle in chapter 15.

Zero Point Motion and the Quantum Jitters

A small vessel, perhaps a centimeter in size, is filled with atoms — helium atoms, which are very nonreactive — and then heated to a high temperature. Thanks to the heat, the particles whiz around, continually bouncing off one another and off the sides of the vessel. The constant bombardment creates pressure on the walls.

By ordinary standards, the atoms move fairly fast: the average velocity is about 1,500 meters per second. Next the gas is cooled. As heat is removed, energy is drained away, and the atoms slow down. Eventually, if we continue to remove heat, the gas will cool to the lowest possible temperature — absolute zero, or approximately minus 459.67 on the Fahrenheit scale. The atoms, having lost all their energy, come to rest, and the pressure on the sides of the vessel disappears.

At least that is what's *supposed* to happen. But that reasoning fails to take into account the Uncertainty Principle.

Consider this: what do we know about the position of an atom in the case at hand? Actually, quite a lot: every atom is confined inside the vessel, and the vessel is only one centimeter in size. Obviously, the uncertainty in position, Δx, is less than one centimeter. Suppose for a moment that all the atoms really do come to rest as all the heat is drained away. Every atom would have zero velocity with no uncertainty. In other words, Δv would be zero. But that's not possible. If it were true, it would mean that the product $m\Delta x\Delta v$ also would be zero, and zero is definitely less than Planck's constant. To put it another way, if the velocity of each atom was zero, its position would be infinitely uncertain. But it's not. The atoms are in the vessel. Thus, even at absolute zero, the atoms cannot completely cease their motion; they will continue to bounce off the sides of the vessel and exert pressure. This is one of the unexpected oddities of Quantum Mechanics.

When a system has been drained of as much energy as possible (when the temperature is absolute zero), physicists say that it is in its *ground state*. The residual fluctuating motion in the ground state is usually called *zero point motion,* but the physicist Brian Greene has coined a more descriptive colloquial name for it. He calls it the "quantum jitters."

Positions of particles are not the only thing that jitter. According to Quantum Mechanics, everything that can jitter does jitter. Another example is the electric and magnetic fields in empty space. Vibrating electric and magnetic fields are all around us, filling space in the form of light waves. Even in a dark room, the electromagnetic field vibrates in the form of infrared waves, microwaves, and radio waves. But what if we darken the room as much as science allows by removing all the photons? The electric and magnetic fields continue to do the quantum jitters. "Empty" space is a violently vibrating, oscillating, jittering environment that can never be quieted.

Before anyone knew about Quantum Mechanics, they knew

about the "thermal jitters," which make everything fluctuate. For example, heating a gas causes the random motion of molecules to increase. Even when empty space is heated, it is filled with jittering electric and magnetic fields. This has nothing to do with Quantum Mechanics and was known in the nineteenth century.

Quantum and thermal jitters resemble each other in some ways but not in others. Thermal jitters are very noticeable. The thermal jitters of molecules and electric and magnetic fields tickle your nerve endings and make you feel warm. They can be very destructive. For example, the energy of the thermal jitters of electromagnetic fields can be transferred to atomic electrons. If the temperature is high enough, electrons can get ejected out of atoms. That same energy can burn or even vaporize you. By contrast, even though quantum jitters can be incredibly energetic, they cause no pain. They do not excite your nerve endings or destroy atoms. Why? It takes energy to ionize an atom (knock its electrons out) or fire your nerve endings. But there is no way to borrow energy from the ground state. The quantum jitters are what's left over when a system has the absolute minimum energy. Though incredibly violent, quantum fluctuations have none of the destructive effects of thermal fluctuations because their energy is "unavailable."

Black Magic

To me the queerest piece of magic in Quantum Mechanics is *interference.* Let's go back to the two-slit experiment that I described at the beginning of this chapter. It has three elements: a source of light, a flat obstruction with two small slits, and a phosphorescent screen that lights up when light falls on it.

Let's begin this experiment by blocking off the left slit. The result is a featureless blob of light on the screen. If we turn the intensity way down, we discover that the blob is really a collection of random flashes caused by individual photons. The flashes are unpredictable, but when there are enough of them, a bloblike pattern emerges.

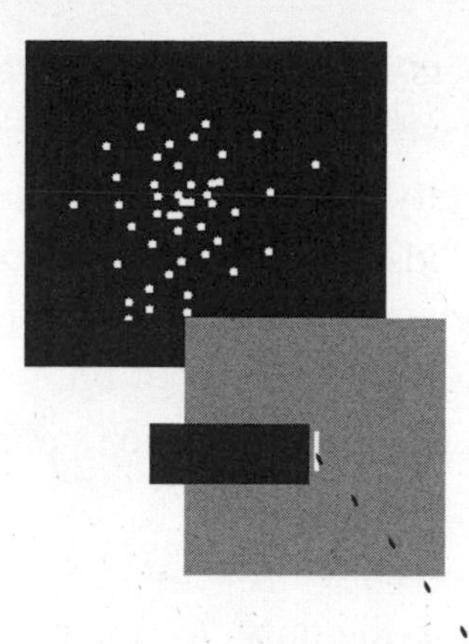

If we open the left slit and block the right slit, the average pattern on the screen looks almost unchanged, except that it shifts very slightly to the left.

The surprise occurs when we open both slits. Instead of just adding the left-slit photons to the right-slit photons and making a more intense but still featureless blob, our action results in a new zebra-stripe pattern.

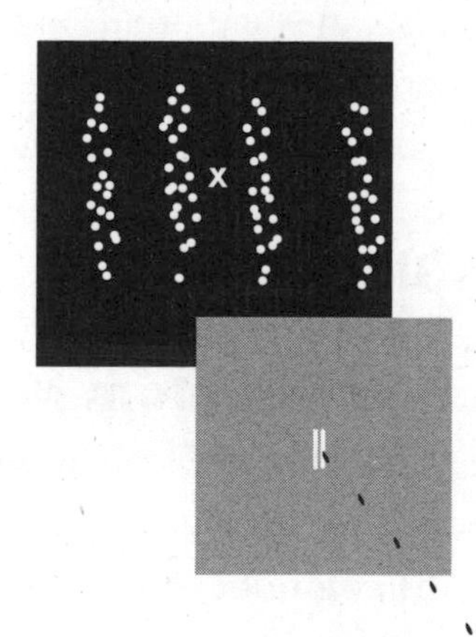

One very odd thing about the new pattern is that there are dark stripes where no photons arrive, *even though the same areas were*

filled with flashes when only one slit was opened. Take the point marked X in the central dark stripe. Photons easily pass through either slit and arrive at X when only one slit is open at a time. You would think that when both slits are open, an even larger number of photons would arrive at X. But opening both slits has the paradoxical effect of shutting off the flow of photons to X. Why does opening both slits make it less probable for a photon to get to destination X?

Imagine a bunch of drunken prisoners staggering around in a dungeon with two doors to the outside. The jailer is careful never to leave one door open, because some prisoners, as drunk as they are, will accidentally find their way out. But he has no qualms about leaving both doors open. Some mysterious magic prevents the drunks from escaping when both doors are open. Of course, this is not what happens to real prisoners, but it is the sort of thing Quantum Mechanics sometimes predicts, not just for photons but for all particles.

The effect seems bizarre when light is thought of as particles, but it is commonplace with waves. The two waves, emanating from the two slits, reinforce each other at some points and cancel each other at others. In the wave theory of light, the dark stripes are due to cancellation, otherwise known as *destructive interference.* The only problem is that light sometimes really does appear to be particles.

The Quantum in Quantum Mechanics

An electromagnetic wave is an example of an oscillation. The electric and magnetic fields at every point of space vibrate with a frequency that depends on the color of the radiation. There are lots of other oscillations in nature. Here are some common examples.

- A clock pendulum. The pendulum swings back and forth, with one full swing taking about one second. The frequency of such a pendulum is one hertz, or one cycle per second.

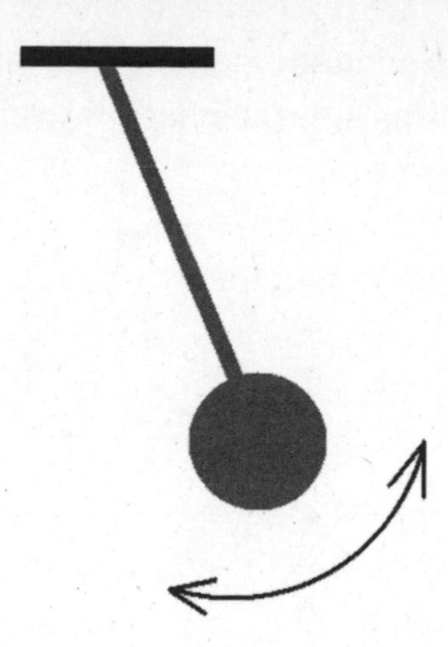

- A weight hanging from the ceiling by a spring. If the spring is very stiff, the frequency can be several hertz.

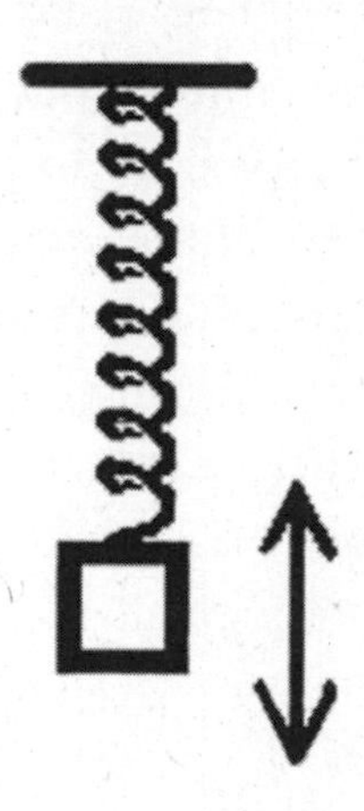

- A vibrating tuning fork or a violin string. Either can achieve a few hundred hertz.
- The electric current in a circuit. This can oscillate at much higher frequencies.

Systems that oscillate are called, not surprisingly, *oscillators*. Oscillators all have energy, at least if they are oscillating, and in classical physics the energy can be any amount. By that I mean that you can smoothly ramp up the energy, gradually if you like, to any desired value. A graph showing how the energy increases as you ramp it up would look like this:

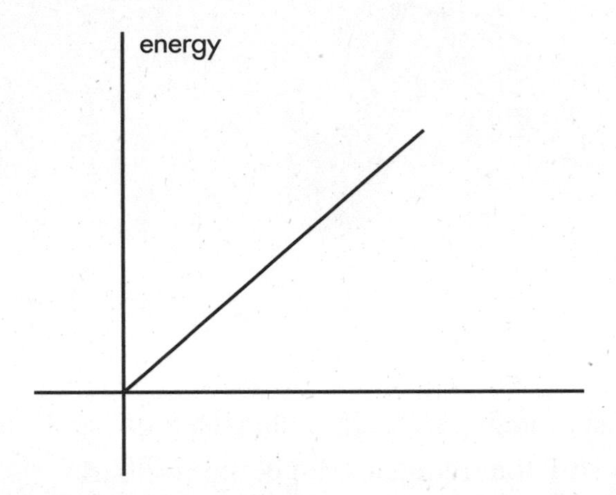

But in Quantum Mechanics, it turns out that the energy comes in little, indivisible steps. When you try to gradually increase the energy of an oscillator, the result is a staircase instead of a smooth ramp. The energy can be increased only in multiples of a unit called an *energy quantum.*

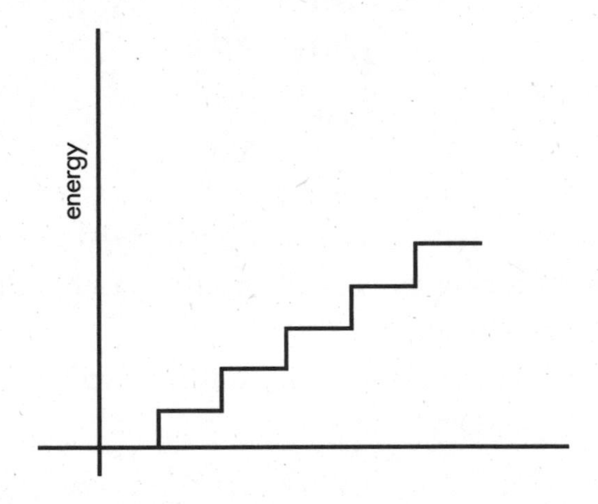

What is the size of the quantum unit? That depends on the frequency of the oscillation. The rule is exactly the same as the one that Planck and Einstein discovered for light quanta: the quantum of energy, E, is the frequency of the oscillator, f, multiplied by Planck's constant, h.

$$E = hf$$

For ordinary oscillators such as a pendulum, the frequency is not very great, and the step height (energy quantum) is extremely small. In that case, the staircase graph is made of such tiny steps that it looks like a smooth ramp. That's why you will never notice the *quantization of energy* in ordinary experience. But electromagnetic waves can have very large frequencies, in which case the steps can be quite high. In fact, as you may have already surmised, increasing the energy of an electromagnetic wave by one step is the same as adding a single photon to a light beam.

To a classically wired brain, the fact that energy can be added only in indivisible quanta seems illogical, but that is what Quantum Mechanics implies.

Quantum Field Theory

Laplace's eighteenth-century vision of the world was a bleak one: particles, nothing but particles, moving in the unalterable orbits demanded by Newton's despotic equations. I wish I could report that today's physics provides a warmer, fuzzier image of reality, but I'm afraid that it doesn't. It's still just particles, but with a modern twist. The iron rule of determinism has been replaced by the arbitrary rule of quantum randomness.

The new mathematical framework that replaced Newton's laws of motion is called Quantum Field Theory, and according to its dictates, all of the natural world consists of elementary particles traveling from one point to another, colliding, splitting, and rejoining. It is a vast network of world lines connecting events (space-time points). The mathematics of this giant spiderweb of lines and points is not easily explained in lay language, but the central points are fairly clear.

In classical physics, particles move from one point of space-time to another along definite trajectories. Quantum Mechanics introduces uncertainty to their motion. Nevertheless, we can think of particles as traveling between space-time points, albeit along uncertain trajectories. These fuzzy trajectories are called *propagators*. We usually represent each propagator as a line between space-time events, but that's only because we have no way of drawing the uncertain motion of real quantum particles.

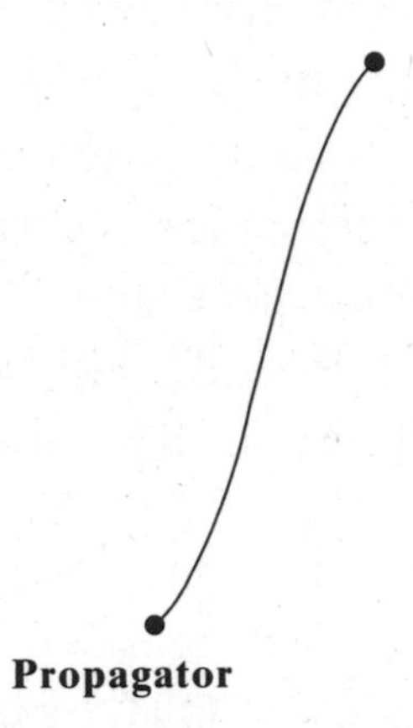

Propagator

Next come the interactions, which tell us how particles behave when they meet. The basic interaction process is called a *vertex*. A vertex is like a fork in a road; a particle proceeds along its world line until it comes to the fork, but then, instead of choosing one road or the other, the particle splits and turns into two particles, one for each branch. The best known example of a vertex is the emission of a photon by a charged particle. A single electron, spontaneously, without any warning, suddenly splits into an electron and a photon.[8]

8. Intuitively, we imagine that when something splits, each part is somehow less than the original. This is an idea inherited from common experience. The splitting of an electron into another electron, and an additional photon, shows how misleading our intuitions can be.

(Photon world lines have traditionally been drawn either as wavy or as dotted lines.)

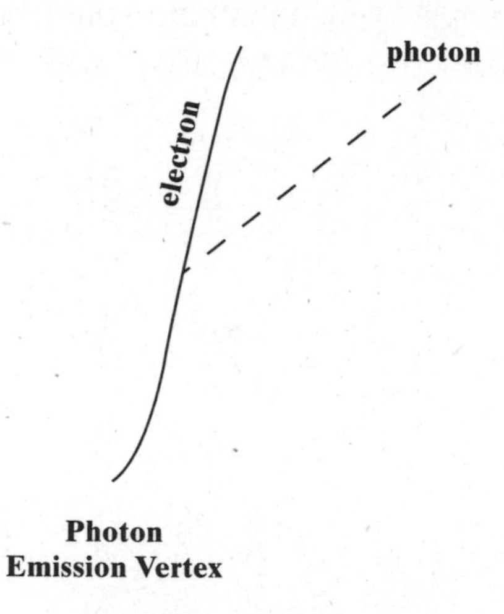

Photon
Emission Vertex

That's the basic process that produces light: jittery electrons splitting off photons.

There are many other kinds of vertices involving other particles. There are also particles called gluons, which are found in the atomic nucleus. A gluon has the capacity to split into two gluons.

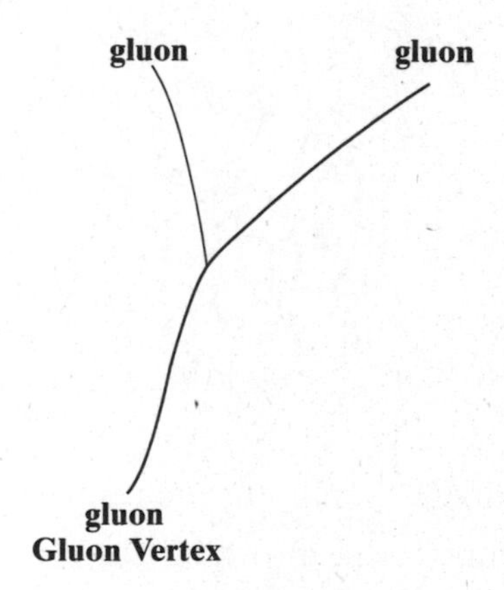

Gluon Vertex

Anything that can happen forward can also happen in reverse. That means that particles can come together and coalesce. For example, two gluons can come together and merge into a single gluon.

Richard Feynman taught us how to combine propagators and vertices to form more complicated processes. For example, there is a Feynman diagram showing a photon hopping from one electron to another, which describes how electrons collide and scatter.

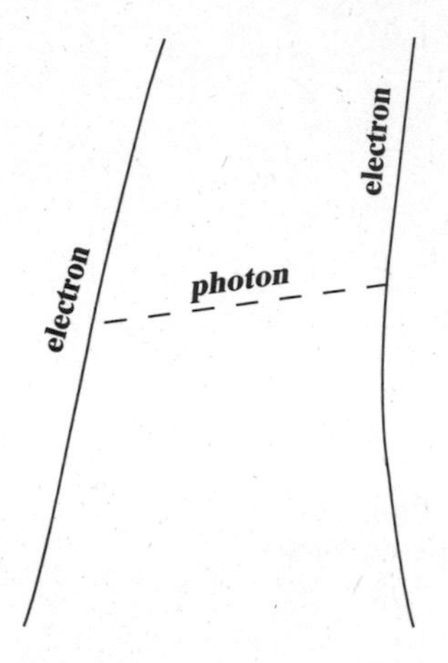

Another diagram shows how gluons form a complicated, sticky, stringy material that holds quarks together in the nucleus.

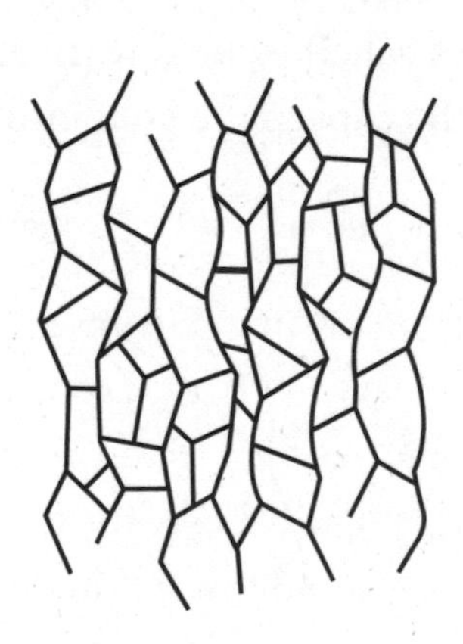

Newton's mechanics seeks to answer the age-old question of predicting the future given an initial starting point, including the positions and velocities of a set of particles. Quantum Field Theory asks the question differently: given an initial set of particles moving in a certain way, what is the probability of different outcomes?

But simply saying that nature is probabilistic (instead of deterministic) is not the full story. Laplace, although he would have disliked the idea, would have understood a world with a bit of randomness. He might have reasoned this way: the behavior of particles is not deterministic, but instead there is a positive probability[9] for each distinct route leading from the past (two electrons) to the future (two electrons plus a photon). Then, following the usual rules of probability theory, Laplace would add up all the various probabilities to get the final grand-total probability. Such reasoning would make perfect sense to Laplace's classically wired mind, but it's not how things really work. The right prescription is bizarre: don't try to grok it — just accept it.

The correct rule is one of the consequences of the strange new "quantum logic" that was discovered by the great English physicist Paul Dirac immediately following the work of Heisenberg and Schrödinger. Feynman was following Dirac's lead when he gave a mathematical rule that computes a *probability amplitude* for each Feynman diagram. Moreover, you do add up the probability amplitudes for all the diagrams, but not to get the final probability. In fact, probability amplitudes needn't be positive numbers. They can be positive, negative, or even complex numbers.[10]

But the probability amplitude is not the probability. To get the overall probability — say, for two electrons to become two electrons plus one photon — you first add up the probability amplitudes for all the Feynman diagrams. Then, according to Dirac's abstract quantum logic, you take the result and *square it!* The result is always positive, and it's the probability for the particular outcome.

This is the peculiar rule that lies at the heart of quantum weirdness. Laplace would have thought it nonsense, and even Einstein

9. In ordinary probability theory, probabilities are always positive numbers. It's hard to imagine what a negative probability could mean. Try to make sense of the following sentence: "If I flip a coin, the probability to get heads is minus one-third." It's clearly nonsense.

10. A complex number is one that contains the imaginary number i, which is the abstract mathematical symbol for the square root of minus one.

didn't think it made sense. But Quantum Field Theory is an incredibly accurate account of everything we know about elementary particles, including the way they combine to form nuclei, atoms, and molecules. As I said in the introduction, quantum physicists had to rewire themselves with new rules of logic.[11]

Before closing this chapter, I want to come back to the thing that troubled Einstein so deeply. I don't know for sure, but I suspect that it had to do with the ultimate meaningless nature of probabilistic statements. I have always been mystified by what they actually say about the world. As far as I can tell, they don't say anything very definite. I once wrote the following very short story, originally included in John Brockman's book *What We Believe but Cannot Prove,* that illustrates the point. The story, "Conversation with a Slow Student," is about a discussion between a physics professor and a student who just can't get the point. When I wrote the story, I was thinking of myself as the student, not the professor.

Student: Hi Prof. I've got a problem. I decided to do a little probability experiment — you know, coin flipping — and check some of the stuff you taught us. But it didn't work.

Professor: Well I'm glad to hear that you're interested. What did you do?

Student: I flipped this coin 1,000 times. You remember, you taught us that the probability to flip heads is one half. I figured that meant that if I flip 1,000 times I ought to get 500 heads. But it didn't work. I got 513. What's wrong?

Professor: Yeah, but you forgot about the margin of error. If you flip a certain number of times then the margin of error is about the square root of the number of flips. For 1,000 flips

11. I don't really expect the lay reader to fully understand the rule or even why it is so strange. Nevertheless, I hope it will give some flavor of how the rules of Quantum Field Theory work.

the margin of error is about 30. So you were within the margin of error.

Student: Ah, now I get it. Every time I flip 1,000 times I will always get something between 470 and 530 heads. Every single time! Wow, now that's a fact I can count on.

Professor: No, no! What it means is that you will *probably* get between 470 and 530.

Student: You mean I could get 200 heads? Or 850 heads? Or even all heads?

Professor: Probably not.

Student: Maybe the problem is that I didn't make enough flips. Should I go home and try it 1,000,000 times? Will it work better?

Professor: Probably.

Student: Aw come on Prof. Tell me something I can trust. You keep telling me what *probably* means by giving me more *probablies*. Tell me what probability means without using the word probably.

Professor: Hmmm. Well how about this: It means I would be surprised if the answer were outside the margin of error.

Student: My god! You mean all that stuff you taught us about statistical mechanics and Quantum Mechanics and mathematical probability: all it means is that you'd personally be surprised if it didn't work?

Professor: Well, uh . . .

If I were to flip a coin a million times I'd be damn sure I wasn't going to get all heads. I'm not a betting man but I'd be so sure that I'd bet my life or my soul. I'd even go the whole way and

> bet a year's salary. I'm absolutely certain the laws of large numbers – probability theory – will work and protect me. All of science is based on it. But, I can't prove it and I don't really know why it works. That may be the reason why Einstein said, "God doesn't play dice." It probably is.

From time to time, we hear physicists claim that Einstein didn't understand Quantum Mechanics and therefore wasted his time with naive classical theories. I very much doubt that this is true. His arguments against Quantum Mechanics were extremely subtle, culminating in one of the most profound and most cited papers in all of physics.[12] My guess is that Einstein was disturbed by the same thing that bothered the slow student. How could the ultimate theory of reality be about nothing more concrete than our own degree of surprise at the outcome of an experiment?

I have shown you some of the paradoxical, almost illogical, things that Quantum Mechanics forced on a classically wired brain. But I suspect that you are not entirely satisfied. Indeed, I hope you are not. If you are confused, you should be. The only real remedy is a dose of calculus and submersion in a good Quantum Mechanics textbook for a few months. Only a very unusual mutant, or a person brought up in an extremely peculiar family, could naturally have the wiring to understand Quantum Mechanics. Remember, in the end even Einstein couldn't grok it.

12. A. Einstein, B. Podolsky, and N. Rosen. "Can Quantum-Mechanical Description of Physical Reality Be Considered Complete?" *Physical Review* 47 (1935): 777–80.

5

PLANCK INVENTS A BETTER YARDSTICK

One day in the Stanford cafeteria, I noticed a number of students from my "physics for premeds" class studying at a table. "What are you guys studying?" I asked. The answer surprised me. They were memorizing the table of constants, down to the last digit, listed on the front cover of the textbook.[1] The table includes the following constants, plus about twenty more.

h (Planck's constant) = 6.626068×10^{-34} m^{2}kg/s
Avogadro's number = 6.0221415×10^{23}
Electron charge = $1.60217646 \times 10^{-19}$ coulombs
c (speed of light) = 299,792,458 m/s
Diameter of proton = 1.724×10^{-15} m
G (Newton's constant) = 6.6742×10^{-11} m^{3}s^{-2}kg^{-1}

Premeds are trained to memorize huge amounts of material in their other science classes. They are good physics students, but they often try to learn physics the same way they learn physiology. The truth is that physics requires very little memory work. I doubt whether very

1. All of the constants are in standard metric units — meters (m), kilograms (kg), and seconds (s).

many physicists could tell you much more than the rough orders of magnitude of these constants.

That raises an interesting question: why are the constants of nature such awkward numbers? Why can't they be simple numbers, such as 2 or 5 or even 1? Why are they always so small (Planck's constant, electron charge) or so large (Avogadro's number, speed of light)?

The answers have very little to do with physics but a lot to do with biology. Let's take Avogadro's number. What it represents is the number of molecules in a certain quantity of gas. How much gas? The answer is an amount of gas that early-nineteenth-century chemists could easily work with; in other words, an amount of gas that could be contained in beakers or other containers that were more or less of human size. The actual numerical value of Avogadro's number has more connection with the number of molecules in a human being than with any deep principle of physics.[2]

Another example is the diameter of the proton — why is it so small? Once again, the key is human physiology. The numerical value in the table is given in meters, but what is a meter? A meter is the metric version of the English yard, which may have referred to the distance from a man's nose to his outstretched fingertips. Quite likely it was a useful unit to measure cloth or rope. The lesson from the smallness of the proton is merely that it takes a lot of protons to make a man's arm. From a fundamental physics point of view, there is nothing special about the number.

So why don't we change the units to make the numbers easier to remember? Actually, we often do just that. For example, in astronomy the light-year is used as a measure of length. (I hate it when I hear light-year misused as a unit of time, as in "Gee, it's been light-years since I saw you last.") The speed of light is not so big when expressed in units of light-years per second. In fact, it's very small,

2. Well, then, why do humans have so many molecules? Again it has to do with the nature of intelligent life, not fundamental physics. It takes a lot of molecules to make a machine complicated enough to think and ask chemistry questions.

only about 3×10^{-8}. But what if we also changed our units of time from seconds to years? Since it takes light exactly one year to go one light-year, the speed of light is one light-year per year.

The speed of light is one of the most fundamental quantities in physics, so it makes sense to use units in which c equals one. But something like the proton radius is not very fundamental. Protons are complicated objects made of quarks and other particles, so why give them pride of place? It makes better sense to pick constants that control the deepest and most universal laws of physics. There is no dispute about what those laws are.

- The maximum velocity of *any* object in the universe is the speed of light, c. This speed limit is not just a law about light but a law about *everything* in nature.
- *All* objects in the universe attract each other with a force equal to the product of their masses and the Newton constant, G. *All* objects means *all* objects, with no exceptions.
- For *any* object in the universe, the product of the mass and the uncertainties of position and velocity is never smaller than Planck's constant, h.

The italics are there to emphasize the all-embracing character of these laws. They apply to *any* and *all* things — *everything*. These three laws of nature truly deserve to be called universal, far more so than the laws of nuclear physics or the properties of any specific particle such as the proton. It may seem trivial, but one of the most profound insights into the structure of physics occurred in 1900 when Max Planck realized that specific units of length, mass, and time could be chosen in order to make the three basic constants — c, G, and h — all equal to one.

The basic ruler is Planck's unit of length. The Planck length is far smaller than the meter or even the diameter of a proton. In fact, it's about a hundred billion billion times smaller than a proton (in meters, it's about 10^{-35}). Even if the proton were magnified to the size of the solar system, the Planck length would be no bigger than a virus. It is to Planck's everlasting credit that he realized that such impossibly tiny dimensions must play a basic role in any ultimate theory of the physical world. He didn't know what that role would be, but he might have guessed that the smallest building blocks of matter would be "Planck sized."

The unit of time that Planck required to make c, G, and h equal to one was also unimaginably small — namely 10^{-42} seconds, the time it takes light to travel one Planck length.

Finally, there is Planck's unit of mass. Given that the Planck length and the Planck time are so incredibly small (in ordinary, bio-friendly units), it would be natural to expect the Planck unit of mass to be much smaller than the mass of any ordinary object. But there you would be wrong. It turns out that the most basic unit of mass in physics is not terribly small on the biological scale: roughly the mass of ten million bacteria. It's about the same as the mass of the smallest object that can be seen with the naked eye — a dust mote, for example.

These units — the Planck length, time, and mass — have an extraordinary meaning: they are the size, half-life, and mass of the smallest possible black hole. We will return to this point in later chapters.

$E = mc^2$

Take a pot, fill it with ice cubes, seal it tight, and weigh it on a kitchen scale. Then put it on the stove and melt the ice, turning it into hot water. Weigh it again. If you do it carefully, making sure nothing enters or escapes the pot, the final weight will be the same as the original, at least to very high accuracy. But if you could do the measurements to one part in a trillion, you would notice a discrepancy; the hot water would weigh slightly more than the ice. To say it differently, heating adds a few trillionths of a kilogram to the weight.

What's going on here? Well, heat is energy. But according to Einstein, energy is mass, so adding heat to the contents of the pot increases the mass. Einstein's famous equation $E = mc^2$ expresses the fact that mass and energy are the same thing measured in different units. In a sense, it is like converting miles to kilometers; the distance in kilometers is 1.61 times the distance in miles. In the case of mass and energy, the conversion factor is the square of the speed of light.

The physicist's standard unit of energy is the joule. One hundred joules is the energy required to illuminate a 100-watt lightbulb for one second. One joule is the kinetic energy of a one-kilogram weight moving with a speed of one meter per second. Your daily food provides about ten million joules of energy. Meanwhile, the standard international unit of mass is the kilogram – slightly more than the mass of a quart of water.

What $E = mc^2$ tells us is that mass and energy are interchangeable concepts. If a bit of mass can be made to disappear, it will be converted to energy – often in the form of heat, but not necessarily. Imagine that a kilogram of mass disappears and is replaced by heat. To see how much heat, multiply the single kilogram by the very large number c^2. The result is about 10^{17} joules. You could live on that for thirty million years, or you could create a very large nuclear weapon. Fortunately, it is very difficult to convert mass into other forms of energy, but as the Manhattan Project[3] proved, it can be done.

To a physicist, the concepts of mass and energy have become so closely identified that we rarely bother to distinguish them. For example, the mass of the electron is often quoted as a certain number of *electron volts,* the electron volt being a unit of energy useful to atomic physicists.

With that knowledge, let's return to the Planck mass – the mass of a speck of dust – which we might as well also call the Planck en-

3. The development of the atomic bomb at Los Alamos, New Mexico, during World War II.

ergy. Imagine the speck being converted into thermal energy by some new discovery. The energy would be roughly the same as a full tank of gasoline. You could drive across the United States with ten Planck masses.

The unimaginable smallness of Planck-sized objects and the overwhelming difficulties of ever directly observing them are sources of deep frustration to theoretical physicists. Just the fact that we know enough to even ask these questions is a triumph of the human imagination. But it is this remote world to which we have to look for the key to the paradoxes of black holes, for it is Planck-sized *bits of information* that densely "wallpaper" the horizon of a black hole. Indeed, a black hole horizon is the most concentrated form of information that the laws of nature allow. Later we will learn what is meant by the term *information* and its twin concept, *entropy.* Then we will be in a good position to understand what the Black Hole War was all about. But first I want to explain why Quantum Mechanics undermines one of the most solid conclusions of General Relativity: the eternal nature of black holes.

6

IN A BROADWAY BAR

The first conversation I ever had with Richard Feynman was in the West End Café on Broadway in upper Manhattan. The year was 1972. I was a thirty-two-year-old, relatively unknown physicist; Feynman was fifty-three. Even if no longer at the height of his powers, the aging lion was still an awesome figure. Feynman had come to Columbia University to give a lecture on his new Parton Theory. *Parton* was Feynman's term for the hypothetical constituents (parts) of subnuclear particles such as protons, neutrons, and mesons. Today we call them quarks and gluons.

At that time, New York City was a major center of high-energy physics. The focal point was the physics department at Columbia. Physics at Columbia had a glorious and distinguished history. I. I. Rabi, a pioneer in American physics, had established Columbia as one of the world's most prestigious physics institutions, but by 1972 Columbia's reputation was in decline. The theoretical physics program at Yeshiva University's Belfer Graduate School of Science, where I was a professor, was at least as good, but Columbia was Columbia, and Belfer was far less exalted.

Feynman's lecture was anticipated with enormous excitement. He held a very special place in the hearts and minds of physicists. Not only was he one of the greatest theoretical physicists of all time, but he was everyone's hero. Actor, comedian, drummer, bad boy, iconoclast, intellectual giant — he made everything look easy. Everyone else would struggle through hours of complicated calculations to answer some physics problem, but Feynman would explain in twenty seconds why the answer was obvious.

Feynman's ego was gargantuan, but he was great fun to be around. A few years later, he and I became good friends, but in 1972 he was a celebrity, and I was a starstruck stage-door Johnny from the hinterlands north of 181st street. I arrived at Columbia by subway two hours early for the lecture, hoping to have a few words with the great man.

The theoretical physics department was on the ninth floor of Pupin Hall. I figured that Feynman would be hanging out there. The first person I saw was T. D. Lee, the mandarin of Columbia physics. I asked him if Professor Feynman was around. "What do you want?" was Lee's friendly response. "Well, I want to ask him a question about partons." "He's busy." End of conversation.

That would have been the end of the story, except for nature's call. When I entered the men's room, I saw Dick standing in front of a urinal. Sidling up next to him, I said, "Professor Feynman, can I ask you a question?" "Yeah, but let me finish what I'm doing, then we can go into the office they gave me. What's the question about?" Right then and there, I decided that I really didn't have a question about partons, but I could concoct one about black holes. The term *black hole* had been coined by John Wheeler four years earlier. Wheeler had been Feynman's thesis adviser, but Feynman told me that he knew almost nothing about black holes. The very little that I knew I had learned from my friend David Finkelstein, one of the pioneers in black hole physics. In 1958 Dave had written an influential paper explaining that a black hole horizon was a point of no return. Among the few things I knew was that a black hole had a singularity at its center and a horizon surrounding the singularity. Dave also had explained to me why nothing could escape from behind the horizon. The final thing I knew, although I can't remember how I knew it, was that once a black hole formed, it couldn't split or disappear. Two or more black holes could merge and form a bigger black hole, but nothing could ever cause a black hole to split into two or more black holes. In other words, once a black hole formed, there was no way to get rid of it.

At about this time, the young Stephen Hawking was revolutionizing the classical theory of black holes. Among his most important discoveries was the fact that the area of the horizon of a black hole can never decrease. Stephen and his collaborators, James Bardeen and Brandon Carter, had used the General Theory of Relativity to derive a set of laws governing the behavior of black holes. The new laws had an uncanny resemblance to the laws of thermodynamics (laws of heat), although this similarity was presumed to be a coincidence. The rule about the area never decreasing was the analog of the Second Law of Thermodynamics, which states that the entropy of a system can never decrease. I doubt that I knew of this work, or even of the name Stephen Hawking at the time of Feynman's lecture, but Stephen's laws of black hole dynamics would eventually have a momentous effect on my research for more than twenty years.

In any case, the question I wanted to pose to Feynman was whether Quantum Mechanics could cause a black hole to disintegrate by breaking up into smaller black holes. I was imagining something similar to the fragmenting of a very large nucleus into smaller nuclei. I hurriedly explained to Feynman why I thought it ought to happen.

Feynman said that he had never thought about it. What's more, he had come to dislike the subject of quantum gravity. The effects of Quantum Mechanics on gravity, or gravity on Quantum Mechanics, were just too tiny ever to be measured. It wasn't that he thought the subject was intrinsically uninteresting, but without some measurable experimental effects to guide theory, it was hopeless to guess how it really worked. He said that he had thought about it years ago and didn't want to get started thinking about it again. He guessed that it might be five hundred years before quantum gravity would be understood. Anyway, he said, he had to give a lecture in an hour and needed to relax.

The lecture was pure Feynman. His presence filled the stage — a larger-than-life character with a Brooklyn accent and body language to illustrate every point. The audience was spellbound. He showed us how to think about difficult problems of Quantum Field Theory in a simple, intuitive way. Almost everyone else was using another, older method to analyze the problems he was addressing. The older method was harder, but he had found a trick that made it all easy — the parton trick. Feynman waved his magic wand, and all the answers popped out. Ironically, the older method was based on the Feynman diagrams!

The best part of the lecture for me was when T. D. Lee interrupted to ask a question — or, more likely, to make a statement disguised as a question. Feynman had claimed that a certain kind of diagram never occurred in his new method, and that simplified things. It was called a Z-diagram. Lee asked, "Isn't it true that in some theories with vector and spinor fields, Z-diagrams don't always give zero? But I believe it can probably be fixed up." The lecture hall was as silent as a crypt. Feynman looked at the mandarin for five seconds, then said, "Fix it!" Then he went on.

After the lecture, Feynman came over to me and asked, "Hey, what's your name?" He said that he had thought about my question and wanted to talk about it. Did I know a place where we could meet later? That's how we wound up at the West End Café.

We'll come back to the café, but first I need to fill you in on some additional points about gravity and Quantum Mechanics.

The question that I wanted to discuss had to do with the effects of Quantum Mechanics on black holes. The General Theory of Relativity is a classical theory of gravity. When a physicist uses the term *classical,* he doesn't mean that it comes from ancient Greece. It just means that the theory does not include the effects of Quantum Mechanics. Very little was understood about how quantum theory would influence the gravitational field, but what little was known had to do with small disturbances that propagate through space as *gravitational waves*. Feynman had contributed most of what we knew about the quantum theory of these disturbances.

In chapter 4, we learned that God apparently ignored Einstein about playing dice. The point, of course, is that things that are certain in classical physics become uncertain in quantum physics. Quantum Mechanics never tells us what will happen; it tells us the probability that this or that will happen. When, exactly, a radioactive atom will decay is unpredictable, but Quantum Mechanics can tell us that it will probably decay in the next ten seconds.

The Nobel Prize–winning physicist Murray Gell-Mann borrowed this slogan from T. H. White's *The Once and Future King:* "Everything not forbidden is compulsory." In particular, there are many events in classical physics that just cannot happen. In most cases, however, the same events are possible in quantum theory. Instead of being impossible, these events are just very improbable. But no matter how improbable, if you wait long enough, they will eventually happen. Thus, everything not forbidden is compulsory.

A good example is a phenomenon called *tunneling*. Imagine a car parked on a hill with a dip in it.

Let's ignore all irrelevant things such as friction and air resistance. Let's also assume that the driver left the hand brake off so that the car is free to roll. It's clear that if the car is parked at the bottom of the dip, it won't suddenly start to move. Motion in either direction would involve going uphill, and if the car is initially at rest, it won't have the energy to move uphill. If we later found the car rolling down the hill beyond the hump, we would assume that either someone had pushed it or it had gotten the energy to go over the hump in some other way. Spontaneously jumping over the hump would be impossible in classical mechanics.

But remember, everything not forbidden is compulsory. If the car was quantum mechanical (as all cars really are), nothing would prevent it from suddenly appearing on the other side of the hump. That might be very unlikely — and for a large, heavy object such as a car, it would be very, *very* unlikely — but it would not be impossible. Thus, given enough time, it would be compulsory. If we waited long enough, we would find the car rolling down the other side of the hump. This phenomenon is called tunneling because it would be as if the car had passed under the hump, through a tunnel.

For an object as massive as an automobile, the probability that it will tunnel is so small that it would take an enormous amount of time (on average) for the car to appear spontaneously on the other side of the hump. To write a number big enough to express this amount of time would require so many digits that even if each digit were no bigger than a proton and they were solidly packed, the digits would more than fill the universe. However, exactly the same effect can allow an alpha particle (two protons and two neutrons) to tunnel out of a nucleus, or an electron to tunnel across a gap in a circuit.

What I was imagining that day in 1972 is that although classical black holes have a fixed shape, quantum fluctuations can cause the shape of the horizon to jiggle. Normally, the shape of a nonspinning black hole is a perfect sphere, but a quantum fluctuation should be able to deform it briefly to a flattened or an elongated shape. Furthermore, every so often a fluctuation can be so large that the black hole will be deformed almost to a pair of smaller spheres connected by a thin neck. From there it is easy for it to split. Heavy nuclei spontaneously split in this way, so why not a black hole? Classically, it cannot happen, just as the car cannot spontaneously hop over the hump. But is it absolutely forbidden? I could see no reason why it should be. Wait long enough, I reasoned, and the black hole will split into two smaller black holes.

My Idea About How Black Holes Decay

Now back to the West End Café. Nursing a beer, I waited for Feynman in the café for about half an hour. The more I thought about it, the more sense it seemed to make. The black hole could disintegrate by quantum tunneling, first into two pieces, then into four, eight, and eventually a large number of microscopic components. In light of Quantum Mechanics, it made no sense to believe that black holes were permanent.

Feynman entered the café just a minute or two early and came over to where I was sitting. I was in the mood to be a big shot, so I ordered two beers. Before I had a chance to pay, he pulled out his wallet and put down the required amount. I don't know whether he left a tip. I sipped my beer, but I noticed that Feynman's glass never left the table. I began by reviewing my argument and ended by saying that I thought the black hole ought to eventually disintegrate into tiny pieces. What could those tiny pieces be? Although it went unsaid, the only reasonable answer was elementary particles such as photons, electrons, and positrons.

Feynman agreed that there was nothing to prevent this from happening, but he thought I had the wrong picture. I had visualized the black hole first splitting into more or less equal fragments. Each fragment would split in half, until the fragments were microscopic.

The problem was that a gigantic quantum fluctuation would be required for a large black hole to split in half. Feynman felt that there was a more plausible picture in which the horizon would split into a piece almost equal to the original horizon and a second microscopic piece that would fly away. As the process repeated, the large black hole would gradually shrink until there was nothing left. This sounded right. A tiny piece of the horizon breaking off seemed much more likely than the black hole splitting into two large fragments.

Feynman's Idea About
How Black Holes Decay

The conversation lasted for about an hour. I don't recall saying good-bye, nor did we plan to pursue the idea. I had met the lion, and he hadn't disappointed me.

Had we thought more about the problem, we might have realized that gravity would most likely pull the tiny fragments back down to the horizon. Some ejected fragments might collide with falling fragments. The region just above the horizon would be a complicated mess of colliding fragments that might be heated by the repeated collisions. We might even have realized that the region just above the horizon would be a seething mass of particles that formed a hot atmosphere. And we might have realized that this heated mass would behave like any heated object and radiate away its energy as heat radiation. But we didn't. Feynman went back to his partons, and I returned to the problem of what keeps quarks confined to the interior of protons.

Now it is time to tell you exactly what *information* means. Information, entropy, and energy are three inseparable concepts that are the subject of the next chapter.

7

ENERGY AND ENTROPY

Energy

Energy is a shape-shifter. Like mythical shape-shifters that could change from human to animal to plant to rock, energy also can change its form. Kinetic, potential, chemical, electrical, nuclear, and thermal (heat) are some of the many shapes that energy can take. It is constantly morphing from one form to another, but there is one constant: energy is conserved; the sum total of all forms of energy never changes.

Here are some examples of shape-shifting.

- Sisyphus is low on energy. So before pushing his boulder to the top of the hill for the umpteenth time, he stops to refresh himself with a meal of honey. When the boulder reaches the top, the condemned man watches while gravity rolls it back to the bottom for the umpteenth-plus-one time. Poor Sisyphus is doomed to eternally convert chemical energy (honey) to potential energy and then to kinetic energy. But wait — what happens to the kinetic energy of the boulder when it rolls to rest at the bottom of the hill? It is converted to heat. A bit of heat flows into the atmosphere and into the ground. Even Sisyphus is heated by the effort. The Sisyphus cycle of energy conversion is as follows:

chemical→potential→kinetic→thermal

- Water flows over Niagara Falls and picks up speed. The flowing water, laden with kinetic energy, is directed into the mouth of a turbine, where it spins the rotors. Electricity is produced and flows through wires into the grid. Can you chart the shape-shifting? Here it is:

  potential→kinetic→electrical

  In addition, some of the energy is uselessly converted to heat: the water that comes out of the turbine is warmer than the water that went in.
- Einstein proclaimed that mass is energy. What Einstein meant when he said $E = mc^2$ was that every object has some latent energy that can be released if somehow its mass can be changed. For example, a uranium nucleus will eventually break apart into a thorium nucleus and a helium nucleus. The thorium and helium will together have slightly less mass than the original uranium. That little bit of excess mass will morph into the kinetic energy of the thorium and helium nuclei and also a few photons. When the atoms come to rest and the photons are absorbed, the excess energy becomes heat.

Of all the usual forms of energy, heat is the most mysterious. What is it? Is it a substance like water, or is it something more ephemeral? Before the modern molecular theory of heat, early physicists and chemists thought that it was a substance and that it behaved like a fluid. They called it *phlogiston* and imagined that it flowed from hot objects to cool ones, cooling the hot and warming the cool. Indeed, we still speak of the flow of heat.

But heat is not a new substance; it's a form of energy. Shrink yourself to the size of a molecule and look around at the hot water in a bathtub. You can see the molecules randomly moving and colliding in a bustling, chaotic dance. Let the water cool and look around again: the molecules move more slowly. Cool it to the freezing point, and the molecules become stuck in a crystal of solid ice. But even in the ice, the molecules continue to vibrate. They cease moving (ignoring quantum zero point motion) only when all the energy has been drained away. At that point, when the water is at minus 459.67 degrees Fahrenheit, or absolute zero, the temperature cannot be lowered any further. Every molecule is rigidly locked in place, in a perfect crystalline lattice, all confusion and chaotic motion stopped.

The conservation of energy as it shifts shape from heat to other forms is sometimes called the *First Law of Thermodynamics*.

Entropy

It would be a bad idea to park your BMW in the rain forest for five hundred years. When you came back, you'd find a pile of rust. That's entropy increasing. If you left the rust pile for another five hundred years, you could be pretty sure it wouldn't turn back into a working BMW. That, in short, is the Second Law of Thermodynamics: entropy increases. Everyone talks about entropy — poets, philosophers, computer geeks — but what exactly is it? To answer this question, consider the difference between the BMW and the heap of rust more closely. Both are collections of about 10^{28} atoms, mostly iron (and in the case of rust, also oxygen). Imagine that you took those atoms and threw them together randomly. What is the likelihood that they would come together to form a working automobile? It would take a good deal of expertise to say just how unlikely that would be, but I think we can all agree that it would be extremely unlikely. Obviously, it would be far more probable that you'd get a pile of rust than a brand-new automobile. Or even an old rusty one.

If you took the atoms apart and threw them together again and again and again, you'd eventually get a car, but in the meantime you'd get a lot more rust piles. Why is that? What's so special about the car — or the rust pile?

If you imagined all the possible ways that you could assemble the atoms, the overwhelming majority of the arrangements would look like rust heaps. A much smaller fraction would resemble an automobile. But even then, if you looked under the hood, you would most likely find a rust heap. An even tinier fraction of the arrangements would form a working automobile. The entropy of a car and the entropy of a rust heap have something to do with the number of arrangements that we would recognize as rust heaps versus the number that we would recognize as a car. If you shook up the atoms of a car, you would be much more likely to get a pile of rust because there are so many more rust pile arrangements than car arrangements.

Here is another example. An ape banging away at a typewriter will almost always type gibberish. Very rarely will he type a grammatically correct sentence such as "I want to arbitrate my hypotenuse with the semicolon." Even less frequently will he write a meaningful sentence such as "King Canute had warts on his chin." What's more, if you take the letters of a meaningful sentence and shake them up like tiles in a Scrabble game, the result will almost always be gibberish. The reason? There are a lot more nonsensical ways to arrange twenty or thirty letters than there are meaningful ones.

The English alphabet has twenty-six letters, but there are simpler systems of writing. Morse code is a very simple system, using only two symbols: dot and dash. Strictly speaking, there are three symbols — dot, dash, and space — but one could always replace the space with some special series of dots and dashes that is otherwise unlikely to occur. Anyway, ignoring the space, here is the description of King Canute and his warts in Morse code — sixty-five symbols in all.

-.-..-.--.-.-..--...--.....-.--.-.-...---.........-.-...-..-.-.-

How many distinct Morse code messages could be made out of sixty-five dots and/or dashes? All you have to do is multiply 2 together 65 times to get 2^{65}, which is about ten billion billion.

When information is coded in terms of two symbols — they could be dots and dashes, ones and zeros, or any other pair — the symbols are called *bits*. Thus, in Morse code, "King Canute had warts on his chin" is a 65-bit message. If you are going to read the rest of this book, it would be a good idea to memorize the definition of the technical term *bit*. Its meaning is not the same as when you say, "I'll take a bit of cream in my coffee." A bit is a single, irreducible unit of information, like the dots and dashes in Morse code.

Why do we go to the trouble of reducing information to dots and dashes, or zeros and ones? Why not use sequences of 0 1 2 3 4 5 6 7 8 9 or, even better, the letters of the alphabet? Messages would be a lot easier to read, and they would take up a lot less space.

The point is that the alphabet (or the ten ordinary numerals) is a human construct that we learn to recognize and store in our memories. But each letter or numeral already has a great deal of information — for example, in the intricate difference between the letters A and B or the numbers 5 and 8. Telegraphers and computer scientists, who rely only on the simplest mathematical rules, prefer — in fact, they are almost forced — to use the *binary code* of dots and dashes or zeros and ones. Indeed, when Carl Sagan designed a system for sending messages to nonhuman civilizations living on distant solar systems, he used the binary code.

Back to King Canute. How many of the 65-bit messages are *coherent* sentences? I really don't know — perhaps a few billion. But whatever the number is, it is an infinitesimal fraction of 2^{65}. So it's almost certain that if you take the 65 bits, or the 27 letters, of "King Canute had warts on his chin" and scramble them, gibberish will be the result. Leaving out the spaces, here is what I got when I did it with Scrabble tiles:

HTKIDGENCUONNHTSRNISAWACHAI

Suppose you only scramble the letters a little at a time. The sentence will gradually lose its coherence. "King Canut ehad warts on his chin" is still recognizable. So is "Knig Canut ehad warts o his chinn." But gradually the letters will become a meaningless jumble. There are so many more meaningless combinations that the trend toward gibberish is inescapable.

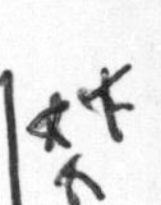

Now I can give you a definition of entropy. *Entropy is a measure of the number of arrangements that conform to some specific recognizable criterion.* If the criterion is that there are 65 bits, then the number of arrangements is 2^{65}.

But entropy is *not* the number of arrangements, in this case 2^{65}. It is just 65 — the number of times you have to multiply 2 together to get the number of arrangements. The mathematical term for the number of times 2 must be multiplied together to get a given number is called its *logarithm*.[1] Thus, 65 is the logarithm of 2^{65}. Entropy is, therefore, the logarithm of the number of arrangements.

Of the 2^{65} possibilities, only a very small fraction are actually meaningful sentences. Let's guess that there are a billion. To make a billion, you have to multiply together about 30 factors of 2. In other words, a billion is about 2^{30}, or, equivalently, 30 is the logarithm of a billion. It follows that the entropy of meaningful sentences is only about 30, a good deal less than 65. Meaningless jumbles of symbols clearly have more entropy than combinations that spell out coher-

1. Strictly speaking, it is the *logarithm to the base 2*. There are other definitions of logarithms. For example, instead of the number of 2s, there is the number of 10s that must be multiplied to get a given number. That would be defined as the *logarithm to the base 10*. Needless to say, you need fewer 10s than 2s to make a given number.

The official physics definition of entropy is the number of times you must multiply the mathematical number e. This "exponential" number is approximately equal to e = 2.71828183. In other words, entropy is the *natural logarithm* or *logarithm to the base e,* while the number of bits (65 in the example) is the *logarithm to the base 2*. The natural logarithm is a bit smaller than the number of bits by a factor of about .7. So for the purists, the entropy of a 65-bit message is .7 × 65, which equals about 45. In this book, I will ignore this difference between bits and entropy.

ent sentences. It's hardly surprising that entropy increases when you scramble the letters.

Suppose that the BMW company improved its quality control to the point where every car that came off the assembly line was identical to every other. In other words, suppose there was one, and only one, atomic arrangement that would be accepted as a true BMW. What would its entropy be? The answer is zero. There would be no uncertainty whatever about any detail when a BMW rolled off the assembly line. Whenever one specifies a unique arrangement, there is no entropy at all.

The Second Law of Thermodynamics, which says that entropy increases, is just a way of saying that as time goes on, we tend to lose track of the details. Imagine that we put a tiny droplet of black ink into a tub of warm water. At first we know precisely where the ink is located. The number of possible configurations of the ink is not too large. But as we watch the ink diffuse through the water, we begin to know less and less about the locations of the individual ink molecules. The number of arrangements that correspond to what we see – namely, a uniform, slightly gray tub of water – has become enormous. We can wait and wait, but we won't see the ink rearrange into a concentrated drop. Entropy increases. That's the Second Law of Thermodynamics. Things tend toward boring uniformity.

Here is yet another example – a bathtub full of hot water. How much do we know about the water in the tub? Assume that it has been sitting in the tub long enough that there is no detectable motion. We can measure the amount of water in the tub (50 gallons), and we can measure the temperature (90 degrees Fahrenheit). But the tub is full of water molecules, and there are obviously a very great number of molecular arrangements that correspond to the given conditions – that is, 50 gallons of water at 90 degrees Fahrenheit. We could know a lot more if only we could measure every atom precisely.

Entropy is a measure of how much information is hidden in the details — details that for one reason or another are too hard to observe. Thus, *entropy is hidden information.* In most cases, the information is hidden because it concerns things that are too small to see and too numerous to keep track of. In the case of the bathwater, these things are the microscopic details of the water molecules — the location and the motion of each of the billion billion billion individual water molecules in the tub.

What happens to the entropy if the water is cooled until its temperature is absolute zero? If we remove every bit of energy from the water, the molecules will arrange themselves in a unique arrangement, the frozen lattice that forms a perfect crystal of ice.

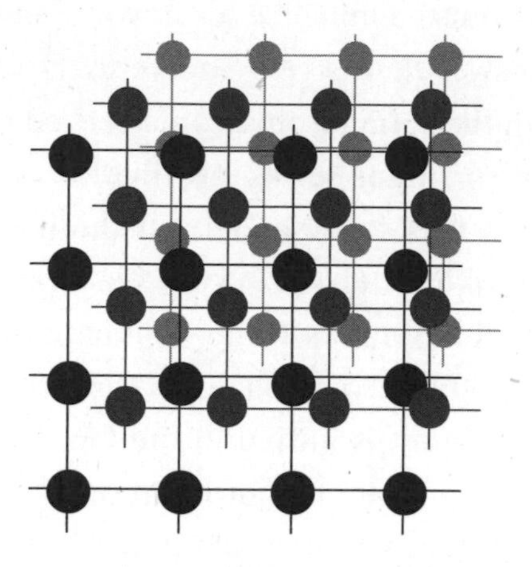

Crystal Lattice

Even if the molecules are too small to see, if you are familiar with the properties of crystals, you can predict the position of every molecule. A perfect crystal, like a perfect BMW, has no entropy at all.

How Many Bits Can You Stuff in a Library?

Ambiguity and subtle nuance in the use of language is often highly valued. Indeed, if words had perfectly precise meanings that could be programmed into a computer, language and literature would be impoverished. But precision in science requires a great degree of linguistic exactitude. The word *information* can mean many things: "I think your information is wrong." "For your information, Mars has two moons." "I have a master's degree in information science." "You can find the information in the Library of Congress." In each of these sentences, the word *information* is being used in a specific way. Only in the last sense of the word does it make sense to ask "Where is the information located?"

Let's pursue this idea of location. If I told you that Grant is buried in Grant's Tomb, we would no doubt agree that I have given you a piece of information. But where is that information? Is it in your head? In my head? Is it somehow too abstract to have a location? Is it diffused throughout the universe for anyone, anywhere to use?

Here is one answer that is very concrete: the information is on the page, stored in the form of physical letters composed of carbon and other molecules. In this sense, information is a concrete thing, almost a substance. It is so concrete that the information in your book and in my book is different information. In your book, it says that Grant is buried in Grant's Tomb. You may suspect that it says the same thing in my book, but you don't know for sure. Just maybe my book says that Grant is buried in the Great Pyramid of Giza. In fact, neither book contains the information. The information that Grant is buried in Grant's Tomb is in Grant's Tomb.

In the sense that physicists use the word, information is made of matter,[2] and it is found somewhere. The information in this book is in a rectangular volume of space about 10 inches by 6 inches by 1 inch — that is, $10 \times 6 \times 1$, or 60 cubic inches.[3] How many bits of information are hidden between the covers of the book? In a line of print, there is room for about 70 characters — letters, punctuation

2. When physicists use the word *matter,* they don't mean only things made of atoms. Other elementary particles, such as photons, neutrinos, and gravitons, also qualify as matter.

3. These dimensions are a rough guess based on the size of the hardcover edition of my previous book. No doubt the actual dimensions of this book are a bit different.

marks, and spaces. At 37 lines per page and 350 pages, that's almost a million characters.

My computer keyboard has about 100 symbols, including lower- and uppercase letters, numbers, and punctuation marks. That means that the number of distinct messages that could be contained in this book is about 100 multiplied by itself a million times — in other words, 100 to the millionth power. That amount, a massive number, is roughly the same as multiplying 2 about 7 million times. The book contains about 7 million bits of information. In other words, if I had written the book in Morse code, it would have taken 7 million dots and dashes. Dividing that by the volume of the book, we find that there are about 120,000 bits per cubic inch. That's the information density in this volume of printed pages.

I once read that the great library in Alexandria contained a trillion bits of information before it burned to the ground. Though not one of the official Seven Wonders of the World, the library was nonetheless one of the greatest marvels of antiquity. Built during the reign of Ptolemy II, it is said to have contained a copy of every important document ever written, in the form of half a million parchment scrolls. No one knows who burned it down, but we can be sure that a lot of priceless information went up in smoke. Just how much? I'd guess that there were about fifty modern pages in an ancient scroll. If those pages were anything like the pages you are reading, a scroll was worth a million bits, give or take a few hundred thousand. At that rate, Ptolemy's library would have contained half a trillion (1 trillion = 10^{12}) bits — close enough to what I read.

The loss of that information is one of the great misfortunes that scholars of the ancient world have to live with today. But it could have been worse. What if every nook and cranny, every available cubic foot, had been filled with books like this one? I don't know exactly how big the great library was, but let's say $200 \times 100 \times 40$ feet, or 800,000 cubic feet — the size of a good-size public building today. That would be 1.4 billion cubic inches.

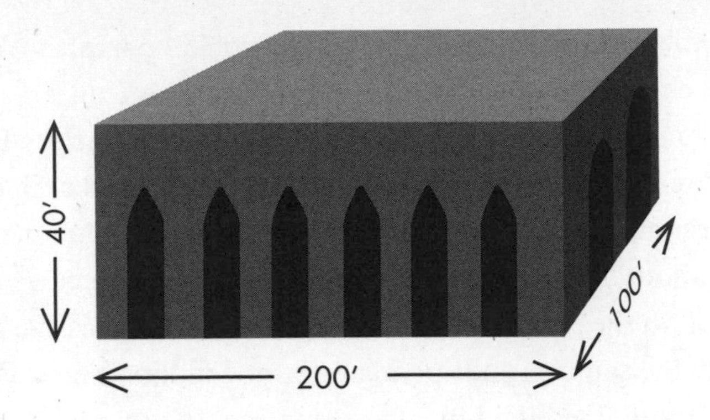

With that knowledge, it's easy to estimate how many bits could be stuffed into the building. At 120,000 bits per cubic inch, the grand total is 1.7×10^{14} bits. Stupendous.

But why stop with books? If each book were shrunk to one-tenth its volume, ten times as many bits could be crammed in. Transferring the contents to microfiche could allow for even more storage. And digitizing each book might allow for even more.

Is there a fundamental physical limitation to the amount of space needed to contain a single bit? Must the physical size of a real bit of data be bigger than an atom, a nucleus, a quark? Can we subdivide space indefinitely and fill it with endless amounts of information? Or is there a limit — not a practical technological limit, but a consequence of a deep law of nature?

The Littlest Bit

Smaller than an atom, smaller than a quark, smaller even than a neutrino, the single bit may be the most fundamental building block. Without any structure, the bit is just there, or not there. John Wheeler believed that all material objects are composed of bits of information, and he expressed the idea with this slogan: *"It from bit."*

John imagined that a bit, being the most basic of all objects, is as small as the smallest possible size — that fundamental quantum of distance discovered by Max Planck more than a century ago. A rough picture that most physicists have in their minds is that space can be divided into tiny Planck-sized cells, similar to a three-dimensional checkerboard. A bit of information can be stored in each cell. The bit can be pictured as a very simple particle. Each cell may contain a particle or not. Another way to think of the cells is that they make up a huge, three-dimensional tic-tac-toe game.

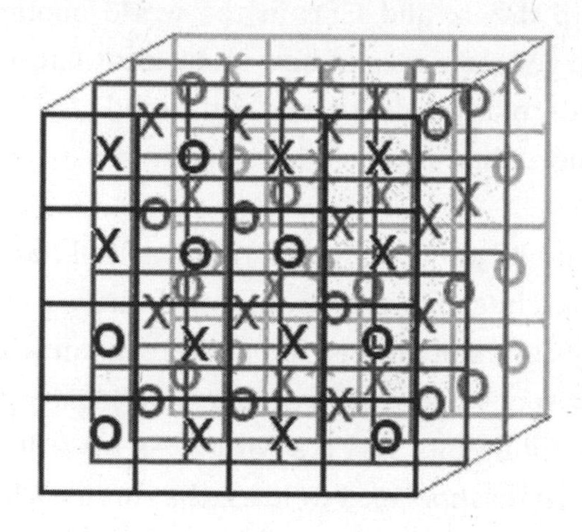

According to Wheeler's "It from bit" philosophy, the physical conditions of the world at any given time can be represented by such a "message." If we knew how to read the code, we would know exactly what was going on in that piece of space. For example, is it what we normally call empty space — a vacuum — or is it a piece of iron or the interior of a nucleus?

Since things in the world change with time — planets move, particles decay, people are born and die — the message in the Os and Xs also must change. At one instant, the pattern might look like the picture above. A little later, it might be rearranged.

In this Wheeler world of information, the laws of physics would consist of rules for how the configuration of bits is updated from instant to instant. Such rules, if correctly constructed, would allow waves of Os and Xs to propagate across the lattice of cells and represent light waves. A large, solid lump of Os might disturb the distribution of Xs and Os in its vicinity, and in this way the gravitational field of a heavy mass might be represented.

Now let's return to the question of how much information could be stuffed into the Alexandria library. All we have to do is divide the volume of the library — 1.4 billion cubic inches — into Planck-sized cells. The answer is about 10^{109} bits.

That's a lot of bits: far more than the entire Internet and all of the books, hard drives, and CDs in the world contain — indeed, vastly more. To get some idea of how much information 10^{109} bits is, let's imagine how many ordinary books it would take to store it. The answer is far more than we could possibly fit into the entire observable universe.

The "It from bit" philosophy, describing a "cellular" world filled with Planck-sized bits of information, is an enticing one. It has influenced physicists on many levels. Richard Feynman was a great advocate of it. He spent a good deal of time building simplified worlds made of space-filling bits. But it's wrong. As we will see, Ptolemy would have been disappointed to learn that his great library could never have held more than a mere 10^{74} bits.[4]

I can more or less picture what a million means: a cube one meter on a side contains a million gumdrops. But what about a billion or a trillion? It's harder to visualize the distinction, even though a trillion is one thousand times bigger than a billion. And numbers like 10^{74} and 10^{109} are far too big to comprehend, except to say that 10^{109} is a lot bigger than 10^{74}. Indeed, 10^{74} — the actual number of bits that could fit in the Alexandria library — is an infinitesimal fraction of the 10^{109} bits that we computed. Why is there such an enormous discrepancy? That is a story for a later chapter, but I will give you a hint here.

Fear and paranoia among kings and princes is an all-too-common theme of history. I have no idea whether Ptolemy suffered from it, but let's imagine how he might have responded to a rumor that secret information was being hidden in his library by his enemies. He

4. As it happens, that is about the number of bits that could be contained in a universe full of printed books.

might have felt himself justified in passing a draconian law that forbade any hidden information. In the case of the Alexandria library, Ptolemy's imaginary law would require every bit of information to be visible from outside the building. To satisfy the law, all of the information had to be written on the exterior walls of the library. The librarian was forbidden to hide a single bit in the interior. Hieroglyphs on the exterior walls – allowed. Roman, Greek, or Arabic writing on the walls – allowed. But scrolls brought into the interior – forbidden. What a waste of space! But it was the law. Under these circumstances, what was the maximum number of bits that Ptolemy could expect to store in his library?

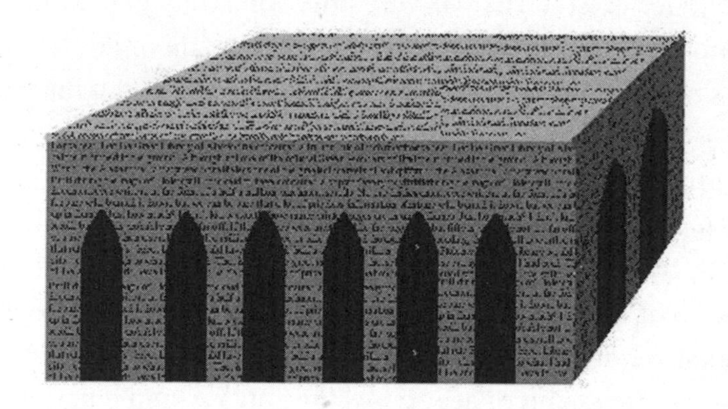

To find the answer, Ptolemy had his servants carefully measure the external dimensions of the building and compute the *area* of the external walls and roof (let's ignore the arches and the floor). They came up with $(200 \times 40) + (200 \times 40) + (100 \times 40) + (100 \times 40) + (200 \times 100)$, which equals 44,000 square feet. Notice that this time, the unit is *square* feet rather than *cubic* feet.

But the king wanted the area measured in Planck units rather than square feet. I'll do it for you. The number of bits that he could plaster on the walls and ceiling was about 10^{74}.

One of the most surprising and outlandish discoveries of modern physics is that in the real world, there is no need for Ptolemy's law. Nature already provides such a law, and even kings can't break it. It is one of the deepest and most profound laws of nature we have discovered: *The maximum amount of information that can be stuffed into a region of space is equal to the area of the region, not the volume.* This strange restriction on filling space with information is the subject of chapter 18.

Entropy and Heat

Heat is the energy of random chaotic motion, and entropy is the amount of hidden microscopic information. Consider the tub of water, now cooled to the coldest possible temperature — absolute zero — at which point every molecule is locked into its precise location in an ice crystal. There is very little ambiguity in the location of each molecule. In fact, anyone who knows the theory of ice crystals could say exactly where each atom lies, even without a microscope. There is no hidden information. The energy, temperature, and entropy are all zero.

Now add a bit of heat by warming the ice. The molecules begin to jiggle, but only a bit. A small amount of information is lost; we lose track, if only a little, of the details. The number of configurations that we might mistakenly confuse with one another is larger than it was. Thus, a bit of heat raises the entropy, and it gets worse as more energy is added. The crystal approaches the melting point, and the molecules begin to slide right past each other. Keeping track of the details quickly becomes prohibitive. In other words, as energy increases, so does entropy.

Energy and entropy are not the same thing. Energy takes many forms, but one of those forms, heat, is joined at the hip with entropy.

More About the Second Law

The First Law of Thermodynamics is the law of energy conservation: you cannot create energy; you cannot destroy it; all you can do is change its form. The Second Law is even more discouraging: ignorance always increases.

Imagine a scene in which a diver plunges into a swimming pool from a springboard:

potential energy→kinetic energy→heat

He quickly comes to rest, and the original potential energy is converted to a slight increase in the thermal energy (heat) of the water. With that slight increase, there is a slight increase in entropy.

The diver would like to repeat the performance, but he's lazy and doesn't want to climb the ladder to the diving board again. He knows that energy can never disappear, so why not just wait until some of the heat in the pool gets converted back to potential energy — *his* potential energy? Nothing about the conservation of energy would prevent him from being shot up to the diving board while the pool cools a bit: the reverse of the dive. Not only would he wind up on the board, but the entropy of the pool would decrease, implying a surprising decrease in ignorance.

Unfortunately, our wet friend finished only half his thermodynamics course — the first half. In the second half, he would have learned what we all know: Entropy *always* increases. Energy *always* degrades. The change of potential, kinetic, chemical, and other forms of energy into heat always favors more heat and less of those organized, nonchaotic forms of energy. This is the Second Law: the total entropy of the world always increases.

It's for this reason that a car will screech to a halt when the brakes are applied, but applying the breaks to a stationary car will not get it moving. The random heat of the ground and the air cannot be turned into the more organized kinetic energy of a moving vehicle. It's also the reason that the heat in the sea cannot be tapped to solve the world's energy problems. On the whole, organized energy degrades to heat, not the other way around.

Heat, entropy, information — what do these practical, utilitarian concepts have to do with black holes and the foundations of physics? The answer is everything. In the next chapter, we will see that black holes are fundamentally reservoirs of hidden information. Indeed, they are the most densely packed information storage containers in nature. And that may be the best definition of a black hole. Let's find out how Jacob Bekenstein and Stephen Hawking came to realize this central fact.

8

WHEELER'S BOYS, OR HOW MUCH INFORMATION CAN YOU STUFF IN A BLACK HOLE?

In 1972, while I was talking to Richard Feynman in the West End Café, a Princeton graduate student named Jacob Bekenstein was asking himself this question: what do heat, entropy, and information have to do with black holes? At that time, Princeton was the world's center for the study of gravitational physics. That may have had something to do with the fact that Einstein had lived there for more than two decades, though by 1972 he had been dead for seventeen years. The Princeton professor who had inspired many brilliant young physicists to study gravity and think about black holes was John Archibald Wheeler – one of the great visionaries of modern physics. Among the many famous physicists who were profoundly influenced by Wheeler during this period were Charles Misner, Kip Thorne, Claudio Teitelboim, and Jacob Bekenstein. Wheeler, who earlier had been Feynman's Ph.D. adviser, was a disciple of Einstein. Like the great man himself, he believed that the key to the laws of nature lay in the theory of gravity. But unlike Einstein, Wheeler, who had worked with Niels Bohr, was also a believer in Quantum Mechanics. Thus, Princeton was a center of research not only on gravity but also on quantum gravity.

At that time, the theory of gravity was a relatively unpopular backwater of theoretical physics. Elementary particle physicists were making giant strides in the reductionist march toward ever tinier structures. Atoms had long given way to nuclei, and nuclei to quarks. Neutrinos were finding their rightful role as equal partners with electrons, and new particles such as the charmed quark had been hypothesized and were within a year or two of experimental discovery. The radioactivity of nuclei was finally being mastered, and the Standard Model of elementary particles was about to be declared. Elementary particle physicists, myself included, thought they had better things to do than waste their time on gravity. There were exceptions, such as Steven Weinberg, but most considered the subject to be frivolous.

In retrospect, this disdain for gravity was a very myopic view. Why was it that the aggressive leaders of physics — the bold frontiersmen of the subject — were so uncurious about gravity? The answer is that they could see no way that gravity could possibly have significance for the way elementary particles interact with one another. Imagine that we had a switch that allowed us to turn off the electrical force between an atomic nucleus and the electrons, thereby leaving only gravitational attraction to hold the electrons in orbit. What would happen to an atom when we flipped the switch? The atom would immediately expand, because the force that holds it together would have diminished. How big would a typical atom become? A good deal bigger than the entire observable universe!

And what would happen if we left the electrical force switched on but turned off gravity? The Earth would fly away from the Sun, but the change in an individual atom would be far too small to make any difference. Quantitatively, the gravitational force between two electrons in an atom is roughly a million billion billion billion billion times weaker than the electrical forces.

Such was the intellectual environment when John Wheeler boldly set out to explore the ocean of ignorance that separated the conventional world of elementary particles from Einstein's theory of gravity. Wheeler himself was a walking enigma. In appearance he looked and sounded like a buttoned-down businessman. He could easily have fit into the boardrooms of the most conservative corporations in America. In fact, his politics were conservative. The cold war was far from over, and John was a staunch anticommunist. And yet, throughout the 1960s and 1970s, in an age of unprecedented campus activism, he was deeply beloved by his students. Claudio Teitelboim, who today is the most eminent Latin American physicist, was one of John's students.[1] Claudio, the scion of a famous Chilean left-wing political family, was one of John's many disciples who later achieved fame. The family was politically allied with Salvador Allende; Claudio himself was a fearless, outspoken foe of the dictatorial Pinochet regime. But despite their political differences, John and Claudio have had an extraordinary friendship, based on love and mutual respect for each other's opinions.

The first time I met Wheeler was in 1961. I was an undergraduate at City College of New York, with a somewhat unorthodox academic record. One of my instructors, Harry Soodak – a cigar-chomping, cussing professor from the same Jewish, left-wing, working-class background that I came from – took me down to Princeton to meet him. The hope was that Wheeler would be impressed and that I would be admitted as a graduate student, despite my lack of an undergraduate degree. At the time, I was working as a plumber in the South Bronx, and my mother thought I should dress properly for the meeting. To my mother, that meant that I should show solidarity with my social class and dress in my work clothes. These days, my plumber in Palo Alto dresses about the same way that I do when I lecture at Stanford University. But in 1961 my

1. Claudio's life has been filled with dramatic events. One of his most exciting adventures occurred about two years ago, when he discovered that his father was Álvaro Bunster, the patriarch of a heroic antifascist family. As a headline in a prominent Chilean newspaper put it, "Famous Chilean Physicist Who Searched for the Origin of the Universe Discovers His Own Origin." As a consequence, Claudio changed his last name to Bunster.

plumbing costume was the same as my father's and that of all his South Bronx plumber buddies – Li'l Abner bib overalls, a blue flannel work shirt, and heavy steel-toed work shoes. I also sported a watch cap to keep the dirt and grime out of my hair.

When Harry picked me up for the drive to Princeton, he did a double take. The big cigar fell from his mouth, and he told me to go back upstairs and change. He said that John Wheeler was not that kind of guy.

When I walked into the great professor's office, I saw what Harry meant. The only way I can describe the man who greeted me is to say that he looked Republican. What the hell was I doing in this Wasp's nest of a university?

Two hours later, I was completely enthralled. John was enthusiastically describing his vision of how space and time would become a wild, jittery, foamy world of quantum fluctuations when viewed through a tremendously powerful microscope. He told me that the most profound and exciting problem of physics was to unify Einstein's two great theories – General Relativity and Quantum Mechanics. He explained that only at the Planck distance would elementary particles reveal their true nature, and it would be all about geometry – quantum geometry. To a young aspiring physicist, the stuffy businessman exterior had morphed into an idealistic visionary. I wanted more than anything to follow this man into battle.

Was John Wheeler really as much of a conservative as he appeared? I don't really know. But he certainly was not a prudish moralizer. Once, while John and my wife, Anne, and I were having a drink in a Valparaiso café on the beach, he got up to take a walk, saying he wanted to check out the South American girls in their bikinis. At the time, he was in his late eighties.

In any case, I never did get to be one of Wheeler's boys; Princeton did not admit me. So I went off to Cornell, where physics was far tamer. It would be many years before I felt the same thrill I had in 1961.

Sometime around 1967, Wheeler became very interested in the gravitationally collapsed objects that Karl Schwarzschild had described in 1917. At that time, they were called black stars or dark stars. But that didn't capture the essence of the objects — the fact that they are deep holes in space whose gravitational attraction is irresistible. Wheeler began calling them black holes. At first the name was blackballed by the preeminent American physics journal *Physical Review*. Today the reason seems hilarious: the term *black hole* was deemed obscene! But John fought it through the editorial board, and black holes they became.[2]

Amusingly, John's next coinage was the saying "Black holes have no hair." I don't know whether *Physical Review* was again outraged, but the terminology stuck. Wheeler wasn't trying to provoke the journal editors. Instead, he was making a very serious point about the properties of black hole horizons. What he meant by "hair" was observable features, perhaps bumps or other irregularities. Wheeler was pointing out that the horizon of a black hole was as smooth and featureless as the smoothest bald pate — actually, much smoother. When a black hole formed — say, by a star collapsing — the horizon very quickly settled down to a perfectly regular, featureless sphere. Apart from their mass and rotational speed, every black hole was exactly like every other. Or so it was thought.

Jacob Bekenstein, an Israeli, is a small, quiet man. But his gentle, scholarly appearance belies his intellectual boldness. In 1972 he was one of Wheeler's graduate students interested in black holes. But he was not interested in these objects as astronomical bodies that might someday be seen through telescopes. Bekenstein's passion was the foundations of physics — the basic underlying principles — and he had a sense that black holes had something profound to say about the laws of nature. He was particularly interested in how black holes might fit together with the principles of Quantum Mechanics and thermodynamics that had so preoccupied Einstein. In fact, Bekenstein's style of doing physics was much like Einstein's; both were masters of the thought experiment. With very little math-

2. I first heard this story from the eminent general relativist Werner Israel.

ematics, but with a lot of deep thinking about the principles of physics and how they apply to imaginary (but possible) physical circumstances, both men were able to draw far-reaching conclusions that would profoundly affect the future of physics.

Here is Bekenstein's question in a nutshell. Imagine yourself orbiting a black hole. In your possession is a container of hot gas — gas with a good deal of entropy. You throw the container of entropy into the black hole. According to the standard wisdom, the container would simply disappear behind the horizon. For all practical purposes, that entropy would completely vanish from the observable universe. According to the prevailing view, the featureless, bald horizon couldn't possibly hide any information. So it would seem that the entropy of the world had decreased, in contradiction with the Second Law of Thermodynamics, which says that entropy never decreases. Could it be that easy to violate so deep a principle as the Second Law? Einstein would have been appalled.

Bekenstein concluded that the Second Law is too deeply rooted in the rules of physics to be so easily violated. Instead, he made a radical new proposal: black holes themselves must have entropy. He argued that when you count up all the entropy in the universe — the missing information in stars, interstellar gas, the atmosphere of the planets, and all the bathtubs of warm water — you must include a certain amount of entropy for each black hole. Furthermore, the bigger the black hole is, the bigger its entropy. With that idea, Bekenstein could rescue the Second Law. Einstein would no doubt have approved.

Here is how Bekenstein thought about it. Entropy always goes together with energy. Entropy has to do with the number of arrangements of something, and that something, in all cases, has energy. Even the ink on a page is made of massive atoms, which, according to Einstein, have energy, because mass is a form of energy. One might say that entropy counts the possible ways of arranging bits of energy.

When Bekenstein, in his imagination, tossed a container of hot gas into a black hole, he was adding to the black hole's energy. In turn, this meant an increase in the black hole's mass and size. If, as Bekenstein guessed, black holes have entropy, which increases with their mass, there was a chance of rescuing the Second Law. The entropy of the black hole would increase more than enough to compensate for what was lost.

Before explaining how Bekenstein guessed the formula for black hole entropy, I will explain why it was such a shocking idea — so shocking that according to Stephen Hawking, he initially dismissed it as nonsense.[3]

Entropy counts alternative arrangements, but arrangements of what? If a black hole horizon is as featureless as the smoothest conceivable bald head, what is there to count? According to this logic, a black hole should have zero entropy. John Wheeler's claim that "black holes have no hair" seemed to directly contradict Jacob Bekenstein's theory.

How to reconcile teacher and student? Let me give you an example that will help you understand. The print on a page decorated with various shades of gray is really composed of tiny black and white dots. Let's suppose we have a million black dots and a million white dots to work with. One possible pattern would be to divide the page in half, either vertically or horizontally. We can make one half black and the other half white. There are only four ways to do this.

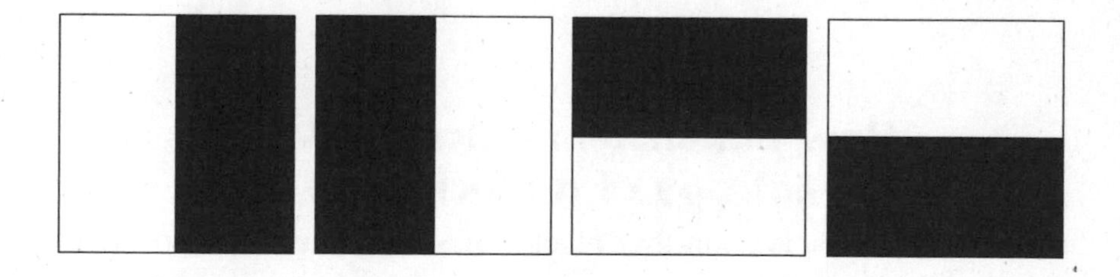

3. You can read all about his initial skepticism in his book *A Brief History of Time*.

We see a strong pattern and sharp distinctions, but only a very few arrangements. Strong patterns and sharp distinctions typically mean small entropy.

But now let's go to the other extreme and randomly distribute the equal number of black and white pixels over the same square. What we see is a more or less uniform gray. If the pixels are really small, the gray will look extremely uniform. There is an enormous number of ways to rearrange the black and white dots that we would not notice without a magnifying glass.

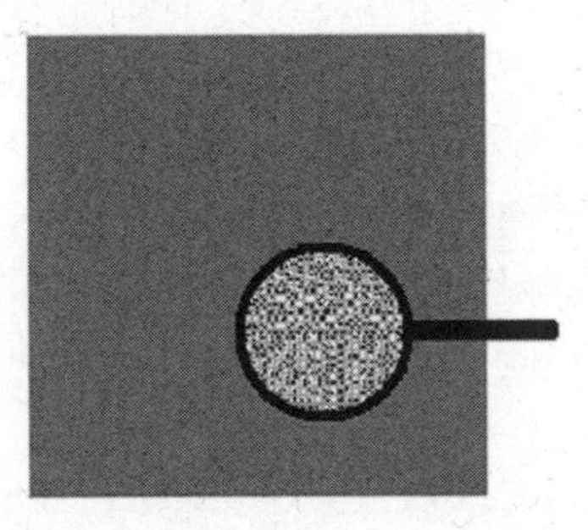

In this case, we see that large entropy often goes together with a uniform, "bald" appearance.

The combination of apparent uniformity and large entropy indicates something important. It implies that the system — whatever it is — must be made of a very large number of microscopic objects that (a) are too small to see and (b) can be rearranged in many ways without changing the basic appearance of the system.

How Bekenstein Calculated the Entropy of a Black Hole

Bekenstein's observation that black holes must have entropy — in other words, that despite their apparent baldness, they hold hidden information — is one of those simple but profound observations that, at a stroke, change the course of physics. When I began to write books for a general audience, I was strongly advised to keep the equations down to only one: $E = mc^2$. I was informed that for every additional equation that I included, I would sell ten thousand fewer books. Frankly, that goes against my experience. People like to be challenged; they just don't like to be bored. So after a good deal of soul-searching, I decided to take the risk. Bekenstein's argument is one of such extraordinary simplicity and beauty that I feel not to include it in this book would be a sad dumbing down of the material. However, I have taken pains to explain the results so that the less mathematically inclined reader can safely skip over the few simple equations without losing the essence.

Bekenstein didn't directly ask how many bits could be hidden inside a black hole of a given size. Instead, he asked how the size of a black hole would change if a single bit of information was dropped into it. This is similar to asking how much the water in a bathtub will rise if a single drop of water is added. Even better, how much will the water rise if a single atom is added?

That raised another question: how to add a single bit? Should Bekenstein throw in a single dot, printed on a scrap of paper? Obviously not; that dot is composed of a huge number of atoms, and so is the paper. There is far more information in that dot than a single bit. The best strategy would be to throw in one elementary particle.

Suppose, for example, that a single photon falls into a black hole. Even one photon can carry more than a single bit of information. In particular, there is a good deal of information in knowing exactly where the photon enters the horizon. Here Bekenstein made clever use of Heisenberg's concept of *uncertainty.* He argued that the location of the photon should be as uncertain as possible, provided only that it gets into the black hole. The existence of such an "uncertain photon" would convey only a single bit of information — namely, it's there, somewhere in the black hole.

Recall from chapter 4 that the resolving power of a light beam is no better than its wavelength. Now in this particular case, Bekenstein did not want to resolve a spot on the horizon; he wanted it to be as fuzzy as possible. The trick was to use a photon of such long wavelength that it would be spread out over the entire horizon. In other words, if the horizon has Schwarzschild radius R_s, the photon should have about the same wavelength. Even longer wavelengths might seem like an option, but they would just bounce off the black hole without getting trapped.

Bekenstein suspected that an extra bit added to the black hole would make it grow by a tiny increment, similar to the way that adding an extra rubber molecule to a balloon would increase its size. But calculating the growth required a few intermediate steps. Let me outline them first.

1. To begin with, we need to know how much the energy of the black hole increases when we add a single bit of information. That amount, of course, is the energy of the photon that carries the bit. So determining the energy of the photon is the first step.

2. Next, we need to determine how much the mass of the black hole changes when the extra bit is added. To do this, we recall Einstein's most famous equation:

   $$E = mc^2$$

   But we read it backward. It tells us the change in mass in terms of the added energy.

3. Once the change in mass is known, we can compute the change in the Schwarzschild radius, using the same formula that Michell, Laplace, and Schwarzschild worked out (see chapter 2):

   $$R_s = 2MG/c^2$$

4. Finally, we must determine the increase in the area of the horizon. For that, we need the formula for the area of a sphere:

$$\text{Area of horizon} = 4\,\pi R_s^{\,2}$$

We begin with the energy of the one-bit photon. As I explained earlier, the photon should have a long enough wavelength that its location is uncertain within the black hole. That means the wavelength should be R_s. According to Einstein, a photon of wavelength R_s has an energy, E, given by the following formula:[4]

$$E = hc\,/\,R_s$$

In this formula, h is Planck's constant, and c is the speed of light. The implication is that dropping a single bit of information into the black hole will increase its energy by an amount hc / R_s.

The next step is to compute how much the mass of the black hole changes. To convert energy to mass, you have to divide by c^2, which means that the mass of the black hole will increase by an amount $h / R_s c$.

$$\text{Change in mass} = h\,/\,R_s c$$

Let's plug in some numbers to see how much a single bit would add to the mass of a solar-mass black hole.

Planck's constant, h	6.6×10^{-34}
Schwarzschild radius of black hole, R_s	3,000 meters (= 2 miles)
Speed of light, c	3×10^{8}
Newton's constant, G	6.7×10^{-11}

Thus, one bit of information added to a solar-mass black hole will add an astoundingly small amount of mass:

4. The frequency, f, of the photon with wavelength R_s is C/R_s. Using the Einstein-Planck formula $E = hf$ tells us that the photon energy is hC/R_s.

$$\text{Increase in mass} = 10^{-45} \text{ kilograms}$$

But still, as they say, "that ain't nothing."

Let's proceed to step three – using the connection between mass and radius to calculate the change in R_s. In algebraic symbols, the answer is as follows:

$$\text{Increase in } R_s = 2hG / (R_s c^3)$$

For the solar-mass black hole, R_s is about 3,000 meters. If we plug in all the numbers, we will find that the radius will increase by 10^{-72} meters. This is not only vastly smaller than a proton, but it is also vastly smaller than the Planck length (10^{-35} meters). With such a small change, you might wonder why we even bother to calculate it, but it would be a mistake to ignore it.

The final step is figuring out how much the area of the horizon will change. For the solar-mass black hole, the increase in the horizon area is about 10^{-70} square meters. That's very small, but again, "that ain't nothing." And not only ain't it nothing, it's something very special: 10^{-70} square meters just happens to be *one square Planck unit.*

Is that an accident? What would happen if we tried it for an Earth-mass black hole (a black hole as big as a cranberry) or a black hole a billion times more massive than the Sun? Try it out, either with numbers or with equations. Whatever the size of the original black hole, this is the rule:

> *Adding one bit of information will increase the area of the horizon of any black hole by one Planck unit of area, or one square Planck unit.*

Somehow, hidden in the principles of Quantum Mechanics and the General Theory of Relativity, there is a mysterious connection between indivisible bits of information and Planck-sized bits of area.

When I explained all of this to my premed physics class at Stanford, someone in the back of the room let out a long, low whistle, then said, "Cooool." It *is* cool, but it's also profound and probably holds the key to the puzzle of quantum gravity.

Now imagine building up the black hole bit by bit, just as you might fill a bathtub atom by atom. Each time you add a bit of information, the area of the horizon increases by one Planck unit. By the time the black hole is completed, the area of the horizon will be equal to the total number of bits of information hidden in the black hole. That was Bekenstein's great achievement, all summarized in this slogan:

The entropy of a black hole, measured in bits, is proportional to the area of its horizon, measured in Planck units.

Or, more succinctly:

Information equals area.

It almost seems that the horizon is densely covered with incompressible bits of information, more or less the same way that a tabletop could be densely covered with coins.

Adding an extra coin to the crowd would increase its area by the area of one coin. Bits, coins, it's the same principle.

The only problem with this picture is that there are no coins at the horizon. If there were, Alice would have discovered them when she fell into the black hole. According to the General Theory of Relativity, to the freely falling Alice, the horizon is an invisible point of no return. The possibility that she will encounter anything like a table full of coins is in direct conflict with Einstein's Equivalence Principle.

This tension — the apparent inconsistency between *the horizon as a surface, packed densely with material bits,* and *the horizon as a mere point of no return* — was the casus belli of the Black Hole War.

Another point has puzzled physicists since Bekenstein's discovery: why is the entropy proportional to the area of the horizon and not the volume of the interior of the black hole? It seems as if there is a lot of wasted space inside. In fact, a black hole sounds an awful lot like Ptolemy's library. We will return to that point in chapter 18, where we will see that all the world is a hologram.

Although Bekenstein had the right idea — black hole entropy is indeed proportional to area — his argument was not perfectly precise, and he knew it. He didn't say that the entropy is *equal* to the area measured in Planck units. Because of a number of uncertainties in his calculation, all he could say was that the entropy of a black hole is *about equal* (or proportional) to the area. In physics, *about* is a pretty slippery word. Is it twice the area or a quarter of the area? Bekenstein's arguments, though brilliant, were not powerful enough to fix the exact factor of proportionality.

In the next chapter, we will see how Bekenstein's discovery of black hole entropy led Stephen Hawking to his greatest insight: black holes not only have entropy, as Bekenstein correctly surmised, but they also have temperature. They are not the infinitely cold, dead objects that physicists had thought them to be. Black holes glow with an inner warmth, but in the end that warmth leads to their destruction.

9

BLACK LIGHT

The winter wind is meanest in big cities. It funnels down long corridors between flat-faced buildings and whips around corners, mercilessly flailing unfortunate pedestrians. One very nasty day in 1974, I was out taking a long run through the icy streets of northern Manhattan, icicles of sweat hanging from my long hair. After fifteen miles, I had reached exhaustion, but I was still a regrettable two miles from my warm office. Without my wallet, I didn't even have the necessary twenty cents to take the subway back. But luck was with me. As I stepped off the curb somewhere around Dyckman Street, a car pulled up next to me, and Aage Petersen stuck his head out the window. Aage was a delightful pixie of a Dane who had been Niels Bohr's assistant in Copenhagen before coming to the United States. He loved Quantum Mechanics and lived and breathed Bohr's philosophy.

Once inside the car, Aage asked if I was on my way to Dennis Sciama's lecture at the Belfer School. I wasn't. In fact, I didn't know anything about Sciama or his lecture. Instead, I was thinking about a hot bowl of soup in the university cafeteria. Aage had met Sciama in England and said that he was an enormously amusing Englishman from Cambridge University who could be counted on for lots of good jokes. Aage thought that the lecture had something to do with black holes — some work that Sciama's student had done and that had Cambridge all abuzz. I promised Aage that I would show up.

The Yeshiva University cafeteria was not a place much to my liking. The food was not bad – the soup was kosher (that I couldn't have cared less about), and it was hot (that was important) – but the conversation among the students irritated me: it was almost always about the law. Not the federal law, or the state or city law, or the laws of science; rather the hairsplitting minutia of Talmudic law that animated the young Yeshiva undergraduates: "Would Pepsi be kosher if it were made in a factory that was built on a former pig farm?" "What if the ground was covered with plywood before the factory was built?" That kind of thing. But the hot soup and cold weather encouraged me to dawdle and eavesdrop on some students at the next table. This time the subject of the conversation was an item that even I cared about – toilet paper! The raging Talmudic debate was over the momentous issue of whether toilet paper rollers may be restocked during the Sabbath, or whether one must use the paper straight from an unmounted roll. From various passages in Rabbi Akiva's writings, one faction had surmised that the great man would have insisted on rigorous obedience to certain laws, which forbade restocking the roller. The other faction believed that the incomparable Rambam[1] had made very clear in *The Guide for the Perplexed* that certain necessary tasks were exempt from these Talmudic injunctions, and logical analysis favored the view that toilet paper restocking was one of those tasks. After half an hour, the argument was still raging. Several new, young rabbis-to-be had joined the fray with additional ingenious, almost mathematical, arguments, and I finally tired of the debate.

You may wonder what this has to do with the subject of this book, black holes. Only this: my dalliance in the cafeteria caused me to miss the first forty minutes of Dennis Sciama's brilliant lecture.

1. "Rambam" is the nickname of Rabbi Moses Ben Maimon, who is better known to the non-Jewish world as Maimonides.

Cambridge University, where Sciama was professor of astronomy and cosmology, was one of three places (besides Princeton and Moscow)[2] where the "brightest and the best" were testing their intellects against the profound puzzles of gravity. And as at Princeton, its young intellectual warriors were led by a charismatic, inspirational leader. Sciama's boys were a stellar group of brilliant young physicists, including Brandon Carter, formulator of the Anthropic Principle in cosmology; Sir Martin Rees, the Astronomer Royal of Great Britain, who now occupies Sir Edmond Halley's chair (Halley of cometary fame); Philip Candelas, the current Rouse Ball Professor of Mathematics at Oxford; David Deutsch, one of the inventors of quantum computing; John Barrow, a distinguished Cambridge astronomer; and George Ellis, the well-known cosmologist. Oh, yes, and there was also Stephen Hawking, who now sits in Isaac Newton's chair at Cambridge. It was, in fact, Stephen's work that Dennis was reporting that frigid day in 1974, but at the time, the name Stephen Hawking meant nothing to me.

By the time I arrived at Sciama's lecture, it was two-thirds over. I was immediately sorry that I hadn't gotten there earlier. For one thing, I was not looking forward to going out into the freezing sleet in my running outfit again. It had gotten dark and no doubt even colder by the time Dennis finished. But it was more than the fear of frostbite that made me wish that Sciama was just beginning. As Aage had said, Dennis was an enormously entertaining speaker. The jokes were indeed outstanding, but more to the point, I was fascinated by the single equation on the blackboard.

Usually by the end of a theoretical physics lecture, the blackboard is filled with mathematical symbols. But Sciama was a man of few equations. When I arrived, the blackboard looked about like this:

2. The great gravity center in Moscow was led by the legendary Russian astrophysicist and cosmologist Yakov B. Zeldovich.

$$T = \frac{hc^3}{16\pi^2 GMk}$$

Within five minutes, I had deciphered what the symbols represented. In fact, they were all standard notation for familiar quantities in physics. But I didn't know the context – what the formula described – although I could tell that it was either very deep or very silly. It had only the most fundamental constants of nature in it: Newton's constant, G, which governs the force of gravity, was in the denominator – an odd place to find it; c, the speed of light, indicated that the Special Theory of Relativity was involved; Planck's constant, h, whispered Quantum Mechanics; and then there was k, Boltzmann's constant. It was the last that seemed so out of place. What the hell was it doing there? Boltzmann's constant has to do with heat and the microscopic origin of entropy. What were heat and entropy doing in a quantum gravity formula?

And what about the number 16 and π^2? These were the kind of mathematical numbers that appear in all sorts of equations. No hint there. M was familiar, and Sciama's words reinforced my impression of its meaning. M was mass. Within a couple of minutes, I could tell that it was the mass of a black hole.

Okay, black holes, gravity, and relativity. That made sense, but adding Quantum Mechanics seemed odd. Black holes are enormously heavy – as heavy as the stars that preceded them. But Quantum Mechanics is for small things: atoms, electrons, and photons. Why bring Quantum Mechanics into a discussion of something as heavy as a star?

Most confusing, the left side of the equation represented temperature, T. The temperature of what?

The last fifteen or twenty minutes of Sciama's lecture were enough for me to put the pieces together. One of Dennis's students had discovered something very odd: Quantum Mechanics gives black holes thermal properties — heat — and along with heat comes temperature. The equation on the blackboard was a formula for the temperature of a black hole.

How strange, I thought. What gave Sciama the daffy idea that a dead star, a star that had completely run out of fuel, would have a temperature other than absolute zero?

Looking at the intriguing formula, I saw an interesting correlation: the temperature of a black hole would be inverse to its mass; the bigger the mass, the smaller the temperature. A huge astronomical black hole, as large as a star, would have a tiny temperature, much colder than any object in any laboratory on Earth. But the real surprise, which made me sit up in my chair, was that tiny black holes, if they existed, were extraordinarily hot — hotter than anything we had ever imagined.

Yet Sciama had an even bigger surprise: black holes evaporate! Physicists till then had believed that black holes, like diamonds, were forever. Once formed, no mechanism known to physical science could destroy or eliminate a black hole. The black vacancy in space formed by a dead star would persist — infinitely cold, infinitely silent — for all eternity.

But Sciama told us that, like a drop of water left out in the Sun, little by little, black holes evaporate and eventually disappear. As he explained, electromagnetic heat radiation carries away the mass of a black hole.

To explain why Dennis and his student thought this, I need to fill you in on some things about heat and heat radiation. I'll come back to black holes, but first a digression.

Heat and Temperature

Heat and temperature are among the most familiar concepts in physics. We all have a built-in thermometer and a built-in thermostat. Evolution provided us with a hardwired sense of cold and warmth.

Warmth is heat, and cold is its absence. But what exactly is this stuff called heat? What goes on in a bathtub of hot water that's absent when the bathwater gets cold? If you look carefully through a microscope at little tiny specks of dust or grains of pollen suspended in warm water, you will see the grains stagger around like drunken sailors. The hotter the water, the more agitated the grains appear. It was Albert Einstein[3] who in 1905 first explained that this Brownian Motion is caused by the grains being constantly bombarded by rapidly moving, energetic molecules. Water, like all other materials, is composed of molecules moving hither and thither, banging into one another, into the walls of the container, and into any foreign impurities. When that motion is random and chaotic, we call it heat. For ordinary objects, when you add energy in the form of heat, the result is an increase in the random kinetic energies of the molecules.

Temperature is, of course, related to heat. When the zigzagging molecules hit your skin, they excite nerve endings, and you experience a sense of temperature. The larger the energy of the individual molecules, the more your nerve endings are affected and the hotter you feel. Your skin is just one of the many types of thermometers that can sense and register the chaotic motion of molecules.

So, roughly speaking, the temperature of an object is a measure of the energy of its individual molecules. When an object cools, en-

3. In 1905 Einstein started two revolutions in physics and finished a third. The two new revolutions were, of course, the Special Theory of Relativity and the quantum (or photon) theory of light. The same year, Einstein gave the first definitive evidence for the molecular theory of matter in his paper on Brownian Motion. Physicists such as James Clerk Maxwell and Ludwig Boltzmann had long suspected that heat was the random motion of hypothetical molecules of matter, but it was Einstein who was able to furnish the definitive proof.

ergy is being drained away, and the molecules are slowing down. Eventually, as more and more energy is eliminated, the molecules reach the lowest possible energy state. If we could ignore Quantum Mechanics, this would happen when the molecular motion ceased altogether. At that point, there would be no more energy to drain away, and the object would be at absolute zero. The temperature cannot be lowered any further than that.

Black Holes Are Black Bodies

Most objects reflect at least a little bit of light. The reason that red paint is red is that it reflects red light. More accurately, it reflects a certain combination of wavelengths that the eye and brain perceive as red. Similarly, blue paint reflects a combination that we perceive as blue. Snow is white because the surface of ice crystals reflects all the visible colors equally. (The only difference between snow and a mirrorlike sheet of ice is that the granular structure of the snow scatters the light in all directions and breaks up the reflected image into thousands of tiny fragments.) But some surfaces reflect almost no light at all. Any light that falls on the sooty surface of a blackened pot is absorbed into the layer of soot, heating both the black exterior and eventually the iron itself. These are the objects that our brains perceive as black.

The physicist's term for a totally light-absorbing object is *black body*. By the time Sciama lectured at my university in New York, physicists had long known that black holes are black bodies. Laplace and Michell had suspected it in the eighteenth century, and Schwarzschild's solution of Einstein's equations had proved it. Light, falling onto the horizon of a black hole, is completely absorbed. Black hole horizons are the blackest of the black.

But what no one knew before Hawking's discovery was that black holes have a temperature. Before that, if you had asked a physicist, "What is the temperature of a black hole?" the initial response probably would have been, "Black holes don't have temperatures." You might have responded, "Nonsense. Everything has a temperature." A little thought might then have provoked the answer, "Okay, a black hole has no heat, so it must be at absolute zero — the lowest possible temperature." In fact, before Hawking, all physicists would have claimed that black holes were indeed black bodies, but black bodies at absolute zero.

Now, it's not correct to say that black bodies don't give off any light at all. Take a sooty pot and heat it to a few hundred degrees, and it will glow red. Even hotter, and it will glow orange, then yellow, and eventually it will have a bright bluish white appearance. Curiously, according to the physicists' definition, the Sun is a black body. How odd, you say; the Sun is about as far from being black as anything you can imagine. Indeed, the Sun's surface radiates plenty of light, but it *reflects none*. To a physicist, that makes it a black body.

Cool the hot pot, and it will glow with invisible infrared radiation. Even the coldest objects radiate some electromagnetic radiation as long as they are not at absolute zero.

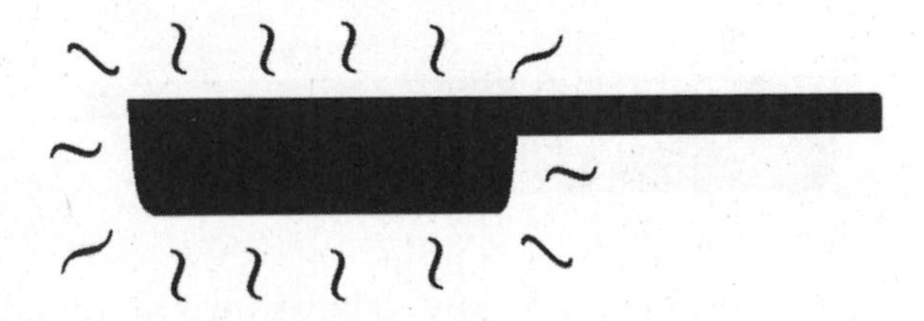

But the radiation emitted by black bodies is not reflected light; it is produced by vibrating or colliding atoms, and unlike reflected light, the color depends on the temperature of the body.

What Dennis Sciama explained was remarkable (and seemed a bit crazy at the time). He said that black holes are black bodies, but they are *not* at absolute zero. Every black hole has a temperature that depends on its mass. The formula was on the blackboard.

He told us one more thing, in some way the most surprising of all. Since a black hole has heat and temperature, it must radiate electromagnetic radiation — photons — in the same way that a hot black pot does. That means that it loses energy. According to Einstein's $E = mc^2$, energy and mass are really the same thing. So if a black hole loses energy, it also loses mass.

That brings us to the punch line of Sciama's story. The size of a black hole — the radius of its horizon — is directly proportional to its mass. If the mass decreases, it follows that the size of the black hole decreases. So as a black hole radiates energy, it shrinks, until it is no bigger that an elementary particle, and then it is gone. According to Sciama, black holes evaporate away, like puddles on a summer day.

Throughout the lecture, or at least the part I witnessed, Sciama made it very clear that he was not the source of these discoveries. It was "Stephen says this" and "Stephen says that." But despite what Dennis said, my impression at the end of the lecture was that the unknown Stephen Hawking was just a lucky student who happened to be at the right place at the right time to hook up with Dennis's research project. It is conventional for a well-known physicist to generously mention the name of a bright student during a lecture. Whether the ideas were brilliant or crackpot, I naturally assumed that they originated with the more senior physicist.

That evening I was firmly disabused of this assumption. Aage, a few other physics faculty members from Belfer, and I took Dennis to dinner at a fine Italian restaurant in Little Italy. Over the meal, Dennis told us all about his remarkable student.

In fact, Stephen was not a student at all. When Dennis spoke of "his student Hawking," he was speaking in the same way that the proud father of a Nobel Prize winner might speak of "my boy." By 1974, Stephen was a rising star in the world of General Relativity. He and Roger Penrose had made major contributions to the subject.

It was my own ignorance that had led me to believe that Stephen was merely the student of a generous thesis adviser.

Over Italian food and good wine, I heard the amazing stranger-than-fiction tale of the young genius who had blossomed only after being diagnosed with a debilitating disease. As a bright but somewhat egotistical graduate student, Stephen contracted Lou Gehrig's disease. The progression of the disease had been rapid, and he was, at the time of the dinner, almost completely paralyzed. Though unable to write equations and barely able to communicate, Stephen was holding his medical nemesis at bay and at the same time exploding with brilliant ideas. The prognosis was grim. Lou Gehrig's disease is a brutal killer, and Stephen would by all accounts be dead in a couple of years. Meanwhile, he was having a ball, joyously (Sciama's word) revolutionizing physics. At that time, Dennis's description of Stephen's courage under adversity seemed an exaggeration. But after knowing Stephen for almost twenty-five years, I would say that it sounded just like him.

Both Stephen and Sciama were unknown quantities to me, and I had no idea whether evaporating black holes were a tall tale, a wild fringe speculation, or genuine. It was possible that I had missed some important parts of the argument while being educated about Jewish toilet paper laws. More likely, Dennis had just reported Stephen's conclusion without providing any technical underpinning. After all, Sciama was not an expert on the advanced methods of Quantum Field Theory that Hawking had used. As I said earlier, he was a man of few equations.

In retrospect, it seems odd that I didn't connect Sciama's lecture with the brief conversation I'd had two years earlier with Richard Feynman at the West End Café. Feynman and I also were speculating about how black holes might eventually disintegrate. But many months would pass before I put the two together.

Stephen's Argument

By his own account, Stephen did not at first believe the strange conclusion that Jacob Bekenstein – the then unknown Princeton student – had come to. How could black holes have entropy? Entropy is associated with ignorance – ignorance of hidden microscopic structure, such as our ignorance of the precise location of the water molecules in a bathtub of warm water. Einstein's theory of gravity and Schwarzschild's black hole solution have nothing to do with microscopic entities. Moreover, it seemed that there was nothing about a black hole to be ignorant of. Schwarzschild's solution of Einstein's equations was unique and precise. For each value of mass and angular momentum, there was one and only one black hole solution. That's what John Wheeler meant when he said, "Black holes have no hair." By the usual logic, a unique configuration (recall the perfect BMW of chapter 7) should have no entropy. Bekenstein's entropy made no sense to Hawking – until he found his own way to think about it.

The key for Hawking was temperature, not entropy. By itself, the existence of entropy doesn't imply that a system has temperature.[4] A third quantity, energy, also comes into the equation. The connection between energy, entropy, and temperature goes back to the origin of thermodynamics[5] in the early nineteenth century. Steam engines were the thing then, and the Frenchman Nicolas Léonard Sadi Carnot was what you might call a steam engineer. He was interested in a very practical question: how to use the heat contained in a given amount of steam to do useful work in the most efficient way – how to get the biggest bang for the buck. In this case, useful work might mean accelerating a locomotive, which would require converting heat energy into the kinetic energy of a large mass of iron.

4. Logically, it is possible to imagine systems that can be arranged in many ways without changing their energy, but this never happens in real-world situations.
5. Thermodynamics is the study of heat.

Heat energy means the disorganized chaotic energy of random molecular motion. By contrast, the kinetic energy of a locomotive is organized into the simultaneous synchronized motion of a huge number of molecules, all moving together. So the problem was how to turn a given amount of disorganized energy into organized energy. The problem was that no one really understood exactly what organized and disorganized energy meant. Carnot was the first to define entropy as a measure of disorganization.

My own introduction to entropy was as an undergraduate mechanical engineering student. Neither I nor any of the other students knew anything about the molecular theory of heat, and I would bet that the professor didn't either. The course, Mechanical Engineering 101: Thermodynamics for Mechanical Engineers, was so confusing that I, who was by far the best student in the class, couldn't make any sense of it. Worst of all was the concept of entropy. We were told that if you heat something a small amount, the change in thermal energy, divided by the temperature, is the change in entropy. Everyone copied it down, but no one understood what it meant. It was as incomprehensible to me as "The change in the number of sausages divided by the onionization is called the floogelweiss."

Part of the problem was that I really didn't understand temperature. According to my professor, temperature is what you measure with a thermometer. "Yes," I might have asked, "but what *is* it?" I am fairly sure the answer would have been, "I told you; it's what you measure with a thermometer."

Defining entropy in terms of temperature is putting the cart before the horse. Although it's true that we all have an innate sense of temperature, the more abstract concepts of energy and entropy are much more basic. The professor should have first explained that entropy is a measure of hidden information and is given in bits. Then he could have gone on to say (correctly):

Temperature is the increase in the energy of a system when you add one bit of entropy.[6]

The energy change when you add one bit? That's exactly what Bekenstein had figured out for a black hole. Apparently, Bekenstein had calculated the temperature of a black hole without realizing it.

Hawking immediately saw what Bekenstein had missed, but the idea that a black hole has a temperature seemed so absurd to Stephen that his first reaction was to dismiss the whole thing as nonsense, entropy along with temperature. Perhaps part of the reason he reacted this way was that the evaporation of a black hole seemed so ridiculous. I don't know exactly what made Stephen rethink things, but he did. Using the sophisticated mathematics of Quantum Field Theory, he found his own way to prove that black holes radiate energy.

The term *Quantum Field Theory* reflects the confusion that was left in the wake of Einstein's discovery of photons. On the one hand, Maxwell had convincingly proved that light is a wavelike disturbance of the electromagnetic field. He and others had thought of space as something that can vibrate, almost like a bowl of Jell-O. This hypothetical Jell-O is called the Luminiferous Ether, and like Jell-O, when it is disturbed by a vibration (in the case of Jell-O, touching it with a vibrating tuning fork would work), waves spread out from the disturbance. Maxwell envisioned oscillating electric charges disturbing the ether and radiating light waves. Einstein's photons confused things for more than twenty years, until Paul Dirac eventually applied the powerful mathematics of Quantum Mechanics to the wavelike vibrations of the electromagnetic field.

The most important consequence of Quantum Field Theory for Hawking was the idea that the electromagnetic field has the "quantum jitters" (see chapter 4), even if no vibrating charges are disturb-

6. Strictly speaking, it's the temperature (measured from absolute zero) times Boltzmann's constant. Boltzmann's constant is nothing but a conversion factor that physicists often set to one by choosing convenient units for temperature.

ing it. In empty space, the electromagnetic field shimmers and vibrates with *vacuum fluctuations.* Why don't we feel the vibrations of empty space? It's not because they are very gentle. In fact, the vibrations of the electromagnetic field in a small region of space are exceedingly violent. But because empty space has less energy than anything else, there is no way for the energy in the vacuum fluctuations to be transferred to our bodies.

There is another kind of jitteriness in nature that *is* very noticeable: thermal jitters. What is the difference between a pot of cold water and a pot of hot water? The temperature, you say. But that's just a way of saying that the hot water feels hot and the cold water feels cold. The real difference is that the hot water has more energy and more entropy — the pot is full of chaotic, randomly moving molecules that are far too complicated to keep track of. This motion has nothing to do with Quantum Mechanics, and it is not subtle. Stick your finger in the pot, and you'll have no trouble detecting thermal fluctuations.

The jittery thermal motion of individual molecules can't be seen because water molecules are too small, but the immediate effect of the thermal jitters is not hard to detect. As I mentioned earlier, grains of pollen suspended in a glass of warm water will move with random, jittery Brownian Motion, which has nothing to do with Quantum Mechanics. It is due to the heat in the water, which causes the water molecules to randomly bombard the grains. When you put your finger in the glass, the same random bombardment on your skin excites the nerve endings and makes the water feel warm. Your skin and nerves absorb a bit of energy from the surrounding heat.

Even in the absence of water, air, or any other substance, heat-sensitive nerves can be excited by the thermal vibrations of black body radiation. In this case, the nerves absorb heat from the surroundings by absorbing photons. But this is possible only if the temperature is above absolute zero. At absolute zero, the quantum jitters of the electric and magnetic field are more subtle and don't have the same obvious effects.

The two kinds of jitters — thermal and quantum — are very different and under ordinary conditions don't get mixed up with each other. Quantum fluctuations are an irreducible part of a vacuum and can't be eliminated, while thermal fluctuations are due to excess energy. The subtleties of quantum fluctuations — why we can't feel them and how they differ from thermal fluctuations — are on the edge of "explainability" in a book trying to avoid complex mathematics, and any analogy or picture that I use will have its logical defects. But some explanation is necessary if you are to grasp what was at stake in the Black Hole War. Just remember Feynman's caveat about explaining quantum phenomena (see page 83).

Quantum Field Theory suggests a way of visualizing the two kinds of fluctuations. Thermal fluctuations are due to the presence of *real photons,* photons that bombard our skin and transfer energy to it. Quantum fluctuations are due to *virtual photon pairs,* which are created, then quickly absorbed back into the vacuum. Here is a Feynman diagram of space-time — time-vertical and space-horizontal — with both real photons and virtual photon pairs.

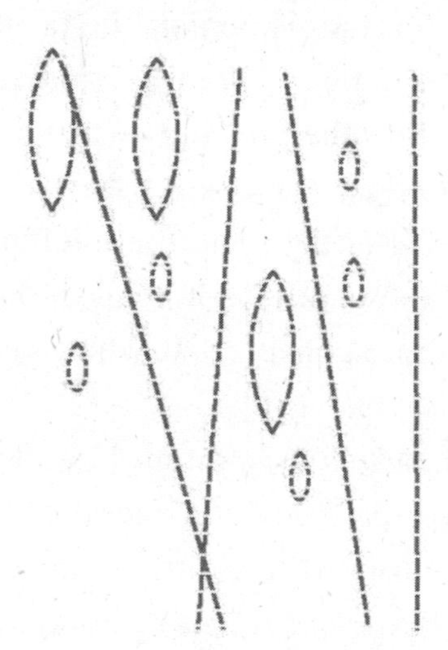

The *real photons* are the unending dashed world lines. Their presence indicates heat and thermal jitters. But if space were at absolute zero, there would be no real photons. Only the microscopic loops of *virtual photons,* rapidly flashing into and out of existence, would remain. The virtual photon pairs are part of the vacuum — what we usually think of as empty space — even when the temperature is at absolute zero.

Under ordinary conditions, there is no confusion between the two kinds of jitters. But a black hole horizon is not ordinary. Near the horizon, the two kinds of fluctuations become confused in a way that no one ever expected. To get some idea of how it happens, imagine Alice freely falling into the black hole in an environment of absolute zero temperature: a perfect vacuum. She is surrounded by virtual photon pairs, but she doesn't notice them. There are no real photons in her vicinity.

Now consider Bob, who hovers above the horizon. For him things are more confusing. Some of the virtual photon pairs — the ones that Alice doesn't notice — may be partly inside the horizon and partly outside. But a particle behind the horizon is irrelevant to Bob. Bob sees a single photon and has no way of recognizing that it belongs to a virtual pair. Believe it or not, such a photon, stranded on the outside while its partner is behind the horizon, will affect Bob and his skin exactly as if it were a real thermal photon. Near the horizon, the separation into thermal and quantum depends on the observer: what Alice detects (or doesn't detect) as quantum jitters, Bob detects as thermal energy. Thermal and quantum fluctuations become two sides of the same coin in regard to a black hole. We will come back to this point in chapter 20, when we learn about Alice's Airplane.

Using the mathematics of Quantum Field Theory, Hawking calculated that the disturbance of the vacuum fluctuations due to a black hole causes photons to be emitted, exactly as if the black hole horizon were a hot black body. These photons are called *Hawking radiation*. Most interesting, a black hole radiates as if its temperature were approximately the same as Bekenstein's argument would give, had Bekenstein made the argument. In fact, Hawking was able to go further than Bekenstein; his methods were so precise that he could calculate the exact temperature and, working backward, the entropy of the black hole. Bekenstein had claimed only that the entropy was proportional to the area of the horizon measured in Planck units. Hawking no longer needed to use the ambiguous term "proportional to." According to his calculation, the entropy of a black hole is precisely one-quarter of the horizon area measured in Planck units.

Incidentally, the equation that Hawking derived for the temperature of a black hole was the one on the blackboard when I walked into Dennis Sciama's lecture.

$$T = \frac{1}{16\pi^2} \times \frac{c^3 h}{GMk}$$

Notice that in Hawking's formula, the mass of the black hole is in the denominator. That means that the larger the mass, the colder the black hole, and conversely, the smaller the mass, the warmer the black hole.

Let's try out the formula for one black hole. Here are the values of the constants.[7]

$$c = 3 \times 10^{8}$$
$$G = 6.7 \times 10^{-11}$$
$$h = 7 \times 10^{-34}$$
$$k = 1.4 \times 10^{-23}$$

Let's take the case of a star with a mass five times that of the Sun that ultimately collapses to form a black hole. Its mass, in kilograms, would be

7. These numbers are all quoted in meters, seconds, kilograms, and degrees Kelvin. The unit of measurement on the Kelvin scale is the same as on the Celsius scale except the temperature is measured from absolute zero instead of from the freezing point of water. Ordinary room temperature is 300 degrees Kelvin.

$$M = 10^{31}$$

If we plug all of these numbers into Hawking's formula, we find that the temperature of the black hole is 10^{-8} degrees Kelvin. That's a very small temperature — about ten billionths of a degree above absolute zero! Nothing in the natural world is that cold. Interstellar, and even intergalactic, space is much warmer than that.

There are black holes of even lower temperature at the centers of galaxies. A billion times more massive than stellar black holes, and a billion times bigger, they are also a billion times colder. But we can also contemplate much smaller black holes. Suppose some cataclysmic event crushed the Earth. The mass of the Earth is about a million times smaller than the mass of a star. The resulting black hole would have the stupendous temperature of about .01 degrees above absolute zero: warmer than the stellar black hole, but still awfully cold — colder than liquid helium and far colder than frozen oxygen. A black hole with the mass of the Moon would get all the way up to 1 degree Kelvin.

But now consider what happens as a black hole emits Hawking radiation and evaporates. As the mass decreases and the black hole shrinks, the temperature rises. In time the black hole will become hot. By the time it has the mass of a large boulder, its temperature will have grown to a billion billion degrees. By the time it has reached the Planck mass, its temperature will have risen to 10^{32} degrees. The only time any place in the universe might have been anywhere near that temperature was at the beginning of the Big Bang.

Hawking's calculation showing how black holes evaporate was more than a brilliant tour de force. I believe that in time, when the repercussions are fully understood, physicists will recognize it as the beginning of a great scientific revolution. It's too early to know exactly how that revolution will play out, but it will touch on the deepest issues: the nature of space and time, the meaning of elementary particles, and the mysteries of the origin of the universe. Physicists incessantly question whether Hawking belongs among the greatest physicists of all time and where he ranks in the hierarchy. In response to those who doubt Hawking's greatness, I will only suggest that they go back and read his 1975 paper "Particle Production by Black Holes."

Yet no matter how great he may be, on at least one occasion, Stephen Hawking lost track of his bits, and that's what started the Black Hole War.

Part II

Surprise Attack

10

HOW STEPHEN LOST HIS BITS AND DIDN'T KNOW WHERE TO FIND THEM

It is impossible as I state it, and therefore I must in some respect have stated it wrong.

—SHERLOCK HOLMES

According to several newspaper articles, the war in Iraq has lasted longer than World War II. What the journalists really mean is that the war has lasted longer than America's involvement in World War II. That war began in the fall of 1939 and didn't end till the summer of 1945. Americans tend to forget that the war was into its third year by the time Pearl Harbor was attacked.

Perhaps I am making the same self-centered mistake by saying that the Black Hole War began in 1983 in Werner Erhard's attic. Stephen's attack actually dates back to 1976, but you can't have a war without two sides. Although the attack was largely ignored, it was a direct frontal assault on one of the most trusted principles of physics: *the law that says that information is never lost,* or, in its short form, *information conservation*. Because it is so central to everything that followed, let's go over the law of information conservation once more.

Information Is Forever

What does it mean for information to be destroyed? In classical physics, the answer is simple. Information is destroyed if the future loses track of the past. Surprisingly, this can happen even with a deterministic law. To illustrate, let's return to the three-sided coin that we played with in chapter 4. The three faces of the coin were called H, T, and F (for heads, tails, and feet). In that chapter, I described two deterministic laws by the following diagrams:

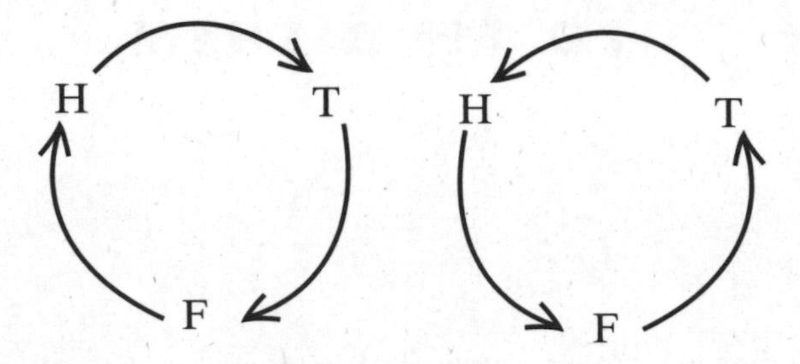

Both laws have the deterministic property that whichever state the coin is in, it is possible to say with absolute certainty both the next state and the previous state. Compare that with a law described by another diagram,

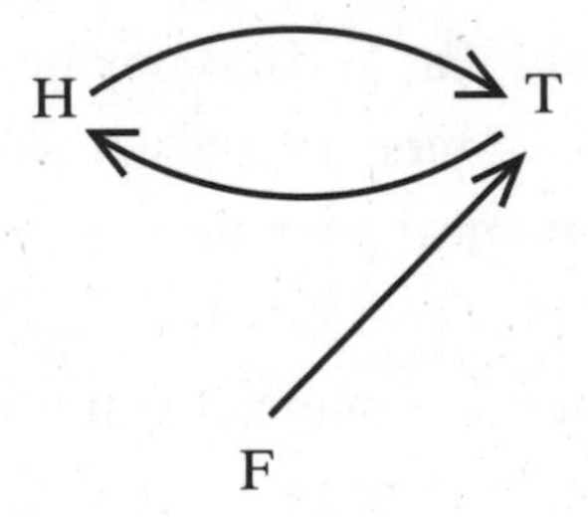

or by the formula

$$H \rightarrow T \quad T \rightarrow H \quad F \rightarrow T$$

In words: If the coin is heads at one instant, it will be tails at the next instant. If it is tails, it will become heads. And if it is feet, it will next be tails. This rule is completely deterministic: wherever you begin, the future is laid out forever by the law. Suppose, for example, that we start with F. The subsequent history is completely determined: F T H T H T H T H T.... If we start with H, the history will be H T H T H T H T H T H T.... And if we start with T, it will be T H T H T H T H T H T....

There is something odd about this law, but what is it exactly? Like other deterministic laws, the future is completely predictable. But when we try to determine the past, things break down. Suppose we find the coin in state H. We can be certain that the previous state was T. So far so good. But let's try to go back another step. There are two states that lead to T, namely H and F. That presents a problem: did we get to T from H or from F? There is no way to know. This is what it means to lose information, but it never happens in classical physics. The mathematical rules that Newton's laws and Maxwell's electromagnetic theory are based on are very clear: every state is followed by a unique state, and every state is preceded by a unique state.

The other way that information can be lost is if there is a degree of randomness in the law. In that case, it is obviously impossible to be certain of the future or the past.

As I explained earlier, Quantum Mechanics has its random element, but there is also a deep sense in which information is never lost. I illustrated this for a photon in chapter 4, but let's do it again, this time using an electron colliding with a stationary target such as a heavy nucleus. The electron comes in from the left, moving in the horizontal direction.

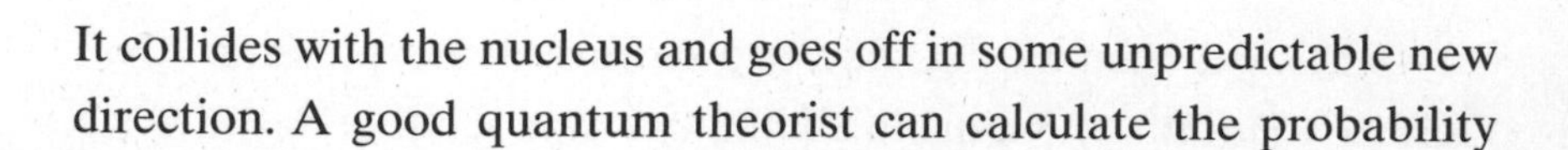

It collides with the nucleus and goes off in some unpredictable new direction. A good quantum theorist can calculate the probability that it will go off in a particular direction, but he cannot predict the direction with certainty.

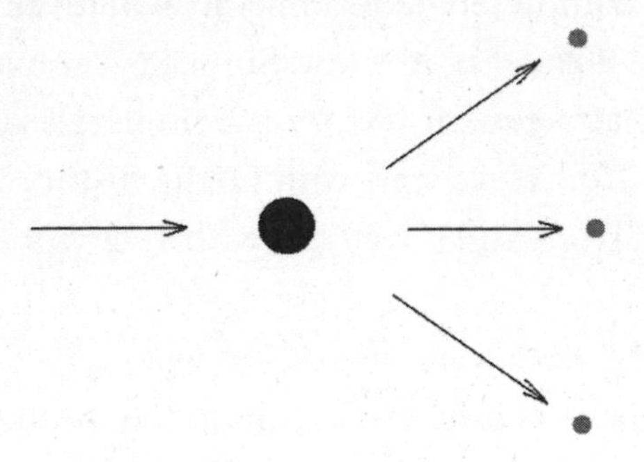

There are two ways to check whether the information regarding initial motion has been retained. Both of them involve running the electron backward with a reversed law.

In the first case, an observer checks where the electron is just before reversing the law. He can do that in various ways, most of them using photons as probes. In the second case, the observer doesn't bother to check; he just reverses the law without intervening in any way with the electron. The results of the two experiments are entirely different. In the first case, after the electron is run backward, it ends up in a random location, moving in an unpredictable direction. In the second case, when no check is made, the electron always ends up reversing its motion back along the horizontal direction. When the observer looks at the electron for the first time after starting the experiment, he will find it moving exactly as it started, except in reverse. It seems that information is lost only when we actively intervene with the electron. In Quantum Mechanics, as long as we don't interfere with a system, the information that it carries is as indestructible as it is in classical physics.

Stephen Attacks

It would have been hard to find two more sullen expressions than Gerard 't Hooft's and mine in all of San Francisco that day in 1983.

High above Franklin Street, in Werner Erhard's attic, war had been declared — a direct attack on our deepest beliefs. Stephen the Bold, Stephen the Daredevil, Stephen the Destroyer had all the heavy weapons, and the angelic/devilish smile showed that he knew it.

In no way was the attack personal. The blitzkrieg was aimed right at the central pillar of physics: the indestructibility of information. Information is often scrambled beyond recognition, but Stephen was arguing that the bits of information that fall into a black hole are permanently lost to the outside world. On the blackboard, he had the diagram to prove it.

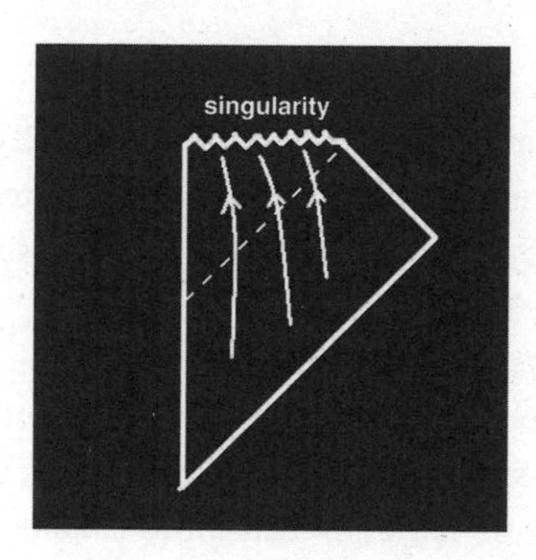

In the course of his own brilliant studies of space-time geometry, Roger Penrose had invented a way to visualize all of space-time on a single blackboard or sheet of paper. Even if space-time were infinite, Penrose would distort it, squeezing it in clever mathematical ways to get the whole thing to fit in a finite area. The Penrose diagram drawn on the blackboard in Werner's mansion showed a black hole with bits of information falling past the horizon. The horizon was shown as a thin diagonal line, and once a bit had passed that line, it could not escape without exceeding the speed of light. The diagram also showed that every such bit was doomed to hit the singularity.

Penrose diagrams are indispensable tools for theoretical physicists, but they take a little training to understand. Here's a more familiar picture that represents the same black hole.

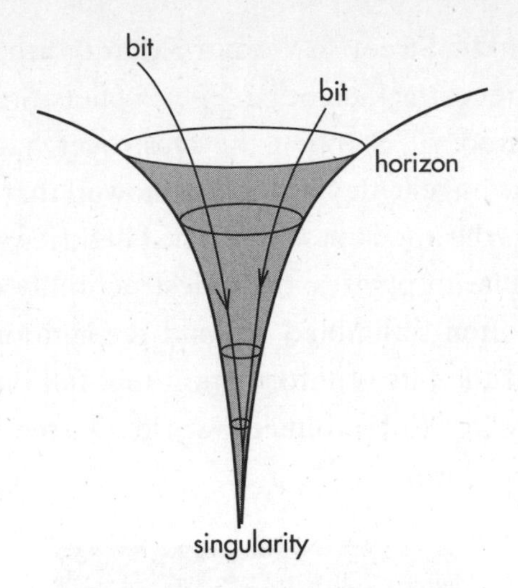

Stephen's point was simple. The bits that fall into a black hole are just like the metaphorical pollywogs in chapter 2 who carelessly fall past the point of no return. Once a bit passes the horizon, there is no way for it to get back to the outside world.

It wasn't the fact that bits of information might be lost behind the horizon that so deeply disturbed 't Hooft and me. Information falling into a black hole is no worse than locking it away in a tightly sealed vault. But something much more sinister was at play here. The possibility of hiding information in a vault would hardly be a cause for alarm, but what if when the door was shut, the vault evaporated right in front of your eyes? That's exactly what Hawking had predicted would happen to the black hole.

By 1983 I had long since made the connection between black hole evaporation and the conversation I had had with Richard Feynman in the West End Café in 1972. The idea that black holes eventually disintegrate into elementary particles didn't bother me at all. But Stephen's claim left me incredulous: *When a black hole evaporates, the trapped bits of information disappear from our universe. Information isn't scrambled. It is irreversibly, and eternally, obliterated.*

Stephen was happily dancing on the grave of Quantum Mechanics, while 't Hooft and I were in complete disarray. To us, such an idea put all the laws of physics at risk. It seemed that combining the General Theory of Relativity with the rules of Quantum Mechanics was going to result in a train wreck.

I don't know whether 't Hooft knew about Stephen's radical idea before the meeting in Werner's attic, but it was the first I had heard of it. Even so, the idea was not new at the time. Stephen had made his arguments several years earlier in a published paper and had carefully done his homework. He had already thought of and dismissed every way that I could think of to escape his "information paradox." Let's go through four of them here.

1. Black Holes Don't Really Evaporate

For most physicists, the conclusion that black holes evaporate was a very big surprise. But the argument for evaporation, though technical, was extremely compelling. By studying quantum fluctuations very near the horizon, Hawking, as well as Bill Unruh, had proved that black holes have a temperature and, like any other warm object, must emit heat radiation (black body radiation). Every so often, a physics paper will appear claiming that black holes don't evaporate. Such papers quickly disappear into the infinite junk heap of fringe ideas.

2. Black Holes Leave Remnants

Although black hole evaporation seemed solid, there was also the fact that as they evaporate, they grow hotter and smaller. At some point, an evaporating black hole will become so hot that it will emit particles of enormously high energy. In its final burst of evaporation, the emitted particles will have energy far beyond anything we have ever experienced. Very little is known about this last gasp. Perhaps the black hole will stop evaporating when it reaches the Planck mass (the mass of a dust mote). At that point, its radius will be the Planck length, and no one can say for sure what happens next. The logical possibility exists that the black hole stops evaporating and leaves a remnant – a miniature information vault – with all the lost information trapped within it. According to this idea, every bit of information that ever fell into the black hole would remain tightly sealed in the infinitesimally small lockbox. The tiny Planckian remnant would have a fantastic property: it would be an infinitesimal particle in which any amount of information could be hidden.

Although the remnant idea was a popular alternative to the destruction of information (in fact, far more popular than the correct idea), it never appealed to me. It seemed contrived to avoid the question. But it wasn't just a matter of taste. A particle that could hide an infinite amount of information would be a particle of infinite entropy. The existence of such infinitely entropic particles would be a thermodynamic disaster: created by thermal fluctuations, they would suck all the heat out of any system. To my way of thinking, remnants were not to be taken seriously.

3. Baby Universes Are Born

Now and then, I receive e-mail messages that all begin the same way: "I am not a scientist, and I don't know much about physics or mathematics, but I think I have the answer to the problem that you and Hawkins" – sometimes it's "Hawkings" and sometimes

"Haskins" – "have been working on." The solution proposed in these messages is almost always *baby universes*. Somewhere deep inside a black hole, a piece of space breaks off and forms a tiny, self-contained universe disconnected from our portion of space-time. (I always think of helium balloons slipping away and disappearing.) The author goes on to argue that all the information that ever fell into the black hole gets trapped in the baby universe. This solves the problem: the information is not destroyed; it's just floating out there in hyperspace or omnispace or metaspace or wherever it is that baby universes go. Eventually, after the black hole evaporates, the rift in space heals, and, being disconnected, the stranded bits of information become totally unobservable.

Baby universes may not be completely silly, especially if we suppose the babies grow up. Our universe *is* expanding. Perhaps each baby universe also expands and eventually matures into a universe with galaxies, stars, planets, dogs, cats, people, and its own black holes. It may even be possible that our own universe originated in this way. But as a solution to the problem of lost information, it simply begs the question. Physics is about observation and experimentation. If baby universes carry off information that becomes unobservable, the results for our world will be exactly the same as if information were destroyed, with all the unfortunate consequences of information destruction.[1]

4. Consider the Bathtub Option

The bathtub option was the least popular argument against Hawking's idea. Black hole experts and general relativists dismissed it as "missing the point." Nevertheless, it was the only possibility that

1. In chapter 1 I briefly mentioned one of the most unfortunate of these consequences: loss of information implies increase of entropy, which in turn means the production of heat. As Banks, Peskin, and I had shown, quantum fluctuations would turn into thermal fluctuations and almost instantaneously heat the world to impossibly large temperatures.

made any sense to me. Imagine drops of ink falling into a tub of water, carrying a message — drip, drip, drop, drip, drop, space, drop, drip.

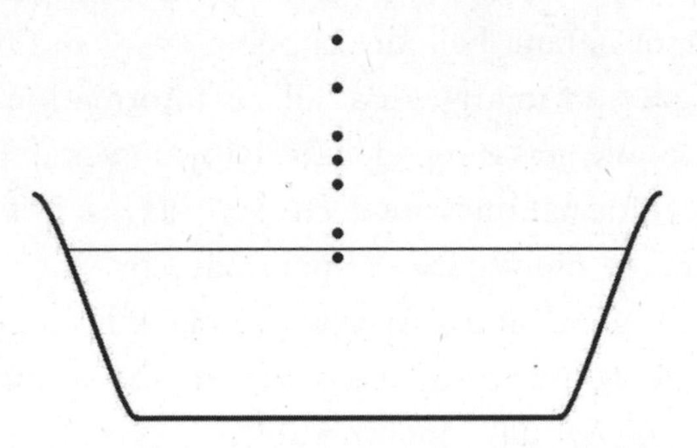

Soon the sharply defined drops begin to dissolve, the message gets harder to read, and the water becomes cloudy.

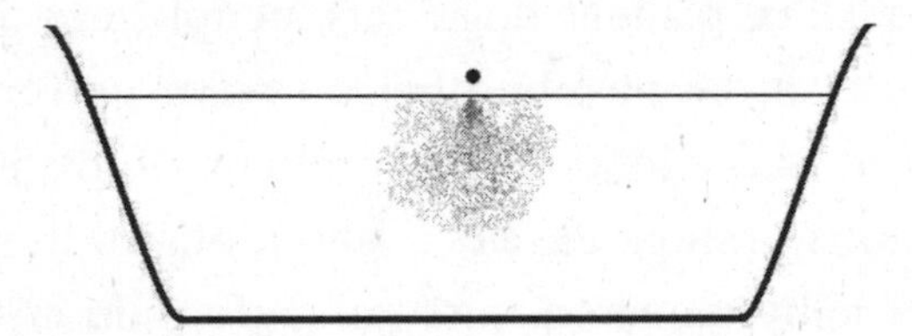

After a few hours, all that's left is a uniform tub of slightly gray water.

Although from a practical point of view, the message is hopelessly scrambled, the principles of Quantum Mechanics ensure that it's still there among the huge number of chaotically moving molecules. But soon the fluid starts to evaporate from the tub. Molecule after molecule escapes into empty space — ink as well as water — eventually leaving the tub dry and empty. The information is gone, but has it been destroyed? Though scrambled far beyond recovery by any practical scheme, not a bit of information has been erased. It's obvious what has happened to it: it's been carried off in the evaporation products, the vaporous molecular cloud escaping into space.

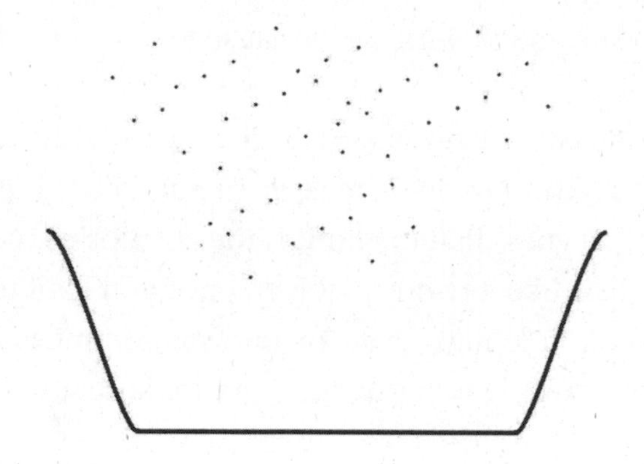

Returning to black holes, what happens to information that previously fell into a black hole when the black hole evaporates? If a black hole is anything like a bathtub, the answer is the same: every bit of information is eventually transferred to the photons and other particles that carry off the energy of the black hole. In other words, the information is stored among the many particles that make up the Hawking radiation. 'T Hooft and I felt certain that this was true, but practically no one who knew much about black holes believed us.

Here is another way to understand Stephen's information paradox. Instead of letting the black hole disappear, as it evaporates we can keep feeding it new things — computers, books, compact discs — at just the right rate to keep it from shrinking. In other words, we resupply the black hole with an endless stream of information to keep it from growing smaller. According to Hawking, even though the black hole wouldn't grow (it would evaporate as it was fed), information would be swallowed up, seemingly without end.

All of this reminds me of my favorite circus act when I was a child. I loved the clowns better than anything else, and of all the clown acts, the one that fascinated me the most was the clown car trick. I don't know how they did it, but a remarkably large number of clowns would squeeze into a very small car. But what if an endless stream of clowns just kept climbing into the car, with no one getting out? That couldn't go on indefinitely, right? The clown capacity of any car is finite, and once the capacity was saturated, something — maybe clowns, maybe sausages — would have to start coming out.

Information is like clowns, and black holes are like clown cars. A black hole of a given size has a maximum number of bits it can hold. By now you can guess that the limit is the entropy of the black hole. If a black hole is like any other object, once you load it to capacity, either the black hole must grow or information must start leaking out. But how can it leak out if the horizon is really a point of no return?

Was Stephen too stupid to see that the Hawking radiation could contain the hidden information? Of course not. Despite his youth, Stephen knew at least as much about black holes as anyone, and far more than I did. He had thought very deeply about the bathtub option and had powerful reasons for rejecting it.

The geometry of Schwarzschild's black hole was thoroughly understood by the mid-1970s. Everyone who was familiar with the subject saw the horizon as a mere point of no return. Just as in the drain hole analogy, Einstein's theory predicted that anyone inadvertently crossing the horizon would notice nothing special — the horizon is a mathematical surface with no physical reality.

The two most important facts that had been drilled into the psyches of relativists were as follows:

- There is no obstruction at the horizon that could prevent an object from passing into the interior of a black hole.

- Nothing, not even a photon — no signal of any type — can return from behind the horizon. To do so would require exceeding the speed of light — an impossibility according to Einstein.

Just to make the point clear, let's return to the infinite lake of chapter 2 and the dangerous drain at the center.

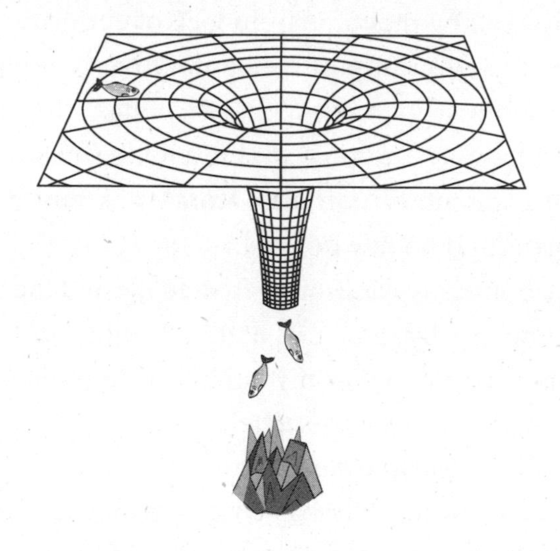

Imagine a bit of information floating downstream. As long as it hasn't crossed the point of no return, the bit can be retrieved. But there is no warning at the point of no return; the bit will float right past it, and once it does so, it can't return without exceeding the speed limit. The bit is lost forever.

The mathematics of the General Theory of Relativity was quite clear about black hole horizons. They were simply unmarked points of no return and presented no obstruction to falling objects.

That was the logic that had been deeply ingrained into the consciousness of all theoretical physicists. It was the reason that Hawking was certain that bits would not only fall through the horizon but also would be permanently lost to the outer world. So when he discovered that black holes evaporate, Stephen reasoned that the information could not escape with the radiation. It would be left behind — but behind *where?* Once the black hole evaporated, there would be nowhere left to hide.

I left Werner's in a foul mood. It was very cold by San Francisco standards, and all I had on was a light jacket. I couldn't remember where I had parked my car, and I was very irritated at my colleagues. Before leaving, I had tried to discuss Stephen's argument with them and been surprised by their apparent lack of curiosity and concern. The group consisted mostly of elementary particle physicists who had no great interest in gravity. Like Feynman, they thought that the Planck scale was so remote that it couldn't possibly affect the properties of elementary particles. Rome was burning, the Huns were at the gate, and no one noticed.

As I drove home, my windshield frosted up, and the traffic was so heavy that it repeatedly came to a halt on Route 101. I simply could not get Stephen's claim out of my head. The combination of stalled traffic and frost allowed me to scribble some diagrams and an equation or two on the windshield, but I couldn't see any way out. Either information was lost and the basic rules of physics would require complete rebuilding, or there was something fundamentally wrong with Einstein's theory of gravity near a black hole horizon.

How did 't Hooft see things? Very clearly, I would say. His resistance to Stephen's claims was unambiguous. I will explain Gerard's viewpoint in the next chapter, but first I will explain the meaning of the S-matrix, his most powerful weapon.

11

THE DUTCH RESISTANCE

Let's take the long view of history — not our own history, but that of some solar system with a central star ten times heavier than the Sun. It wasn't always a solar system; it began as a giant cloud of gas, mostly hydrogen and helium atoms, but also a bit of everything else in the periodic table. In addition, there were some free electrons and ions. In other words, it began as a very diffuse cloud of particles.

Then gravity began to do its work. The cloud started to pull itself together. Under its own weight, it contracted, and as it did so, gravitational potential energy morphed into kinetic energy. The particles began to move with greater speed, while the space between them decreased. As the cloud shrank, it also became hot, eventually hot enough to ignite and become a star. Meanwhile, not all of the gas was trapped by the star; some remained in orbit and condensed into planets, asteroids, comets, and other debris.

Tens of millions of years passed before the star ran out of hydrogen, at which point it had a short life — perhaps a few hundred thousand years — as a red supergiant. Finally, it died in a violent implosion, forming a black hole.

Then slowly, very slowly, the black hole radiated away its mass. Hawking evaporation eroded it, as its energy was emitted into photons and other particles. After some horrendously long time — something like 10^{68} years — the black hole disappeared in a final burst of high-energy particles. By that time, the planets had long since disintegrated into elementary particles.

Particles come in, and particles go out: that's the long view of history. All collisions of elementary particles, including those that take place in laboratories, begin and end the same way — particles approach, and particles recede — and some things happen in between. So in what way is the long stellar history, even if it temporarily involves a black hole, fundamentally different from *any* collision of elementary particles? Gerard 't Hooft's view was that it wasn't different and that this could be the key to explaining why Hawking was wrong.

Collisions of particles — atoms as well as elementary particles — are described by a mathematical object called the S-matrix, where S stands for scattering. The S-matrix is a gigantic table of all the possible inputs and outputs of a collision, as well as some quantities that can be processed into probabilities. It's not a table in some fat book, but a mathematical abstraction.

Consider this: An electron and a proton approach each other along the horizontal axis with a respective speed of 20 percent and 4 percent of the speed of light. What is the probability that they will collide and emerge as a final electron and proton, together with four additional photons? The S-matrix is a mathematical table of such probabilities — strictly speaking, probability amplitudes — that summarizes the quantum history of the collision. 'T Hooft deeply believed, as did I, that an entire stellar history — gas cloud→solar system→red giant→black hole→Hawking radiation — could be summarized by an S-matrix.

One of the most important properties of the S-matrix is *reversibility*. To help you understand the meaning of this term, I will give you a very extreme example. The thought experiment involves colliding two "particles." One of these particles is a bit unusual. Rather than being an elementary particle, it's composed of a huge number of plutonium atoms. In fact, this highly dangerous particle is a nuclear bomb, with a trigger so delicate that a single electron can set it off.

The other particle in the collision is an electron. The initial entry in the S-matrix table is "bomb and electron." What comes out? Smithereens. A slapdash eruption of hot gas atoms, neutrons, photons, and neutrinos. Of course, the real S-matrix would be incredibly complicated. The fragments would have to be listed in detail, along with their velocities and directions, and then a probability amplitude would be assigned to each final output. A tremendously oversimplified version of the S-matrix would look something like this:[1]

Output

Input

	electron, proton, and four photons	* * * *	smithereens	more smithereens
electron and proton	.002 + .321 i			
*				
*		Probability Amplitudes		
*				
*				
*				
electron and bomb			.012 + .002 i	.143

Now let's turn to reversibility. The S-matrix has a property called "having an inverse." Having an inverse is a mathematical way to describe the law that says that information is never lost. The inverse of an S-matrix is an operation that undoes the changes that the S-matrix does. In other words, it is exactly the same thing that I described earlier as *reversing the law*. The inverse of the S-matrix runs everything backward, from output to input. You can also think of it as reversing the motion of all the final particles and following the system

1. The real S-matrix would have an infinite number of inputs and outputs, and each box would have a complex numerical entry.

in reverse, much like running a movie backward. If, after the collision is complete, you apply the inverse (run things backward), the fragments will come together and reassemble themselves into the original bomb, including all its high-precision circuitry and delicate mechanisms. Oh, yes, and there will also be the original electron, now flying away from the bomb. In other words, the S-matrix not only predicts the future from the past but also allows you to reconstruct the past from the future. The S-matrix is a code whose details ensure that no information is ever lost.

But the experiment is very difficult. Just one little mistake — one disturbed photon — will ruin the code. In particular, you must not look at, or otherwise interfere with, even a single particle before reversing the action. If you do, instead of the original bomb and electron, all you will get is more random smithereens.

Gerard 't Hooft fought the Black Hole War under the banner of the S-matrix. His view was straightforward: the formation and subsequent evaporation of a black hole is merely a very complex example of a particle collision. It is not in any fundamental way different from the collision of an electron and a proton in the laboratory. In fact, if one could increase the energy of the colliding electron and proton to prodigious proportions, the collision would create a black hole. The collapse of a gas cloud is only one way of forming a black hole. Given a big enough accelerator, just two colliding particles could create a black hole that would subsequently evaporate.

For Stephen Hawking, the fact that the S-matrix implies information conservation proved that it must be the wrong description of black hole history. His view was that the precise details of the gas cloud — whether it was made of hydrogen, helium, or laughing gas — would go down the drain, past the point of no return, and disappear with the black hole when it evaporated. Whether the original gas had lumps or was smooth, precisely how many particles it contained, and all such details would be lost forever. Reversing all the final particles and letting the whole thing run backward would not reconstruct the original input. According to Hawking, the result of reversing the final radiation would just be more undifferentiated Hawking radiation.

If Hawking was right, the entire episode – particles→black hole→Hawking radiation – could not be described by the usual mathematics of the S-matrix. And so Stephen invented a new concept to replace it. The new code would have an extra degree of randomness that would obliterate the original information. To replace the S-matrix, Stephen invented the "not-S-matrix." He called it the $-matrix and it came to be known as the *Dollar-matrix.*

Like the S-matrix, the Dollar-matrix is a rule for relating what goes in to what comes out. But instead of preserving the differences inherent in distinct starting points, in the case of a black hole, it blurs those distinctions to the point that no matter what goes in – Alice, a baseball, three-day-old pizza – when reversed, exactly the same thing comes out. Throw your computer with all its files into a black hole. Out comes featureless Hawking radiation. If you reverse the action, the S-matrix will spit out the computer, but the $-matrix will dribble out more featureless Hawking radiation. According to Hawking, all memory of the past is lost in the heart of the transient black hole.

It was a frustrating stalemate: Gerard said S-matrix; Stephen said $-matrix. Stephen's arguments were clear and persuasive, but Gerard's faith in the laws of Quantum Mechanics was unshakable.

Perhaps, as some people claim, Gerard and I resisted Stephen's conclusion because we were particle physicists, not relativists. Almost all the methodology of particle physics revolves around the principle that collisions between particles are governed by a reversible S-matrix. But I don't think it was particle physics chauvinism that underlay our refusal to abandon the rules. All hell would break loose in all of physics, not just black hole physics, once the door to information loss was opened. Stephen's challenge had ignited a fuse to a bundle of theoretical dynamite.

Given that, perhaps it's a good time to explain why physicists believe that a bomb explosion can be reversed. It's certainly not possible to try it out in the lab. But let's suppose that it was possible to capture all the outgoing atoms and photons and to turn them around. If it could be done with infinite precision, the laws of physics would lead to a reconstructed bomb. But any tiny mistake, perhaps a single lost photon or even a tiny error in the direction of one photon, would be disastrous. Minute errors have a way of getting magnified. A single sperm missing its target might have changed history if that sperm had belonged to Genghis Khan's father. In billiards, an infinitesimal change in the way the balls are initially stacked or in the direction of the first shot will become magnified after a few collisions and lead to a completely different outcome. And so it is with an exploding bomb or a colliding pair of high-energy particles: a tiny mistake in reversing the motion, and the outcome will be nothing like the original bomb or particles.

So how can we be sure that a perfect reversal of the fragments will reassemble into the bomb? We know it because the fundamental mathematical laws of atomic physics are reversible. Those laws have been tested with incredible precision, in contexts much simpler than bombs. A bomb is nothing but a collection of atoms. It is far too complicated to follow the development of some 10^{27} atoms as the explosion unfolds, but our knowledge of the laws of atoms is very secure.

But what replaces atoms, and the laws of atomic physics, when an exploding bomb is replaced by an evaporating black hole? Although 't Hooft had many brilliant insights into the nature of horizons, he had no clear answer to that question. Oh, he knew that the replacement for atoms had to be the microscopic objects that make up the entropy of the horizon. But what were they, and what were the precise laws governing the way they would move, combine, separate, and recombine? 'T Hooft didn't know. Hawking and most other relativists simply dismissed the idea of such a microscopic underpinning, declaring, "The Second Law of Thermodynamics tells us that physical processes cannot be reversed."

In fact, that is not what the Second Law says. It says only that reversing physics is incredibly hard, and the littlest error will defeat your effort. Moreover, you better know the exact details — the microstructure — or you will fail.

My own view in those early years of the controversy was that the S-matrix, not the $-matrix, was right. But just saying "S, not $" would not be convincing. The best thing to do was to try to discover the mysterious microscopic origin of black hole entropy. Above all, it was necessary to understand what was wrong with Stephen's argument.

12

WHO CARES?

No one is ever going to use Hawking radiation to cure cancer or make a better steam engine. Black holes will never be useful for storing information or swallowing enemy missiles. Even worse, unlike elementary particle physics or intergalactic astronomy — two subjects that also may never have any practical applications — the quantum theory of black hole evaporation will probably never even lead to direct observations or experiments. Why, then, does anyone waste his or her time on it?

Before I tell you why, let me first explain why Hawking radiation is unlikely ever to be observed. Let's grant that in the future, we will be able to get close enough to an astronomical black hole to observe it in some detail. Even then, there is no chance of observing it evaporate for one very simple reason: no astronomical black hole is currently evaporating. Quite the opposite, they are all absorbing energy and growing; even the most isolated black hole is surrounded by heat. The emptiest regions of intergalactic space, cold as they are, are still far warmer than a stellar-mass black hole. Space is filled with black body radiation (photons) left over from the Big Bang. The coldest place in the universe is a sultry three degrees above absolute zero, whereas the warmest black holes are a hundred million times colder.

Heat, thermal energy, always flows from warm to cold, never the other way, so it follows that radiation from the warmer regions of space flows into the colder black holes. Instead of evaporating and shrinking, as they would if the temperature of space was absolute zero, real black holes are constantly absorbing energy and growing.

Space was once much hotter than it is now, and in the future the expansion of the universe will make it colder. Eventually, in a few hundred billion years, it will cool to the point where it is colder than stellar black holes. When that happens, black holes will begin to evaporate. (Will there be anyone around to see it? Who knows, but let's be optimists.) Still, the evaporation will be extremely slow — it will take at least 10^{60} years to detect any change in the mass or size of a black hole — so it is unlikely that anyone will ever detect a black hole shrinking. Finally, even if we effectively had all the time in the world, there would be no hope of unscrambling the information contained in Hawking radiation.

If deciphering the messages in Hawking radiation is so utterly hopeless and there is no practical reason to do so, why has the problem fascinated so many physicists? In a sense, the answer is a very selfish one: we do it to satisfy our curiosity about how the universe works and how the laws of physics fit together.

The truth is, much of physics follows this pattern. On occasion practical questions have led to profound scientific developments. Sadi Carnot, the steam engineer, revolutionized physics while trying to build a better steam engine. But more often, pure curiosity has led to the great paradigm shifts in physics. Curiosity is like an itch; it demands to be scratched. And for a physicist, nothing itches more than a paradox, an incompatibility among the various things that one thinks one knows. Not knowing how something works is bad enough, but finding contradictions among things you thought you knew is unbearable, particularly when a clash of fundamental principles is involved. It's worth recalling a few such clashes and how they drove physics to its most far-reaching conclusions.

The ancient Greek philosophers left a paradoxical legacy, a clash of two incompatible theories governing completely separate worlds of phenomena: the celestial and the terrestrial. *Celestial* referred to the world of heavenly bodies, what we call astronomy. The celestial world was a better, cleaner, more perfect world — a world of perfect, eternal clockwork precision. Indeed, according to Aristotle, all heavenly bodies moved on one of fifty-five perfect concentric crystal spheres.

By contrast, the laws of terrestrial phenomena were deemed to be corrupt. Nothing moved simply on the Earth's grungy surface. A heavy cart would wobble and grind to a halt unless a horse continued to pull it. Lumps of matter would fall gracelessly to the ground, where they would remain. These basic laws governed the four elements: fire rises, air hovers, water falls, and earth sinks to the lowest elevation.

The Greeks were apparently happy with two completely different sets of rules. But Galileo, and to an even greater extent Newton, found this dichotomy intolerable. Galileo's simple thought experiment demolished the idea that there could be two separate sets of laws of nature. He imagined standing on the top of a mountain and throwing a stone, first hard enough that it landed a few yards from his feet; then harder, so it traveled a thousand miles before landing; and then even harder, so that it traveled the whole circumference of the Earth. He realized that the stone would orbit the Earth in a circular orbit. This created a new paradox: how could the laws of terrestrial phenomena be so completely different from the laws of celestial phenomena if a terrestrial stone could become a celestial body?

Newton, who was born in the year of Galileo's death, solved the puzzle. He realized that the same law of gravity that made apples fall from trees also held the Moon in orbit around the Earth, and the Earth in orbit around the Sun. His laws of motion and gravity were the first comprehensive physical laws with universal validity. Did Newton know how useful they would be to future aerospace engineers? It is doubtful that he would have cared. Curiosity, not practicality, drove him.

The next great itch that comes to mind is the one that Ludwig Boltzmann scratched so strenuously. Again a clash of principles: how could a one-way law requiring entropy always to increase coexist with Newton's reversible laws of motion? If, as Laplace believed, the world was made of particles obeying Newton's laws, then it should be possible to run it in reverse. Eventually, Boltzmann solved the problem, first by recognizing that entropy is *hidden microscopic information* and then by realizing that entropy does not *always* increase. Every so often, an unlikely event occurs. You shuffle a random deck, and by pure chance it comes up in perfect numerical order, with hearts following diamonds following clubs following spades. But entropy-decreasing events are the very rare exceptions. Boltzmann solved the paradox by saying that *entropy almost always increases*. Today Boltzmann's statistical vision of entropy is the foundation for the practical information sciences, but to him the puzzle of entropy was just a terrible itch that needed to be scratched.

It is interesting that in the cases of Galileo and Boltzmann, the clashes were not uncovered by surprising new experimental discoveries. The key in each case was the right thought experiment. Galileo's rock-throwing and Boltzmann's time-reversal experiments never had to be carried out; it was enough just to think of them. But the greatest master of thought experiments was Albert Einstein.

Two profoundly disturbing contradictions afflicted physicists at the turn of the twentieth century. The first was a conflict between the principles of Newtonian physics and Maxwell's theory of light. The Principle of Relativity, which we associate so closely with Einstein, really goes back to Newton and, even further, to Galileo. It is a simple statement about the laws of physics as seen from different frames of reference. To illustrate it, imagine a circus performer — a juggler of balls — who boards a train to get to the next town. While on the train, he feels the need for a little practice juggling. Having never tried juggling on a moving train, he wonders, "Will I need to compensate for the motion of the train every time I throw a ball into the air and retrieve it? Let's see. The train is moving west. So whenever I make a catch, I'd better reach east a bit." He tries it with a single ball. Up it goes, east goes his catching hand, and plop goes the ball on the floor. He tries again, this time decreasing the eastward compensation. Plop again.

Now as it happens, the train is of extremely high quality. The rails are so smooth and the suspension system so perfect that the motion of the train is undetectable to the passengers. The juggler laughs and says to himself, "I understand. Without my being aware of it, the train slowed down and came to a stop. Until it starts again, I can practice in the usual way. I'll just revert back to the good old laws of juggling." It works perfectly.

Imagine his surprise when he looks out the window and sees the countryside whizzing past at a good 90 miles per hour. Deeply puzzled, the juggler asks his friend the clown (who happens to be a Harvard physics professor during the off-season) for clarification. Here is what the clown says: "According to the principles of Newtonian mechanics, the laws of motion are the same in all reference frames, as long as they are moving with uniform velocity relative to each other. Thus, the laws of juggling are exactly the same in the reference frame at rest on the ground and the reference frame traveling with the smoothly moving train. It is impossible to detect the motion of the train by any experiment done entirely inside the railroad car. Only by looking out the window are you able to tell that the train is moving with respect to the ground, and even then, you can't tell whether it is the train or the ground that's moving. All motion is relative." Amazed, the juggler picks up his balls and continues his practice.

All motion is relative. The motion of a railroad car with a velocity of 90 miles per hour, the motion of the Earth around the Sun at 30 kilometers per second, and the motion of the solar system around the galaxy at 200 kilometers per second — all of these are undetectable as long as they are smooth.

Smooth? What does that mean? Consider the juggler as the train starts out. Suddenly, it lurches forward. Not only do the balls get jerked backward, but the juggler himself might even get knocked off his feet. When the train comes to a halt, something similar happens. Or suppose the train goes around a sharp bend. Surely, in all these situations, the rules of juggling will require modification. What is the new ingredient? The answer is *acceleration*.

Acceleration means change in velocity. When the railroad car lurches forward, or when it comes to a sudden halt, the velocity changes and acceleration happens. What about when it goes around a bend? It may be less obvious, but it is nonetheless true that the velocity is changing — not the magnitude of the velocity, but its *direction*. To a physicist, any change in velocity is called acceleration, be it in magnitude or direction. Thus, the Principle of Relativity has to be sharpened:

> *The laws of physics are the same in all reference frames that are moving with uniform velocity (without acceleration) relative to each other.*

The Principle of Relativity was first formulated about 250 years before Einstein was born. So why is Einstein so famous? It's because he revealed the apparent conflict between the Principle of Relativity and another principle of physics — a principle that we might call Maxwell's Principle. As discussed in chapters 2 and 4, James Clerk Maxwell discovered the modern theory of electromagnetism — the theory of all electric and magnetic forces in nature. Maxwell's greatest discovery was unraveling the great mystery of light. Light, he argued, consists of waves of electrical and magnetic disturbances moving through space like waves through the sea. But for us the most important thing that Maxwell proved is that light moving through empty space always moves at exactly the same speed: approximately 300,000 kilometers per second.[1] That's what I call Maxwell's Principle:

> *Light moving through empty space, no matter how it was created, always moves at the same velocity.*

But now we have a problem: a serious clash between two principles. Einstein was not the first to worry about the clash between the Principle of Relativity and Maxwell's Principle, but he saw the problem most clearly. And whereas others were troubled by experimental data, Einstein — master of thought experiments — was troubled by an experiment that took place entirely within his head. According to his own recollection, in 1895, at the age of sixteen, Einstein produced the following paradox. Picturing himself riding in a railroad carriage *moving with the speed of light,* he observes a light wave moving alongside him in the same direction. Would he not see the light ray standing still?

There were no helicopters in Einstein's day, but we might imagine him hovering above the sea, moving with exactly the speed of ocean waves. The waves would appear to be standing still. In the same way, the sixteen-year-old reasoned that the passenger in the railway carriage (remember, he is moving with the speed of light) would detect a completely motionless light wave. Somehow, at that early age, Einstein knew enough about Maxwell's theory to realize that what he was imagining was impossible: Maxwell's Principle asserted that all light moves with the same velocity. If the laws of nature are the same in all reference frames, then Maxwell's Principle had better apply in the moving train. Maxwell's Principle and the Principle of Relativity of Galileo and Newton were on a collision course.

Einstein scratched that itch for a decade before he saw the way out. In 1905 he wrote his famous paper "On the Electrodynamics of Moving Bodies," in which he postulated an entirely new theory of

1. When light moves through water or glass, it moves somewhat slower.

space and time — the Special Theory of Relativity. The new theory radically changed the concepts of length and duration (of time) and especially what it means for two events to be simultaneous.

At the same time that he was figuring out Special Relativity, Einstein was puzzling over another paradox. At the turn of the twentieth century, physicists were extremely perplexed by black body radiation. Recall that in chapter 9, I explained that black body radiation is the electromagnetic energy that is produced by a glowing-hot object. Imagine a completely empty closed container at absolute zero. The interior of the vessel is a perfect vacuum. Now let's warm the vessel from the outside. The exterior walls begin to radiate black body radiation, and so do the interior walls. The radiation from the interior walls goes into the closed space inside the vessel and fills it with black body radiation. Electromagnetic waves of all different wavelengths rattle around, bouncing off the interior walls: red light, blue light, infrared, and all the colors of the spectrum.

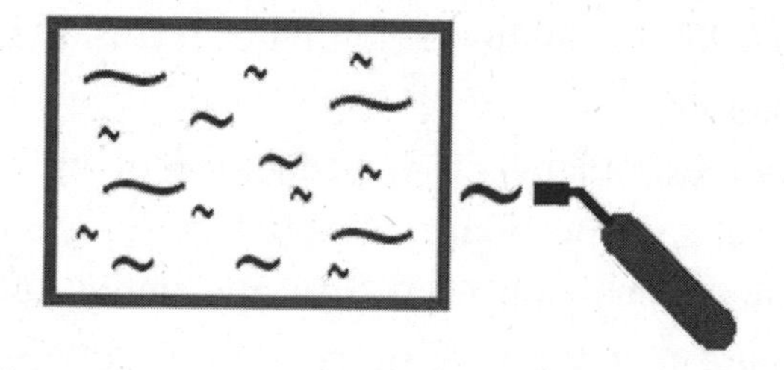

According to classical physics, each wavelength — microwave; infrared; red, orange, yellow, green, blue, and ultraviolet waves — should contribute the same amount of energy. But why stop there? Even shorter wavelengths — X-rays, gamma rays, and tinier and tinier wavelengths — should also contribute the same amount of energy. Since there is no limit to how short a wavelength can be, classical physics predicted an *infinite* amount of energy in the vessel. This was manifest nonsense. All that energy would instantly vaporize the vessel, but what exactly was wrong?

The problem that this led to was so bad that in the late nineteenth century it became known as the *ultraviolet catastrophe*. Again the problem resulted from a clash of deeply believed principles, both of which were very difficult to give up. On the one hand, the wave theory was incredibly successful in explaining the well-known properties of light, such as diffraction, refraction, reflection, and, most impressively, interference. No one was ready to abandon the wave theory, but on the other hand, the principle that each wavelength should have the same energy, called the Equipartition Principle, followed from the most general aspects of the theory of heat: in particular that heat is random motion.

In 1900 Max Planck contributed some important new ideas that got close to resolving the dilemma. But it was Einstein, in 1905, who found the correct answer. With no hesitation, the unknown patent clerk made an incredibly bold move. Light, he said, is not a continuous smear of energy as imagined by Maxwell. It is composed of indivisible particles of energy, or quanta, later called photons. One can only marvel at the sheer arrogance of the young man who told the greatest scientists in the world that everything they knew about light was wrong.

The hypothesis that light is composed of indivisible photons whose energy is proportional to their frequency solved the problem. Applying Boltzmann's statistical mechanics to these photons, Einstein found that the very short wavelengths (high frequency) have less than a single photon. Less than one means none. So the very short wavelengths carry no energy, and the ultraviolet catastrophe ceased to exist. But this was not the end of the discussion. It would take almost thirty years for Werner Heisenberg, Erwin Schrödinger, and Paul Dirac to reconcile Einstein's photons with Maxwell's waves. But Einstein's breakthrough opened the door.

The General Theory of Relativity, Einstein's greatest masterpiece, was also born out of a simple thought experiment about a conflict of principles. The thought experiment was so simple that a child could have performed it. All it involved was the everyday observation that when a train accelerates from rest, the passengers get pushed back against their seats, much as if the railroad car had been tipped upward so that gravity would pull them toward the rear of the car. So how, he asked, can we tell that a reference frame is accelerated? Accelerated relative to what?

Einstein's answer as conveyed by the clown: *we cannot tell.* "What?" the juggler says. "Of course you can. Didn't you just tell me that you get pushed against the back of your seat?" "Yes," answers the clown, "exactly as if someone had tipped the car up so that gravity pulled you back." Einstein seized on that idea: it is impossible to tell acceleration from the effect of gravity. The passenger has no way of knowing whether the train is starting out on its journey or gravity is pulling him back in his seat. Out of paradox and contradiction was born the Equivalence Principle:

> *The effects of gravity and acceleration are indistinguishable from one another. The effect of gravity on any physical system is exactly the same as the effect of acceleration.*

Again and again, we see the same pattern. At the risk of overstating the point, the greatest advances in physics have been uncovered by thought experiments that revealed a clash between deeply held principles. The present is no different from the past in this respect.

The Clash

Let's come back to the original question posed at the beginning of this chapter: why should we care whether information is lost in black hole evaporation?

In the days and weeks after the meeting in Werner Erhard's attic, it began to dawn on me that Stephen Hawking had put his finger on a clash of principles that rivaled the great paradoxes of the past. Something was seriously awry with our most basic concepts of space and time. It was obvious — Hawking had said it himself — that the Equivalence Principle and Quantum Mechanics were on a collision course. The paradox could bring down the whole structure, or reconciling the two could bring deep new insight into both.

For me the clash created an unbearable itch, but it wasn't very contagious. Stephen seemed satisfied with the conclusion that information is lost, and few others seemed to care very much about the paradox at all. Throughout the decade from 1983 to 1993, this complacency bothered me a great deal. I could not understand how everyone — above all Stephen — could fail to see that reconciling the principles of Quantum Mechanics with those of relativity was the great problem of our generation — the great opportunity to match the achievements of Planck, Einstein, Heisenberg, and the other heroes of the past. I felt that Stephen was being dense in failing to see the depth of his own question. It became something of an obsession to convince Stephen and the others (but mostly Stephen) that the key was not to abandon Quantum Mechanics, but to reconcile it with the theory of black holes.

It seems obvious to me — and I'm sure that Stephen, Gerard 't Hooft, John Wheeler, and just about every relativist, string theorist, and cosmologist that I know would agree — that having two incompatible theories of nature is intellectually intolerable and that the General Theory of Relativity must be made compatible with Quantum Mechanics. But theoretical physicists are a contentious bunch.[2]

2. Recently, I was very surprised to find that not everyone agrees. In a review of Brian Greene's book *The Fabric of the Cosmos,* Freeman Dyson made a remarkable declaration: "As a conservative, I do not agree that a division of physics into separate theories for large and small is unacceptable. I am happy with the situation in which we have lived for the last 80 years, with separate theories for the classical world of stars and planets and the quantum world of atoms and electrons." What could Dyson have been thinking? That like the ancients before Galileo, we should accept two *unbridgeable* theories of nature? Is that conservative? Or is it reactionary? To my ear, it just sounds incurious.

13

STALEMATE

When I was young and people — especially at parties or social gatherings — would ask me what I did for a living, I didn't really want to talk about it. It wasn't that I was ashamed or embarrassed. It was just too hard to explain. So to avoid the subject, I would say, "I'm a nuclear physicist, but I can't talk about it." It worked in the sixties and seventies, but not now that the cold war is over.

I still have a little trouble with the question, though for a different reason: I don't know exactly what the answer is. The obvious reply, "I'm a theoretical physicist," usually leads to another question, "What kind of physics do you do?" That's the point where I get stuck. I could say that I am an elementary particle physicist, but I've also worked on big things such as black holes and the whole universe. I could say that I'm a high energy physicist, but I sometimes work on the lowest energies — even the properties of empty space. There is just no good name for what I and most of my friends are interested in. Being called a string theorist irritates me; I don't like being pigeonholed so narrowly. I'd like to say that I work on the fundamental laws of nature, but that sounds pretentious. So usually my answer is that I am a theoretical physicist and I've worked on many things.

In fact, before the early 1980s, most of what I worked on could legitimately be called elementary particle physics. Around that time, however, the field became somewhat stagnant. The Standard Model of particle physics was a done deal, and its most interesting variants had been worked out. It was just a matter of time — a lot of time — until accelerators could be built to test these variants. So the truth is that I was a bit bored and decided to see what I could figure out about quantum gravity. After a few months, I began to worry that Feynman was right — quantum gravity was too remote; there seemed to be no way to make any progress. It wasn't even clear to me what the problems were. John Wheeler, in his inimitable way, had said, "The question is — what is the question?" and I certainly didn't see the answer. I was on the verge of returning to conventional particle physics when all of a sudden, Stephen dropped the bomb that answered Wheeler's query: the question is, How do we rescue physics from the anarchy of information loss?

If particle physics at the time was stagnant, so was the quantum theory of black holes, and it stayed that way for about nine years. Even Hawking published nothing about black holes during the years 1983 to 1989. I have been able to find only eight journal articles from that entire period that address the question of information loss in black holes. I wrote one of them, and 't Hooft wrote all the rest, largely expressing his faith in the S-matrix rather than Hawking's $-matrix.

The reason that I published almost nothing about black holes for the nine years after 1983 was that I simply could find no way to solve the conundrum. During that time, I found myself going in circles, asking the same questions over and over, and coming up against the same impenetrable obstacles. Hawking's logic was so clear: the horizon is merely a point of no return, and nothing that crosses it can get back out. The reasoning was persuasive, but the conclusion was absurd.

Here is how I explained the problem in a lecture to a group of amateur physics and astronomy enthusiasts in San Francisco sometime in 1988.[1]

Paradox of the Very Big Black Hole: A San Francisco Lecture

I want to draw your attention to a serious conflict of principles that was first described about thirteen years ago by Stephen Hawking. My reason for bringing it up is that it indicates a very serious crisis that needs to be resolved before we can hope to understand the most profound questions of physics and cosmology. These questions involve gravity on the one hand and quantum theory on the other.

Why, you may ask, will we ever need to mix these two realms of experience together? After all, gravity deals with the very large and the very heavy, while Quantum Mechanics governs the world of the very small and light. Nothing is both heavy and light at the same time, so how can both theories be important in the same context?

Let's start with elementary particles. As you all know, the gravitational force between electrons and the atomic nucleus is incredibly weak by comparison with the electrical forces that hold the atom together. The same is true, but even more so, for the nuclear forces that hold quarks together in a proton. In fact, the gravitational force is about a million billion billion billion billion times weaker than the usual forces. So it

1. What follows is an approximate reconstruction of the lecture based on notes that I still have. I have taken some liberties in order to replace the equations with words. The story "Don't Forget to Take Your Antigravity Pills" was intended for a popular science magazine. It never made it to the final draft, but in a shortened form, it was part of the San Francisco lecture.

is clear that gravity plays no important role in atomic or nuclear physics — but what about elementary particles?

Ordinarily, we think of particles such as the electron as infinitely small points of space. But that can't be the whole truth. The reason is that elementary particles have so many properties that make them different from one another. Some are electrically charged, while others are not. Quarks have properties with names such as *baryon number, isospin,* and the misnamed *color.* Particles spin like toy tops about an axis. It's unreasonable to think that a mere point could possess so much structure and variety. Most of us particle physicists believe that if we could examine particles down to some incredibly small size-scale, we would begin to see the hidden machinery that makes them tick.

If it's really true that electrons and their various cousins are not infinitely small, they must have some size. But all we really know by direct observation (smashing them together) is that they are no bigger than about one ten-thousandth the size of the atomic nucleus.

But extraordinary things are happening. In recent years, we have been accumulating indirect evidence that the machinery in the interior of particles is not much bigger, nor is it much smaller, than the Planck length. Now the Planck length has a very dramatic meaning for a theoretical physicist. We are used to thinking of gravity as so much weaker than electrical and subnuclear forces that it is completely irrelevant to the behavior of elementary particles, but this is not so when bits of matter approach one another to within the Planck length. By that point, gravity has become not only as strong as the other forces but even stronger.

What all of this means is that down at the bottom of the world — distances so small that even electrons are complicated structures — gravity may be the most important force holding those particles together. So you see, gravity and

Quantum Mechanics may well come together at the Planck scale and explain the properties of electrons, quarks, photons, and all their friends. We elementary particle physicists had better get quantum gravity straight.

Cosmologists, too, can avoid a quantum theory of gravity for only so long. As we follow the universe back into the past, we know that it was more densely packed with particles. Today [1988] the photons that make up the CMB[2] are about a centimeter apart, but when they were first emitted, they were a thousand times closer. As we trace backward, the particles get packed together like sardines in an ever smaller can. It seems likely that at the time of the Big Bang, they may have been no farther apart than the Planck length. If so, the particles were so close together that the most important forces between them were gravitational. In other words, the same quantum gravity forces that hold the key to elementary particles may also be the primary forces responsible for the Big Bang.

So, granted that quantum gravity is important to our future (and our past), what do we know about it? Not a lot other than that quantum theory and gravity clash in a big way, especially about black holes. That's a good thing, because it means that we have a chance of learning important things by resolving that clash. Today I'm going to tell you a short story that illustrates the problem — not the solution — just the problem.

Don't Forget to Take Your Antigravity Pills

Year 8,419,677,599

Long ago the Earth escaped from its orbit around the now dead star Sol. After wandering for countless generations, it found its place orbiting a giant black hole somewhere in the

2. Cosmic microwave background — the radiation that was originally emitted by the Big Bang.

Coma Supercluster. The entire planet has been governed by the same corporation ever since the late twenty-first century, when a bloodless coup left all power in the hands of the pharmaceutical industry.

"Yes, Count Geritol, what is it now? You've been promising action for the last five years. Are you wasting my time with another 'progress' report?"

"Please, Your Royal Highness, this lowly, worthless worm begs your royal pardon for his unpardonable stupidity, but this time I have truly good news. We've caught him!"

His Royal Highness, Emperor Merck LLXXXVI, frowns for a moment. Then turning his enormous bald head toward the count — Minister of Bogus Information Generation and Antirational Science Enforcement — he pins the count to the wall with his piercing gaze. "Fool. Who did you catch? A codfish?"

"No, Excellency. It's the heretic, Great One. We caught that equation-solving son of a filthy physicist — the one who has been infecting our people with ugly rumors that antigravity pills are fraudulent. Right now he is chained to the wall in the antechamber. Shall I bring him in?" The count bends his weasel-face into a sycophantic grin. "I bet he could use some Valium right now. Ha-ha."

A smile briefly flickers across His Royal Highness's visage. "Bring the dog in."

The prisoner, ragged and bruised, but unrepentant, is brutally flung to the floor at Geritol's feet. "What is your name, dog, and who is your family?" Rising to his feet and defiantly brushing the dust off his tunic, the prisoner looks his persecutor straight in the eye and proudly answers, "My name is Steve."[3]

3. By the late twentieth century, a good fraction of the world's great physicists were named Steve. Steve Weinberg, Steve Hawking, Steve Shenker, Steve Giddings, and Steve Chu were among the many Steves of physics. In the late twenty-first century, those who aspired to be parents of great physicists started naming their children (female as well as male) Steve.

With a long, defiant pause — too long for the count's comfort — he continues. "I descend from an ancient line that traces its roots back to the Black Hole War. My ancestor was Stephen the Bold of Cambridge."

The emperor's features momentarily cloud over with uncertainty, but recovering his poise, he smiles. "Well, Steve — I presume that Dr. is the appropriate title — now look where your ancient lineage has gotten you. Your existence offends me. The only question is how to eliminate your presence."

Later, as the artificial Sun sets in the west, Steve is brought his last meal. As if to mock him, the emperor has sent the choicest morsels from his own table, along with a message of "sympathy." With head hung low, the glum prison guard (Steve was well liked by the prison staff) reads the message. To the guard, it seems the worst possible news. "Tomorrow at first hour, you, your family, and all your heretical friends are to be put aboard a small habitable planet and then dropped headlong into the abyss — the giant maw of dark fire and heat that surrounds the black hole. First you will feel uncomfortably warm. Soon your flesh will cook and your blood will boil. Your bits will be scrambled until they evaporate, irreversibly scattered to the heavens." For no apparent reason, Steve's features relax in a faint smile. "An odd reaction to bad news," the prison guard thinks.

The emperor and the count rise early the next day. The emperor is friendly, almost jovial. "Today will be amusing. Don't you think so, Count?" "Oh, yes, Your Excellency. I have announced the execution. The people will find it very entertaining to watch through their telescopes as the heretic's blood begins to boil."

Anxious for the emperor's approval, the toady count suggests a last quick check on the black hole's temperature. "Yes,

Minister, let's do that. From this distance, the horizon looks cold, but let's drop a thermometer down to the surface, on a cable, and record the temperature near the horizon. Of course, it has been done many times, but I'll enjoy seeing the mercury rise." And so a small rocket is made ready to carry the thermometer away from the Earth. Once past the Earth's gravitational pull, the thermometer falls toward the horizon, pulling the cable behind it.

Down goes the thermometer until the cable tightens. "Warm, but not hot. Lower it a bit more, Count," the emperor orders. A bit more cable slowly peels off the reel. Through the telescope, the emperor watches the mercury rise – past the boiling point of water, past the evaporation point of glass and mercury – until the thermometer is vaporized. "Hot enough for you, Your Highness?" the count asks. "You mean hot enough for Steve, Count. Yes, I think the climate will be perfect. Let's go. It's time to start the execution."

A short time later, a second rocket, this one big enough for two hundred people, is readied to transport the unfortunate rational-science heretics to a small but hospitable satellite. Steve's wife, sobbing with despair, steadies herself by grimly hanging on to his arm. The physicist longs to explain the truth, but it's still too soon. The emperor's guards are all around them.

A few hours later, the count himself presses the button that activates the giant rockets that blast the small blue-green satellite from its orbit around the Earth. With two hundred frightened passengers (the guards no longer with them), the colony begins its plunge toward the dark fire.

"I can see them, Count," the emperor observes. "The heat is beginning to affect them. They are becoming lethargic and slow. Very slooooow." The dome of the observatory is large, and the eyepiece of the telescope is located in a most precarious position. The count smiles, pops an antigravity pill, and offers one to the emperor. "For safety, Your Highness. The fall from here would be most unpleasant." His Excellency swallows the pill and looks into the eyepiece again. "I can still see them. But, look, they are beginning to fall into the stretched horizon. Now my loyal subjects will see how my enemies are scrambled. Look, their individual bits are gradually merging into the hot, dense soup. And one by one, they are being carried off by photons. Let's count them and make sure they have been completely vaporized."

They watch as one by one the photons are recorded and analyzed in the telescope's giant bank of computers.

"Ha," says the count. "It's just as the principles of Quantum Mechanics predict. Every bit of information is accounted for. But it is scrambled far beyond recognition. No one is going to put Humpty-Dumpty together again."

The emperor throws his arm over the count's shoulders and says, "Congratulations, Count. A most profitable morning's work." But the careless gesture affects their balance. Two hundred feet from the floor below, the count suddenly wonders if the rumors about antigravity pills might not be true after all.

Steve studies his notebook with concentration. Then he looks up with glee and hugs his wife. "My dear, we will soon be safely past the horizon." Ms. Steve and the others are clearly puzzled as Steve goes on. "The Equivalence Principle is our salvation," he explains. "There is no danger at the horizon. It is merely a harmless point of no return." He adds, "Fortunately, we will be in free fall, and our acceleration will exactly cancel out the effects of the black hole's gravity. We'll feel nothing as we sail through the horizon." His wife is still skeptical: "Well, even if the horizon is harmless, I've heard terrible tales of an inescapable singularity inside the black hole. Won't it crush and grind us to bits?" "Yes, that's true," he answers. "But this black hole is so big that it will be about a million years before our planet gets anywhere near the singularity."

And so they sail happily through the horizon — at least if you believe in the Equivalence Principle.

The End

There are many things wrong with this story, apart from its literary merit. Among other things, if a black hole were large enough that Steve and his followers could survive for many years before arriving at the singularity,[4] it would take equally many years for the count's thermometer to fall to its destination. Far worse, the time for the black hole to emit the bits of information that Steve and his followers originally comprised would be incredibly long — much longer than the age of the universe. But if we ignore such numerical details, the basic logic of the story makes sense.

Or does it?

Was Steve immolated at the horizon? The count and emperor counted every bit, and all were there in the evaporation products, "just as the principles of Quantum Mechanics predict." So Steve was destroyed as he approached the horizon. But the story also claims that Steve safely passed to the other side with no harm done to him or his family — just as the Equivalence Principle predicts.

Evidently, we have a clash of principles. Quantum Mechanics implies that all objects encounter a superhot region just above the horizon, where the extreme temperature turns

4. Which, being behind the horizon, the emperor and the count could never see.

all matter into disassociated photons and radiates them back out of the black hole like light from the Sun. In the end, every bit of information carried by falling matter must be accounted for in these photons.

But it seems that the Equivalence Principle has a different and contradictory tale to tell.

Seminar Interrupted

Let me interrupt the flow of the 1988 seminar to clarify some points, which many of the physics enthusiasts in the audience knew, but you may not. First of all, why did the Equivalence Principle give the exiles confidence that the horizon was a safe environment? A thought experiment I mentioned in chapter 2 helps. Imagine life in an elevator, but in a world where gravity is much stronger than on the surface of the Earth. If the elevator is stationary, the passengers feel the full force of gravity on the bottoms of their feet and in every other part of their squashed bodies. Suppose that the elevator begins to ascend. The upward acceleration makes things worse. According to the Equivalence Principle, the acceleration adds an additional component to the gravity experienced by the passengers.

But what if the elevator cable snaps and the elevator begins to accelerate downward? Then the elevator and the passengers are in free fall. The effects of gravity and downward acceleration exactly cancel each other, and the passengers cannot tell that they're in a powerful gravitational field — at least not until they hit the ground and experience a violent upward acceleration.

In the same way, the exiles on their freely falling planet should not experience any effects of the black hole's gravity at the horizon. They are like the freely drifting pollywogs in chapter 2 as they unknowingly drift past the point of no return.

The second point is less familiar. As I have explained, the Hawking temperature of a large black hole is extremely small. Then why do the count and the emperor detect such a high temperature near the horizon when they lower their thermometer? To understand this point, we need to know what happens to a photon as it moves up out of a powerful gravitational field. But let's start with something more familiar — a stone thrown vertically upward from the surface of the Earth. If it's not thrown with enough initial velocity, it will fall back to the surface. But given enough initial kinetic energy, the stone will escape the confines of the Earth. However, even if the stone manages to escape, it will be moving with a lot less kinetic energy than when it started. Or, said another way, the stone had a lot more kinetic energy when it started than when it finally escaped.

Photons all move with the speed of light, but that doesn't mean they all have the same kinetic energy. In fact, they are much like the stone. When they rise out of a gravitational field, they lose energy; the stronger the gravity they have to overcome, the greater the energy loss. By the time a gamma ray rises from near the horizon, its energy is so depleted that it is now a very low-energy radio wave. Conversely, a radio wave observed far from the black hole must have been a high-energy gamma ray when it left the horizon.

Now consider the count and the emperor far above the black hole. The Hawking temperature is so low that the radio wave photons have very little energy. But with a little thought, the count and the emperor would have realized that the same photons must have been superhigh-energy gamma rays when they were emitted near the horizon. That's the same thing as saying it was hotter down there. In fact, gravity is so strong at the horizon of a black hole that photons from that region would have to have enormous energy to escape. Seen from far away, the black hole may be very cold, but up close the thermometer would be bombarded by ferociously energetic photons. That's why the executioners were certain that their victims would be vaporized at the horizon.

Seminar Resumed

So it appears that we have derived a contradiction. One set of principles — General Relativity and the Equivalence Principle — says that information sails uninterrupted, down through the horizon. The other, Quantum Mechanics, brings us to the opposite conclusion: the in-falling bits, though badly scrambled, are eventually returned in the form of photons and other particles.

Now, you may ask, how do we know that the bits, after falling through the horizon but before hitting the singularity, can't make their way back out into the Hawking radiation? The answer is obvious: they would have to exceed the speed of light to do so.

I have presented you with a powerful paradox and told you why it may be very important to the future of physics. But I have given you no hint of a possible way out of the dilemma. That's because I don't know the resolution. But I do have prejudices, so let me tell you what they are.

I don't believe that we will have to give up either the principles of Quantum Mechanics or those of the General Theory. In particular, I, like Gerard 't Hooft, believe that there is no information loss in black hole evaporation. Somehow we are missing a very deep point about information and how it is located in space.

That lecture in San Francisco was the first of a great many similar lectures that I gave in physics departments and physics conferences on at least five continents. I had decided that even if I couldn't solve the puzzle, I would become a proselytizer for its importance.

I recall one lecture particularly well. It was at the University of Texas, one of the premier physics departments in the United States. The audience included a number of extremely accomplished physicists, including Steven Weinberg, Willy Fischler, Joe Polchinski,

Bryce DeWitt, and Claudio Teitelboim, all of whom have made major contributions to the theory of gravity. I was quite interested in their views, so at the end of the lecture, I took a poll of the audience. If my memory serves me right, Fischler, DeWitt, and Teitelboim held the minority view, that information is not lost. Polchinski was convinced by Hawking's arguments and voted with the majority. Weinberg abstained. The overall vote was about three to one in favor of Hawking, but there was a noticeable reluctance on the part of the audience members to commit themselves.

During the stalemate, Stephen and I crossed paths a number of times. Of all those encounters, the one that stands out most took place in Aspen.

14

SKIRMISH AT ASPEN

Before the summer of 1964, I had never seen a hill higher than mighty Mount Minnewaska (which is all of 3,000 feet tall) in the Catskill Mountains. Aspen, Colorado, was a strange and magical mountain kingdom for me when I first laid eyes on it as a twenty-four-year-old graduate student. The tall, snow-covered peaks surrounding the town gave it a wild, otherworldly feel, especially for a city boy like me. Though already a popular ski town, Aspen still had a bit of the frontier flavor of the colorful silver-mining days of the late nineteenth century. The streets were unpaved, and in June the tourists were so sparse that you could camp almost anywhere outside town. It was a place of quirky characters. In any of the local bars, you might sit down between a genuine American cowboy and a rough, unshaved mountaineer. Or you might find yourself wedged between a grubby fisherman and a Polish sheepherder. You could also strike up a conversation with a member of the power elite of American business, the concertmistress of the Berkeley student orchestra, or a theoretical physicist.

Nestled at the west end of town, between Aspen Mountain to the south and Red Mountain to the north, is a group of low buildings surrounded by a large, grassy lawn. On summer days, you can see about a dozen physicists seated at picnic tables, arguing, debating, and just enjoying the fine weather. The main building of the Aspen

Institute for Theoretical Physics is not much to look at, but right behind it, in a pleasant outdoor space, is a blackboard shaded by an awning. This is where the real action takes place, where some of the world's greatest theoretical physicists meet for seminars to discuss their latest brainstorms.

In 1964 I was the only student at the center — I believe I was the only student in the entire two-year history of the institute — but the truth is that I wasn't there because of any talent I had in physics. Rushing down from the nearby Continental Divide, the Roaring Fork River runs through town. The water is turbulent, swift, very cold, and, most important for me that summer, full of silver: not the metallic silver of silver mines, but the living silver of wild rainbow trout. My adviser, Peter, was a fly fisherman, and when he found out that I could fly-fish, he invited me to join him for the summer in Aspen.

When I was a small boy, my father had taught me to fish on quieter eastern trout rivers, the legendary Beaverkill River and Esopus Creek in the Catskills. There the pools were quiet, and you could wade in up to your chest. Often you could see not only your fly but also the brown trout as it struck. But on the Roaring Fork in June, a sane fisherman would have to stay on the edge and make his best guess where the fly was. Although it took some time for me to master the technique, I caught a lot of rainbows that summer, but I learned no physics.

I'm not so fond of Aspen now. The glitterati have replaced the cowboys, to no advantage as far as I am concerned. Over the years, I've gone back a few times for the physics but not the fishing. Sometime around 1990, as I was passing through town on my way to Boulder, I stopped to give a lecture.

By that time, black holes and the puzzle of information loss were starting to appear on the radar screen. The general consensus was that Hawking was right, but a few people in addition to 't Hooft and me were questioning it. The inimitable Sidney Coleman was among them.

Sidney was a colorful character and a hero to an entire generation of physicists. With a mustache, droopy eyes, and unkempt long hair, he always reminded me of Einstein. His mind was incredibly fast, and his ability to get quickly to the heart of the matter, especially when the question involved difficult subtleties, was legendary. Sidney was a kindly man but not known to suffer fools gladly. More than one well-known seminar speaker at Harvard (where Sidney was a distinguished senior professor) had walked away with his tail between his legs after being mercilessly interrogated by Coleman. In Aspen that day, his presence meant that the seminar speaker would be held to a high standard.

By sheer coincidence, another familiar face was in the audience. As I entered the outdoor seminar space and headed toward the blackboard, the familiar high-tech wheelchair rolled in, and Stephen Hawking took his place in the front row. As everyone knew, my purpose would be to undermine Stephen's arguments about information loss. My strategy was to first outline the nature of the problem by repeating Stephen's logic. That would take about half of the allotted hour. Then I would explain why I believed that the logic couldn't be correct. But I also wanted to add an extra ingredient to Stephen's argument to make it even stronger. The stronger Stephen's case, the more it would imply a major paradigm shift if it was eventually proved wrong.

In explaining Stephen's logic, I wanted to fill a hole that apparently no one had thought of. Here was the idea: Imagine that the region just outside the horizon is occupied by a lot of tiny, invisible Xerox machines. When any information — a written document, for example — falls to the horizon, the Xerox machines copy the information, leaving two precisely identical versions. One of the copies continues, undisturbed, through the horizon and into the interior of the black hole, eventually to be destroyed at the singularity. But the fate of the second copy is more complex. First it is thoroughly scrambled or shuffled until it is unrecognizable without the shuffling code. Then it is radiated back out amid the Hawking radiation.

The photocopying of information just before it crosses the horizon would seem to solve the problem. Think first of the observers who hover far outside the black hole. They see every bit of information returned in the Hawking radiation. Thus, they conclude that there is no reason to change the rules of Quantum Mechanics. In blunter terms, they conclude that Hawking's ideas about the destruction of information are wrong.

What about the freely falling observer? The instant after passing the horizon, he looks around and sees that nothing has happened. His bits are still with him, assembled into the same person, accompanied by whatever fell with him. The horizon from this point of view is nothing but a harmless point of no return, and Einstein's Equivalence Principle is perfectly respected.

Could it really be that the horizon of a black hole is covered with perfectly faithful, miniature (perhaps Planck-sized) copying devices? It seemed a tempting idea. If correct, it would explain Stephen's paradox in a simple, logical way: no information would ever be lost in a black hole, and future physicists could go on using the usual principles of Quantum Mechanics. Quantum Xerox machines at the horizon of every black hole would bring the Black Hole War to a sudden end.

Sidney was impressed. He turned around in his chair to face the audience. Then, as only Sidney could, he explained what I had said in terms that were even clearer than those I had used. But Stephen said nothing. Slumped in his wheelchair, he had a wide smile on his face. It was clear that he knew something that Sidney did not. In fact, both Stephen and I were aware that my explanation was a straw man that I had created just to knock it down.

Stephen and I both knew that a perfect copier of quantum information contradicted the principles of Quantum Mechanics. In a world governed by the mathematical rules laid down by Heisenberg and Dirac, a perfect copying machine was impossible. I had a name for this principle: the *No-Quantum-Xerox Principle*. In the modern field of physics known as quantum information theory, the same idea is called the *No-Cloning Principle.*

Triumphantly, I looked at Coleman and said, "Sidney, quantum Xerox machines are impossible," expecting that he would instantly catch on. But for once, his rapid-fire brain was slow. I had to explain the point in detail. The explanation that I gave to Sidney and the others in the seminar filled the blackboard with mathematical equations and took all of the remaining time allocated for the seminar. Here is a simpler version.

Imagine a machine with an input port and two output ports. Any system, in any of its possible quantum states, can be inserted into the input port. For example, an electron can be loaded into the copier. The machine takes the input and ejects two identical electrons. The outputs are identical not only to each other but also to the original input.

A Quantum Xerox Machine

One electron with wave function goes in.
Two identical electrons come out.

If such a machine could be built, it would give us a way to beat Heisenberg's unbeatable Uncertainty Principle. Suppose we want to know both the position and the velocity of an electron. All we have to do is copy it and then measure the position of one clone and the velocity of the other. But, of course, that's impossible given the principles of Quantum Mechanics.

By the end of the hour, I had successfully defended Stephen's paradox and explained the No-Quantum-Xerox Principle, but I had left myself no time to explain my own point of view. Just as the seminar was breaking up, Stephen's disembodied mechanical voice loudly crowed, "So now you agree with me!" There was a mischievous twinkle in his eyes.

Clearly, I had lost that battle. I had been defeated by my own friendly fire, by lack of time, and especially by Stephen's quick wit. On my way out of Aspen that evening, I stopped at Difficult Creek and took out my fly rod. But my favorite pool was full of noisy kids floating on tire tubes.

PART III

Counterattack

15

THE BATTLE OF SANTA BARBARA

It was a Friday afternoon in 1993, and everyone else had gone home. John, Lárus, and I were sitting in my Stanford office shooting the breeze, drinking coffee that Lárus had brewed. Icelanders brew the most powerful coffee on Earth. According to Lárus, it has something to do with their late-night drinking habits.

Lárus Thorlacius, a tall Icelandic Viking (he claims that he is descended not from Norse warriors, but from Irish slaves), was a Stanford postdoc who had just received his Ph.D. from Princeton. John Uglum, a Texan and a Republican (not of the religious variety, but an Ayn Rand Libertarian), was my graduate student. Whatever our political and cultural differences — I'm a liberal Jew from the South Bronx — we were buddies, and we did a lot of male bonding: sitting around drinking coffee (occasionally something stronger), arguing politics, and talking about black holes. (A little later, Amanda Peet, a student from New Zealand, would expand our little "band of brothers" to three brothers and a sister.)

By 1993 black holes had not only arrived on physicists' radar screens; they had become the focal point. Part of the reason was a provocative paper that had been written about a year and a half earlier by four well-known American theoretical physicists. Curt Callan, a Princeton aristocrat, had been a leader in the field of elementary particle physics and an influential member of the American science establishment since the 1960s. (He had been Lárus's Ph.D. adviser.) Andy Strominger and Steve Giddings were younger, up-and-coming faculty members at the University of California, Santa Barbara (UCSB). At the time, the thing that distinguished them in my mind was that Giddings wore shorts and Strominger wore suspenders. The University of Chicago's Jeff Harvey was (and still is) a great physicist, a talented composer (see the end of chapter 24), and a stand-up comic. Collectively, they were known as CGHS, and the simplified version of black holes that they had written about were called CGHS black holes. Their joint paper had created a short-lived sensation, partly because the authors claimed to have finally solved the problem of information loss in black hole evaporation.

What made the CGHS theory so simple — in retrospect, deceptively simple — was that it described a universe with only a single dimension of space. Their world was even simpler than Flatland,[1] Edwin Abbott's fictional world of two dimensions. CGHS envisioned a universe of creatures that lived on an infinitely thin line. These creatures were as simple as could be: nothing more than single elementary particles. At one end of this one-dimensional universe sat a massive black hole, heavy enough and concentrated enough to trap anything that got too close.

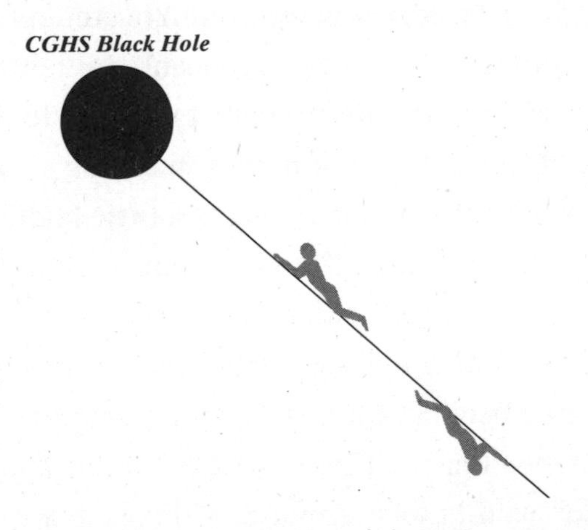

1. See Edwin A. Abbott, *Flatland: A Romance of Many Dimensions* (1884).

The paper that CGHS had written was an extremely elegant mathematical analysis of Hawking radiation, but somewhere in the analysis, they had made a mistake, claiming that Quantum Mechanics eliminated the singularity, and with it the horizon. Lárus and I, along with our colleague Jorge Russo, were among several people who pointed out the error. That made us experts on CGHS black holes. (There was even a particular version of the CGHS theory called the RST — Russo, Susskind, and Thorlacius — model.)

Now, the reason that John, Lárus, and I were sitting around after hours that Friday was that there was an upcoming conference devoted specifically to the puzzles and paradoxes of black holes. The conference was to take place a couple of weeks later in Santa Barbara, home to the Institute for Theoretical Physics (ITP) at UCSB.[2] Just how good a physics institution was the ITP? The short answer is really good. By 1993 it had become an active center of black hole research.

James Hartle was the most senior of the black hole theorists on the UCSB faculty. Jim was a very distinguished elder statesman who had done groundbreaking work with Stephen Hawking on quantum gravity long before it had become popular. But there were four younger members in the physics department there who were all destined to play big roles in the Black Hole War. All four were in their midthirties and extremely active. You've already met Steve Giddings and Andy Strominger (the G and S of CGHS). Although both were friends of mine whose physics I admired, they proved to be very exasperating foes for the next two years. They often drove me to distraction with their stubborn attachment to wrongheaded ideas. Eventually, however, they more than redeemed themselves.

Gary Horowitz was the third of the young UCSB faculty. Gary is a General Relativity expert — a *relativist* — who was by then making a name for himself as a brilliant leader in the field. He had also worked closely with Hawking and knew as much about black holes

2. Today the ITP is known as the KITP, the Kavli Institute for Theoretical Physics.

as anyone. Finally, Joe Polchinski had recently migrated to Santa Barbara from the University of Texas. Joe and I had worked together on a number of research projects, and I knew him well. Although I always found him a very pleasant person, full of genuine good humor, I also was awed by his intellectual power and speed and his sheer brilliance. Since the earliest days of our friendship — Joe must have been about twenty-five, and I was forty — I had had no doubt that he was fated to become one of the greatest theoretical physicists of the era. I wasn't disappointed.

These extraordinary young physicists worked closely together. Sometimes the subject was black holes, sometimes String Theory. The enormous talent of this small, close-knit group made them a very potent force in theoretical physics. It also made Santa Barbara one of the most exciting places (if not *the* most exciting place) for a theoretical physicist to hang out. There was no question that a conference in Santa Barbara devoted to the puzzles of black holes would be an important event.

The conference was probably organized to celebrate the excitement that the CGHS paper had created. It was hoped that the technical mathematics that CGHS had invented would hold the key to what by then was being called the *information paradox*. I had been asked by the conference organizers to report on the work that Lárus, Jorge, and I had done at Stanford, so there we were, late that Friday, discussing what I would say.

Maybe it was the super-caffeinated coffee, a surge of testosterone, or just our Three Musketeers camaraderie, but I said to John and Lárus, "Damn it, I don't want to talk about CGHS or RST. It's a dead end.[3] I want us to do something that will really shake things up. Let's go way out on a limb and say something very bold that will really get their attention."

3. In retrospect, I think the CGHS theory taught us a great deal. More than anything that came earlier, it gave a crystal clear, mathematical formulation of the contradiction that Hawking had exposed. It certainly had a very large influence on my own thinking.

The three of us had been looking for a way out of Stephen's paradoxical conclusion for some time, and an idea was beginning to solidify. It was not much more than a foggy notion — it didn't even have a name — but it was time for action.

"I think that the three of us should gather up the loose strands of our half-baked idea and, even if we can't prove it, try to make it more precise. Just the act of naming a new concept can sometimes bring clarity. I propose that we write a paper on Black Hole Complementarity, and I'll announce the new idea at the Santa Barbara meeting."

The story "Don't Forget to Take Your Antigravity Pills" (see chapter 13) is a good place to begin explaining what I had in mind. Like Akira Kurosawa's film *Rashômon,* it's a tale seen through the eyes of different participants — a tale with completely contradictory conclusions. In one version — that of the emperor and the count — Steve, the persecuted physicist, was annihilated by the incredibly hot environment surrounding the horizon. According to Steve, the story had a different and happier ending. Obviously, one or the other, if not both, had to be wrong; Steve could not have both survived and been killed at the horizon.

"The point about Black Hole Complementarity," I explained to my colleagues, "is that as crazy as it sounds, both stories *are* equally true."

My two friends were puzzled. I no longer remember exactly what I told them next, but it must have gone something like this: Everyone who remained outside the black hole — the count, the emperor, and the emperor's loyal citizens — all saw the same thing:[4] Steve was heated, vaporized, and turned into Hawking radiation. What's more, it all happened just before he reached the horizon.

How could we make sense of this? The only way consistent with the laws of physics would be to assume that some kind of superheated layer exists just above the horizon, perhaps no more than a

4. I am using "saw" in a somewhat general sense. The observers outside the black hole could detect the energy and even the individual bits of information that Steve's body comprised in the form of Hawking radiation.

Planck length thick. I admitted to John and Lárus that I didn't know precisely what this layer was made of, but I explained that the entropy of a black hole means that the layer must be composed of tiny objects, very likely no bigger than the Planck length. The hot layer would absorb anything that fell onto the horizon, just like drops of ink dissolving in water. I remember referring to the unknown tiny objects as *horizon-atoms,* but of course I didn't mean ordinary atoms. I knew as much about these atoms as physicists in the nineteenth century knew about ordinary atoms: only that they exist.

This hot layer of stuff needed a name. Astrophysicists had already coined the name that I eventually settled on. They had used the idea of an imaginary membrane covering the black hole just above its horizon to analyze certain electrical properties of black holes. The astrophysicists had called this imaginary surface the *stretched horizon,* but I was proposing a real layer of stuff, located a Planck length above the horizon, not an imaginary surface. What's more, I claimed that any experiment — lowering a thermometer and measuring the temperature, for example — would confirm the existence of horizon-atoms.[5]

I liked the sound of "stretched horizon" and adopted it for my own purposes. Today the stretched horizon is a standard concept in black hole physics. It means the thin layer of hot microscopic "degrees of freedom" located about one Planck distance above the horizon.

5. Physicists had known since the 1970s that a thermometer lowered to the vicinity of a horizon would report a high temperature. Bill Unruh, the conceiver of dumb holes, had discovered this fact while a student of John Wheeler.

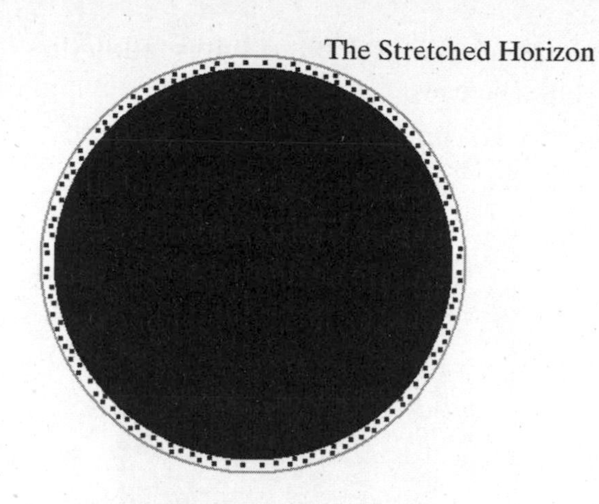

The stretched horizon helps us understand how a black hole evaporates. Every so often, one of the energetic horizon-atoms gets hit a little harder than usual and is ejected from the surface out into space. You could almost think of the stretched horizon as a thin, hot layer of atmosphere. In this sense, the description of black hole evaporation would closely parallel the way the Earth's atmosphere gradually evaporates into outer space. What's more, since a black hole loses mass as it evaporates, it must also shrink.

But this was only half the story — the half seen from a vantage point outside the black hole. By itself, this half of the story was hardly radical. Stuff falls into hot soup. Hot soup evaporates. Bits of information get carried out in the evaporation products. All very ordinary. If I had been talking about anything but a black hole, it would have been an unremarkable explanation.

What about the view from the inside, or more precisely, the view seen by a freely falling observer? We could call it Steve's version, which would seem to contradict the account from outside (the emperor's and the count's version).

I proposed two postulates.

1. To any observer who remains outside a black hole, the stretched horizon appears to be a hot layer of horizon-atoms that absorb, scramble, and eventually emit (in the form of

Hawking radiation) every bit of information that falls onto the black hole.

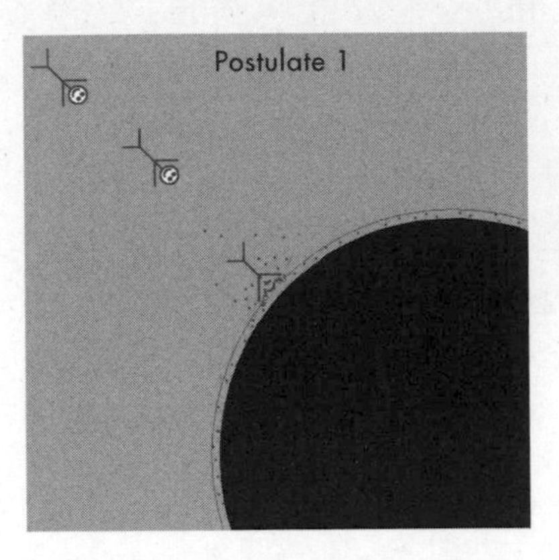

2. To a freely falling observer, the horizon appears to be absolutely empty space. Those falling observers detect nothing special at the horizon, although for them it is a point of no return. They only encounter a destructive environment much later, when they eventually approach the singularity.

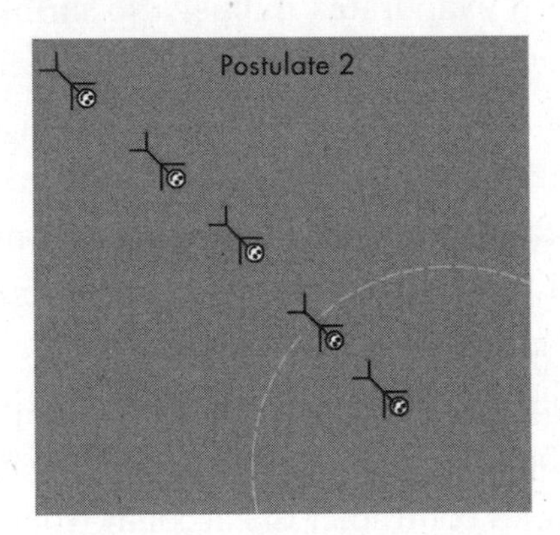

It was superfluous to add a third postulate, but I did so anyway.

3. Postulates 1 and 2 are both true, and the apparent contradiction is not real.

Lárus was skeptical. How, he asked, is it possible for two contradictory stories to both be true? There is a logical contradiction in saying that the in-falling Steve was killed at the horizon and, in the same breath, that he survived for another million years. Basic logic says that a thing and its opposite cannot both be true. In fact, I was asking the same question myself.

On the second floor of the Stanford physics department, there used to be a display of a hologram. Light reflecting off a two-dimensional film with a random pattern of tiny dark and light spots would focus in space and form a floating three-dimensional image of a very sexy young woman who would wink at you as you walked past.

You could walk partway around the fictitious image and see it from various angles. Lárus, John, and I would make a special point of passing the hologram every so often. Now I joked to Lárus that the surface of the black hole — the horizon — must be a hologram, a two-dimensional film of all the three-dimensional stuff inside the black hole. Lárus didn't buy it. Neither did I, not at that point anyway. In fact, I really didn't see the point of my own remark.

But I had been thinking about it for a while and had a more serious answer. Physics is an experimental and observational science; when all the mental pictures are stripped away, what's left is a collection of experimental data, together with mathematical equations that summarize that data. A genuine contradiction doesn't mean a discrepancy between two mental pictures. Mental pictures have more to do with limitations imposed by our evolutionary past than with the actual realities that we are trying to understand. A genuine contradiction occurs only when experiments lead to contradictory results. For example, if two identical thermometers are plunged into a pot of hot water and they each give a different temperature, we would not accept the results; we would know that something is wrong with one of the thermometers. Mental pictures are valuable in physics, but if they seem to lead to a contradiction where there is none in the data, the picture is not the right one.

Could we expose a genuine contradiction if we postulated that both black hole stories were true — Steve's and the count's? To detect a contradiction, two observers would have to come together at the end of the experiment and compare notes. But if one observation was made behind the horizon and the other observer never crossed it, then by the very definition of a horizon, they couldn't come together and compare data. So there was no real contradiction — only a bad mental picture.

John asked how Hawking would respond. My answer, which proved quite accurate, was, "Oh, Stephen will smile."

Complementarity

The word *complementarity* was introduced into physics by the legendary father figure of Quantum Mechanics, Niels Bohr. Bohr and Einstein were friends, but they disagreed incessantly about the paradoxes and apparent contradictions of Quantum Mechanics. Einstein was the true father of Quantum Mechanics, but he came to loathe the subject. Indeed, he used all of his unrivaled intellectual powers to try to poke holes in the logical foundations of it. Time after time, Einstein thought he had discovered a contradiction, and time after time, Bohr countered his blow with his own weapon — complementarity.

It was no accident that I used *complementarity* in describing how the paradoxes of quantum black holes would be resolved. In the 1920s, Quantum Mechanics was rife with apparent contradictions. One of them was the unresolved argument about light: was it waves, or was it particles? Sometimes it seemed as if light behaved one way, and other times the opposite way. To say that light was both things — waves and particles — was nonsensical. How would we know when to use particle equations and when to use wave equations?

Another puzzle: We think of particles as tiny objects that occupy a position in space. But particles can travel from one point to another. To describe their motion, we have to specify how fast and in what direction they move. Almost by definition, a particle is a thing that has a position and a velocity. But no! With a logic that seemed to defy logic, Heisenberg's Uncertainty Principle insisted that position and velocity cannot both be specified. More nonsense.

Something very strange was afoot. It seemed that reason was being flushed down the toilet. Of course, there was no real contradiction in the experimental data; every experiment gave a definite result, a reading on a dial, a number. But there was something very wrong with the mental picture. The model of reality that's wired into our brains couldn't capture the true character of light or the uncertain way that particles move.

My own view about the paradoxes of black holes was the same as Bohr's view about the paradoxes of Quantum Mechanics. In physics, a contradiction is only a contradiction if it leads to inconsistent experimental results. Bohr was also a stickler for the precise use of words. When used in an imprecise way, words sometimes lead to the appearance of a contradiction where there is none.

Complementarity is about the misuse of a simple three-letter word: a-n-d. "Light is waves, *and* light is particles." "A particle has a position *and* a velocity." In effect, Bohr said, get rid of *and* — and replace it with *or:* "Light is waves, *or* light is particles." "A particle has a position *or* a velocity."

What Bohr meant was that in certain experiments, light behaves like a collection of particles, while in other experiments, it behaves like a wave. There is no experiment where it behaves like both. If you measure some wave characteristic — say, the value of the electrical field along the wave — you will get an answer. If you measure a particle property, such as the location of the photons in a very low-intensity light beam, you will also get an answer. But don't try to measure a wave property at the same time that you measure a particle property. One thing gets in the way of the other. You can measure a wave property *or* a particle property. Bohr said that neither waves nor particles are complete descriptions of light, but that they *complement* each other.

Exactly the same is true of position and velocity. Some experiments are sensitive to the position of an electron — for example, the point at which an electron hits a TV screen and illuminates it. Other experiments are sensitive to its velocity — for example, how much the trajectory of an electron bends when the electron passes near a magnet. But no experiment can be sensitive to the precise position *and* the velocity of the electron.

Heisenberg's Microscope

But *why* can't we measure the position and velocity of a particle at the same time? Determining the velocity of an object is really just measuring its position at two successive instants and seeing how far it moves in between. If it's possible to measure a particle's position once, it surely can be done twice. It seems contradictory to think that the position and velocity can't both be measured. On the face of it, Heisenberg seemed to be speaking nonsense.

Heisenberg's strategy was a brilliant example of the kind of thinking that makes complementarity so compelling. Like Einstein, he became a thought experimenter. How, he asked, would one actually go about trying to measure both the position *and* the velocity of an electron?

First, he understood that he would have to measure the position at two different times in order to deduce the velocity. Furthermore, he would have to measure the position without disturbing the electron's motion, or the disturbance might invalidate the measurement of the original velocity.

The most direct way of measuring an object's location is to look at it. In other words, shine light on the object, and from the reflected light, deduce the position. In fact, our eyes and brains have specially built-in circuitry to determine the location of an object from the image on the eye's retina. This is one of those hardwired physics abilities that evolution provided.

Heisenberg imagined looking at the electron under a microscope.

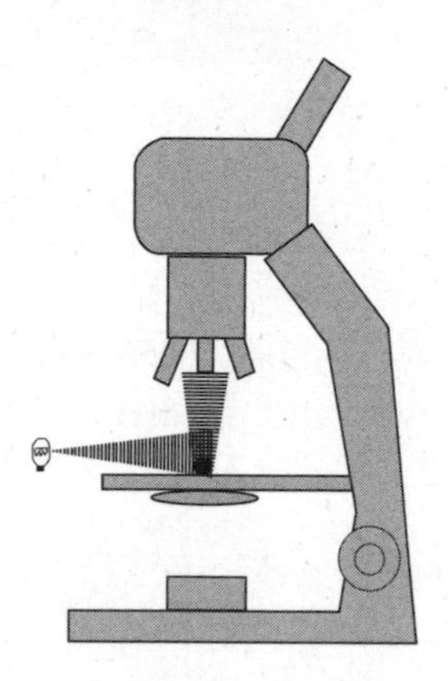

The idea was to very gently strike the electron with a light beam — gently so as not to kick the electron and change its velocity — and then focus the beam to form an image. But Heisenberg found himself trapped by the properties of light. First of all, the scattering of light by a single electron is a job for the particle theory of electromagnetic radiation. Heisenberg could be no more gentle with the electron than hitting it with a single photon. Moreover, it would have to be a very gentle photon — one of very low energy. A collision with an energetic photon would create just the sort of unwanted sharp kick he wanted to avoid.

All images made with waves are inherently fuzzy, and the longer the wavelength, the fuzzier the image. Radio waves have the longest wavelengths, from thirty centimeters up. Radio waves make excellent images of astronomical objects, but if you tried to make a radio portrait of a face, you would just get a blur.

Microwaves are the next-smaller wavelength. A portrait made by focusing ten-centimeter microwaves would still be too fuzzy to make out any features. But as the wavelength got down to a couple of centimeters, a nose, eyes, and mouth would begin to appear.

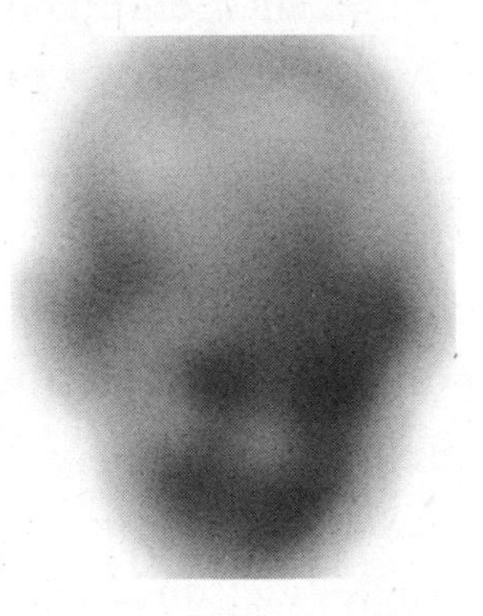

The rule is simple: you can't get better focusing power than the wavelength of the waves that are forming the image. Facial features are a few centimeters in size and will become focused when the wavelength gets that small. By the time the wavelength is a tenth of a centimeter, a face will be fairly distinct, although you might miss a small pimple.

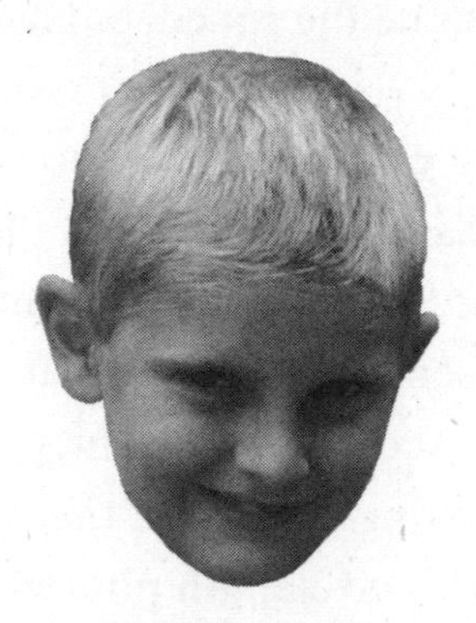

Suppose Heisenberg wanted to take a sharp enough image of the electron to see its location with an accuracy of a micron.[6] He would have to use light with a wavelength smaller than a micron.

Now to spring the trap. Recall from chapter 4 that the shorter the wavelength of a photon, the higher its energy. For example, the energy of a single radio wave photon would be so small that it would have almost no effect on an atom. By contrast, the energy of a one-micron photon would be enough to excite an atom by kicking an electron "upstairs" to a more energetic quantum orbit. An ultraviolet photon, with a wavelength ten times smaller, would be energetic enough to knock the electron clear out of the atom. So Heisenberg was trapped. If he wanted to determine the electron's location with great precision, there would be a cost. He would have to hit it with a very energetic photon, which in turn would "kick" the electron and alter its motion in a random way. If he used a gentle photon with little energy, the best he could do would be to get a very fuzzy idea of the electron's location. It was a real Catch-22.

You might wonder if it is *ever* possible to measure the velocity of an electron. The answer is yes. What you must do is measure its location twice, but with very poor precision. For example, you could use

6. A micron is one-millionth of a meter. It's about the size of a very small bacterium.

a long-wavelength photon to get a very fuzzy image, then do it again at a much later time. By measuring the two fuzzy images, it is possible to accurately determine the speed, but at great cost to the accuracy of the position.

Nothing Heisenberg could think of would allow him to determine both the location and the velocity of the electron at the same time. I imagine that he, and certainly his mentor Bohr, began to wonder if it made any sense to suppose that an electron had both a position and a velocity. According to Bohr's philosophy, one can describe an electron as having a position that can be measured accurately using a very short-wavelength photon, *or* an electron can be described as having a velocity that can be measured using long-wavelength photons, but not both. The measurement of one property precludes the measurement of the other. Bohr expressed this by saying that the two kinds of knowledge — position and velocity — were complementary aspects of the electron. Of course, there was nothing special about the electron in Heisenberg's argument; it could have been a proton, an atom, or a bowling ball.

The story of the count, the emperor, and Steve seems to suffer from a contradiction, but the contradiction is only apparent. Looking for a bit of information inside the horizon and at the same time looking for it outside the horizon preclude each other the same way that position and velocity measurements preclude each other. No one can be both behind the horizon and in front of it at the same time. At least that was the claim that I wanted to make in Santa Barbara.

Santa Barbara

Black holes are real. The universe is full of them, and they are some of the most spectacular and violent astronomical objects. But at the 1993 meeting in Santa Barbara, most of the physicists were not particularly interested in astronomical black holes. Thought experiments, not telescope observations, were the focus of concern. The information paradox had finally surfaced in a big way.

The conference was not huge — maybe a hundred participants at most. When I walked into the auditorium, I saw a lot of people I knew. Stephen was in his wheelchair off to the side. Jacob Bekenstein, whom I had never met, was sitting near the middle of the audience. The local crew — Steve Giddings, Joe Polchinski, Andy Strominger, and Gary Horowitz — were in plain view. They would play major roles in the coming revolution, but at the time they were the enemy, the confused foot soldiers of the army of information losers. Gerard 't Hooft was seated right up in the front row, ready for battle.

Hawking's Lecture

Here is what I remember of Hawking's lecture. Stephen sat slumped in his wheelchair, head too heavy to hold straight, while the rest of us waited in hushed expectation. He was on the right side of the stage, from which he could see the large projector screen at the front of the room and also scan the audience. By now Stephen had lost the capacity to speak with his own vocal cords. His electronic voice delivered a prerecorded message, while an assistant operated the slide projector from the rear. The projector was synchronized with the recorded message. I wondered why he had to be there at all.

Despite its robotic sound, his voice was rich with personality. His smile communicated supreme confidence and assurance. There is a mystery to Stephen's performances: how does the presence of his motionless, frail body breathe so much life into an otherwise lifeless event? With barely a flicker of motion, Stephen's face conveys a magnetism and charisma that few men have.

The lecture itself was not memorable, at least not for the content. Stephen spoke about what he was expected to speak about — what I had not wanted to speak about — the CGHS theory and how CGHS had blown it (he generously credited RST for discovering the error). His main message was that if you did the CGHS mathematics properly, the results supported his own theory that information cannot be radiated out from a black hole. To Stephen the lesson of CGHS was that the mathematics of the theory simply proved his point. To me the lesson was that not only was the mental picture defective, but the mathematical foundations of quantum gravity, at least as embodied in CGHS, were inconsistent.

The most unusual thing about Stephen's lectures is the question-and-answer period that follows. One of the conference organizers gets up on the stage and asks for questions from the audience. Usually the queries are technical, and sometimes they are long-winded, designed to show that the questioner knows his stuff. But then the auditorium becomes deadly quiet. One hundred acolytes turn into mute monks in a bizarrely silent cathedral. Stephen is composing his answer. The method by which he communicates with the outside world is amazing. He cannot speak or lift a hand to do sign language. His muscles are so atrophied that they can barely exert any force. He has neither the strength nor the coordination to type on a keyboard. If I remember correctly, at the time he communicated by exerting a feeble pressure on a control stick.[7]

He has a small computer screen attached to the arm of his wheelchair, and a series of electronic words and letters flash across the screen, more or less in continuous succession. Stephen picks them off one by one and stores them in the computer to form a sentence or two. This can take up to ten minutes. Meanwhile, as the oracle is composing his answer, the room is as silent as a crypt. All conversation ceases while the suspense and anticipation build. Then finally the answer: it can be no more than yes or no, or perhaps a sentence or two.

I have seen this take place in a room with one hundred physicists as well as in a small stadium with five thousand spectators, including a South American president, the chief of staff of the military, and several senior generals. My reaction to the extraordinary silence has ranged from amusement to serious indignation (why is my time be-

7. Today it is even more difficult: the control stick has been replaced by a sensor that detects minute motions of Stephen's cheek muscle.

ing wasted on this farce?). I always feel like making noise, maybe just talking to my neighbor, but I never do.

What is it about Stephen that commands the rapt attention that a holy man, about to reveal the deepest secrets of God and the universe, might receive? Hawking is an arrogant man, very full of himself, extremely self-centered. Then again, the same is true of half the people I know, including myself. I think the answer to this question is partly the magic and mystery of the disembodied intellect who navigates the universe in his wheelchair. But part of it is that theoretical physics is a small world composed of people who have known one another for years. For many of us, it is an extended family, and Stephen is a beloved and deeply respected part of that family, even if he sometimes engenders frustration and annoyance. We are all very aware that he has no way to communicate except through the tedious and lengthy process that he uses. Because we value his point of view, we sit quietly, waiting. I also think it is likely that Stephen's concentration is so intense during the composing process that he's not even aware of the strange silence around him.

As I said, the lecture was not memorable. Stephen made the usual claim: Information goes into a black hole and never comes out. By the time the black hole evaporates, it is completely obliterated.

Gerard 't Hooft followed immediately on his heels. He is also a man of great charisma who is vastly admired in the physics community. Gerard has a terrific stage presence and commands enormous authority. Although he is not always easy to understand, he has none of the oracular mystery of Hawking. He is a rather straightforward and sensible Dutchman.

Gerard's presentations are always fun. He likes to use his body to illustrate points, and he knows how to produce spectacular graphics. After all these years, I still recall a video he made to illustrate the horizon of a black hole. A sphere was tiled with pixels that were randomly either black or white. As the video was played, the pixels began to flicker from black to white and vice versa. The picture looked like white noise on a malfunctioning TV screen. It was quite evident that 't Hooft had ideas that were similar to my own about the existence of an active layer of rapidly changing horizon-atoms that make up the entropy of a black hole. (I waited for him to steal my thunder and give his own version of Black Hole Complementarity, but if he was thinking about it, he didn't quite spell it out.)

'T Hooft is an extremely deep and original thinker. As with many very original people, he is often unclear to others. After his black hole lecture, it was obvious that he had lost the audience. Not that he had bored the audience — far from it — but they couldn't understand his logic. Remember, the horizon of a black hole was supposed to be empty space, not a defective TV screen.

Overall, I doubt that either man had changed any minds about the fate of information in a black hole. No one polled the audience, but I would guess that at that point, there was a two-to-one bias favoring Hawking.

What I found remarkable about most of the rest of the conference was the stubborn refusal to entertain the right solution of the paradox. Most of the lectures mentioned the three possible solutions:

1. The information comes out in the Hawking radiation.
2. The information is lost.
3. The information eventually resides in some sort of tiny black hole remnant that remains after evaporation. (Usually the remnant was no bigger than the Planck size and no heavier than the Planck mass.)

Lecture after lecture repeated these three possibilities and immediately dismissed the first. The broad consensus of the speakers was that either information is lost, as Hawking advocated, or some tiny remnants are capable of hiding indefinitely large amounts of information. There may also have been some baby universe advocates, but I don't remember. Almost no one, with the exception of 't Hooft and a couple of others, expressed confidence in the usual laws of information and entropy.

Don Page came closest to expressing such confidence. Page is an amiable bear of a man, an Alaskan with a huge appetite. Hyperkinetic, loud, and enthusiastic to the extreme, Don is a walking contradiction, at least to me. He is an outstanding physicist and a profound thinker. His understanding of Quantum Field Theory, probability theory, information, black holes, and the general foundations of scientific knowledge are extremely impressive. He is also an Evangelical Christian. He once spent more than an hour explaining to me, using mathematical arguments, that the probability that Jesus was the Son of God exceeded 96 percent. But his physics and mathematics are free of ideology and are brilliant. His work has had a deep impact not only on my own thinking about black holes but also on the entire field.

In his lecture, Don repeated the mantra of the three possibilities, but he seemed much less willing than the others to dismiss the first one. My feeling was that he really believed that black holes behave like all other objects in nature, respecting the usual laws that require information to leak out during evaporation. But he also couldn't see how to reconcile this with the Equivalence Principle. It is remarkable how resistant physicists at the time were to the possibility that information leaks out in the Hawking radiation in the same way that it escapes from an evaporating pot of water.

Black Hole Complementarity

The Black Hole War had reached a stalemate. Neither side seemed able to get any traction. In fact, the fog of war was so dense that it was hard to see two sides. Apart from Hawking and 't Hooft, the impression I had was of a staggering multitude of shell-shocked troops in serious confusion.

My own talk was scheduled for later that day. I felt very much like Sherlock Holmes when he said to Watson, "When you have eliminated all that is impossible, whatever remains, however improbable, must be the truth." When I finally rose to speak, I felt that everything had been eliminated except one possibility — a possibility that, on the face of it, sounded so improbable as to seem ludicrous. Nevertheless, despite the absurdity of Black Hole Complementarity, it had to be right. All the alternatives belonged to the impossible.

"I don't care if you agree with what I say. I only want you to remember that I said it." Those were my two opening sentences; after fourteen years, I still remember them. Then, in technical physics jargon, I outlined the two contradictory outcomes of what in essence was Steve's story. "Obviously, at least one of the endings must be wrong since they say opposite things." There was a great deal of affirmative head nodding. But then I continued, "Nevertheless, I have come to tell you the impossible: Neither story is false. They are both true — in complementary ways."

After explaining how Bohr had used the term *complementarity,* I argued that in the case of a black hole, the experimenter is faced with a choice: to remain outside the black hole and record data from the safe side of the horizon, or to jump into the black hole and do observations from the inside. "You can't do both," I insisted.[8]

Imagine that a package is delivered to your house. A friend passing by observes that the mailman was unable to deliver the package and therefore carried it back to the mail truck. Meanwhile, you (who are in the house) answer the door and take the package from the mailman's hands. I think I would be quite justified in claiming that the observations cannot both be true. Someone is confused.

Why is a black hole different? I suggested that we follow the package story a little further. Translated from technical jargon and mathematical symbols, the story continued more or less as follows: Later the same day, you leave the house and meet your friend at a café. She says, "I was walking by your house earlier, and I saw the

8. The language I used was the usual technical mathematics that theoretical physicists use to communicate, but I was attacking a mental picture drawn from previous experience, not a mathematical formula. I might as well have used pictures.

postman try to deliver a package. But no one answered the door, so he took the package back to the truck." "No, you're mistaken," you say. "He did deliver the package. It was a new dress that I ordered from a catalog." Evidently, a contradiction has been exposed. Both observers know that something is inconsistent. In fact, it is not even essential that you physically leave the house to reveal the contradiction. Having the same conversation over the phone would expose the same contradiction.

But a black hole horizon is fundamentally different from the entryway to your house. You might say it's a one-way door: you can go in, but you can't go out. From the very definition of a horizon, no message can pass from inside the horizon to outside. The observer outside the horizon is permanently cut off from anyone or anything on the inside not by thick walls, but by the basic laws of physics. The very last step that leads to the contradiction – combining the two allegedly inconsistent observations into a single observation – is physically impossible.

I would have liked to add some philosophical remarks about how evolution has created a mental picture that guides our actions when it comes to caves, tents, houses, and doors, but that misleads us when it comes to black holes and horizons. Yet those remarks would have been ignored. Physicists want facts, equations, and data – not philosophy and evolutionary pop psychology.

Stephen was smiling as I delivered the message, but I rather doubted that he approved.

Next I used the analogy of ink drops falling into a pot of water to illustrate how the stretched horizon can absorb then scramble information, and finally, just like water evaporates from a pot, the information is eventually carried off in the Hawking radiation. To anyone outside the black hole, it is all rather ordinary – black holes and bathtubs are not so different, or so I claimed.

The audience was restless; some tentative hands were raised in objection. They knew how information evaporated from bathtubs, but something was missing: What about someone who falls into the black hole? When he reaches the stretched horizon, does he suddenly get wet? Wouldn't that violate the Equivalence Principle?

So I went on to the other half of the story: "To anyone who falls into the black hole, the horizon appears to be perfectly ordinary empty space. No stretched horizon, no incredibly hot microscopic objects, no boiling seething stretched horizon, nothing out of the ordinary: just empty space." I further explained why no contradiction could ever be detected.

I'm not sure whether Stephen was still smiling or not. And as I learned later, most of the relativists in the audience thought I had lost my marbles.

Even during the lecture, it was obvious that I had caught the audience's attention. Gerard, who can be a thorny character, was sitting in the front row shaking his head and frowning. I knew that of everyone present, he best understood what I was saying. I also knew that he agreed. But he wanted it said in his own particular way.

I was most interested in the reactions of the Santa Barbara people — Giddings, Horowitz, Strominger, and especially Polchinski. I couldn't get any impression from the stage, but I later found out that they had not been at all moved by my arguments.

There were two sympathetic listeners. After my lecture, at lunch in the college cafeteria, John Preskill and Don Page came over and sat down next to me. Hyperkinetic Don had a tray heaped with the most astonishing amount of food, including three giant desserts. (It was pretty clear where his energy came from.) Don can talk loudly and frenetically, but he is also a very good listener, and that day he was in listening mode. I already knew that he liked the idea that black holes are more or less like ordinary objects when it comes to information. He had publicly said so in his own energetic talk.

By comparison, John Preskill is more reserved, though not at all stuffy. A wiry man with a wry sense of humor, John is about the same age as Joe Polchinski and was at the time a professor at the California Institute of Technology. Cal Tech had been the home of two of the greatest physicists of the century, Murray Gell-Mann and

Dick Feynman. John himself was a widely admired physicist with a reputation as an exceedingly straight shooter. Like Sidney Coleman, John was one of those people whose clarity of thought gave him a special moral authority. My conversations with John had always been rewarding. The one that unfolded that day was revelatory. But before I can explain it, I have to tell you a little more about Black Hole Complementarity.

Looking at the Horizon with Heisenberg's Microscope

A single hydrogen atom falls into a gigantic black hole. First the naive picture: the tiny atom follows a trajectory and passes through the horizon, perfectly intact. In classical physics, the atom crosses the horizon at a very well-defined point — a dot no bigger than the atom itself. That seems right because, according to the Equivalence Principle, nothing violent is supposed to happen as the dot of hydrogen passes the point of no return.

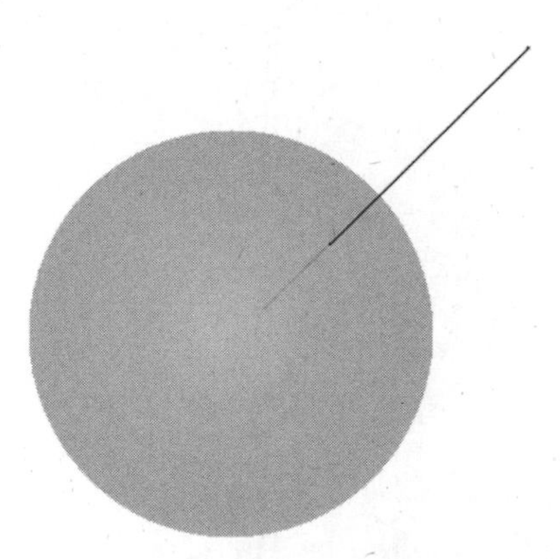

But that's too naive. According to Black Hole Complementarity, the observer watching from outside sees the atom enter a very hot layer (the stretched horizon), just like a particle falling into a pot of hot water. As it falls into the layer of hot stuff, it gets bombarded on every side by violently energetic degrees of freedom. First the atom is hit from the left, then from above, then from the left again, then from the right. It staggers around like a drunken sailor. This Brownian Motion is aptly called a *random walk.*

Brownian Motion

An atom can be expected to do exactly the same thing when it falls into the hot degrees of freedom that make up a stretched horizon — stagger around over the whole horizon.

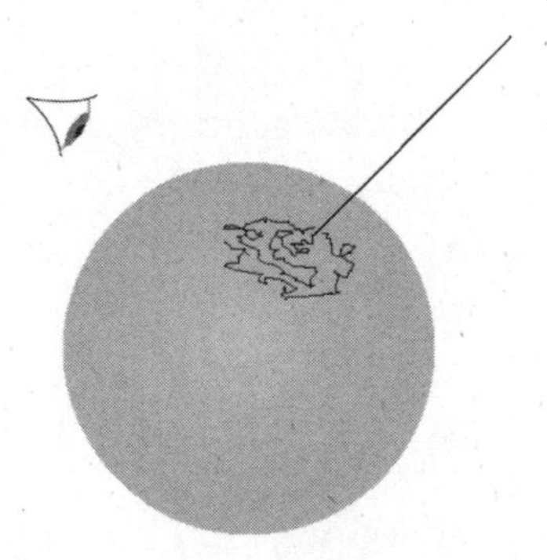

But even that's a little too simple. The stretched horizon is so hot that the atom will get blasted apart – *ionized* is the technical term – and the electron and proton will separately stagger over the horizon. Even the electrons and quarks may be torn apart into more basic elements. Notice that all of this is supposed to happen just *before* the atom crosses the horizon. I think it was Don, during his third dessert, who pointedly asked whether this spelled trouble for complementarity. There seemed to be two descriptions of the atom, even *before* it crosses the horizon. In one, the atom is ionized as it staggers around the whole horizon. But in the other, the atom falls completely undisturbed, straight toward a point on the horizon. Why couldn't someone on the outside observe the atom and see that nothing violent is happening to it? That would falsify Black Hole Complementarity once and for all.

As I began to explain, it quickly became obvious that John Preskill had thought about the same question and had come to the same conclusion as I had. We both began by noting that the atom couldn't be ionized until it reached a point where the temperature near the horizon was about 100,000 degrees. That only happens very close to the horizon, at about a millionth of a centimeter away. That's where we would need to observe the electron. That doesn't sound very hard; a millionth of a centimeter is not terribly small.

What would Heisenberg do? The answer, of course, is that he would get out his microscope and illuminate the atom with the appropriate wavelength of light. In this case, to resolve the atom when it gets to within a millionth of a centimeter of the horizon, he would need photons of wavelength 10^{-6} centimeter. Now here's the usual catch: a photon of such small wavelength has a lot of energy; in fact, it has enough energy that when it hits the atom, *it will ionize it*. In other words, any attempt to prove that the atom was not ionized by the hot stretched horizon will backfire by ionizing the atom. Going even further, we argued that any attempt to see whether the electron and the proton take a random walk over the horizon would blast the particles and scatter them all over the horizon.

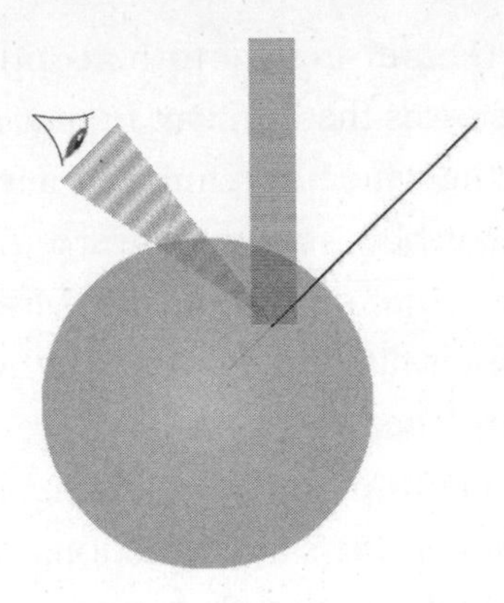

My memory of the discussion is not perfect, but I do recall Don getting very animated and saying, at the top of his considerably robust voice, that I hadn't been kidding when I'd called it Complementarity. It was exactly the kind of thing that Bohr and Heisenberg had talked about. In fact, experimentally disproving Black Hole Complementarity was very much like disproving the Uncertainty Principle — the experiment itself created the kind of uncertainty that it was designed to disprove.

We talked about what would happen as the atom got even closer to the horizon. Heisenberg's microscope would have to use even higher-energy quanta. Eventually, to follow the atom to within a Planck length of the horizon, we would have to hit it with photons of energy even greater than the Planck energy. No one knew anything about what such collisions would be like. No accelerator in the world has ever accelerated particles to anywhere near the Planck energy. John turned this idea into a principle:

> *Any theoretical proof that Black Hole Complementarity leads to an observable contradiction will inevitably depend on unwarranted assumptions about "physics beyond the Planck scale" — in other words, assumptions about nature in a realm far beyond our experience.*

Then Preskill raised a question that worried me. Suppose a bit of information were dropped into a black hole. According to my point of view, someone on the outside could collect the Hawking radiation and eventually recover that bit. But suppose that after collecting the bit, he jumped into the black hole, carrying the bit with him. Wouldn't there be two copies of the bit on the inside? It would be as if after receiving your package from the mailman, you stayed home, and your friend entered your house. Wouldn't there be a contradiction when observers met and compared notes on the inside?

John's question jolted me. I had never thought of that possibility. If someone on the inside discovered two copies of the same bit, that would be a violation of the No-Quantum-Xerox Principle. Here was the most serious challenge to Black Hole Complementarity that I had come across. The answer, although I wouldn't understand it for several weeks, was partly given by Preskill himself. He speculated that perhaps the two replicas could not meet up before they crashed into the singularity. Physics in the vicinity of the singularity is deep into the mysterious terra incognita of quantum gravity. That would allow us to wiggle out of the problem. As it happened, Don Page's ideas would also play a central role in defusing Preskill's initial bomb.

The discussion ended abruptly when someone announced that the next lecture was about to begin. I think it may have been the last lecture of the meeting, and I have no idea who gave it or what it was about. I was too worried about John's question to concentrate. But before the final closing of the conference, an announcement by one of the organizers brought me out of my thoughts. Joe Polchinski got up and said that he would take a poll. The question was this: "Do you think that information is lost when black holes evaporate as Hawking maintains? Or do you think it comes back out as 't Hooft and Susskind claim?" I suspect that before the meeting, the vote would have leaned heavily toward Hawking's view. I was extremely curious to see whether the people at the meeting had been swayed at all.

The participants were asked to cast a vote for one of the usual three candidates, plus a fourth. Here are the options, paraphrased.

1. Hawking's option: information that falls into a black hole is irretrievably lost.
2. 'T Hooft and Susskind's option: information dribbles back out among the photons and other particles in the Hawking radiation.
3. Information becomes trapped in tiny Planck-sized remnants.
4. Something else.

With each show of hands, Joe recorded the result on the whiteboard at the front of the lecture hall. Someone photographed the board for posterity. Here it is, courtesy of Joe.

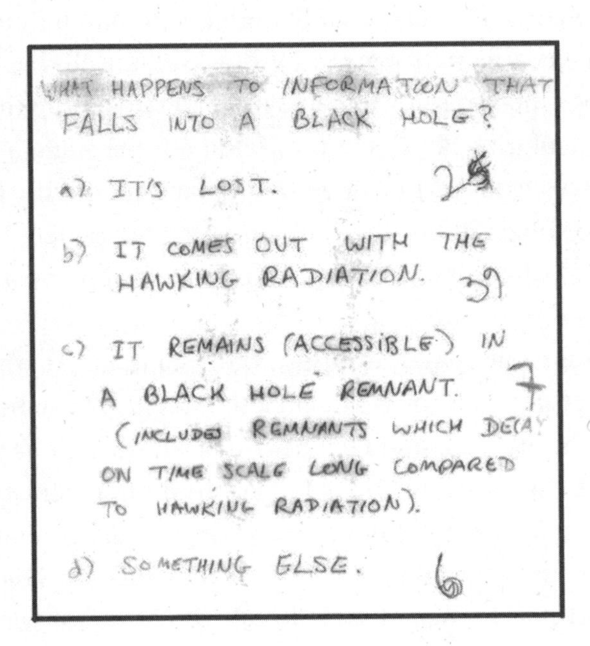

The final results:

- 25 votes for information lost
- 39 votes for information coming out with the Hawking radiation
- 7 votes for remnants
- 6 votes for something else

The momentary victory — 39 votes for what in effect was the principle of Black Hole Complementarity, versus only 38 for all the others combined — was not as satisfying as it might seem. What would a real victory be — 45 to 32, 60 to 17? Did it really matter what the majority thought? Science, unlike politics, is not supposed to be ruled by popular opinion.

Shortly before the Santa Barbara conference, I had read Thomas Kuhn's book *The Structure of Scientific Revolutions.* Generally, like most physicists, I am not very interested when philosophers opine about how science works, but Kuhn's ideas seemed right on target; they managed to put into focus my own fuzzy thoughts about the way physics had advanced in the past and, more to the point, how I hoped it was progressing in 1993. Kuhn's view was that the normal progress of science — the experimental gathering of data, the interpretation of that data by the use of theoretical models, the solving of equations — is every so often punctuated by major paradigm shifts. A paradigm shift is nothing less than the replacement of one worldview with another. Whole new ways of thinking about problems rise to take the place of previous conceptual frameworks. Darwin's principle of natural selection was a paradigm shift; the shifts from space and time to space-time and then to a flexible, elastic space-time were paradigm shifts; so, of course, was the replacement of classical determinism by the logic of Quantum Mechanics.

Scientific paradigm shifts are not like artistic or political paradigm shifts. Changes of opinion in art and policy really are just that — changes of opinion. By contrast, there will never be a swing from Newton's laws of motion back to Aristotelian mechanics. I very much doubt that we will change our minds about the superiority of the General Theory of Relativity over Newton's theory of gravity when it comes to making accurate predictions about the solar system. Progress — the progression of paradigms — in science is real.

Of course, science is a human affair, and during the painful struggles for new paradigms, opinions and emotions can be just as volatile as in any other human endeavor. But somehow, when all the hot opinions have been filtered through the scientific method, some small kernels of truth are left. They may be improved upon, but as a rule they are not reversed.

I felt that the Black Hole War was a classic struggle for a new paradigm. The fact that Black Hole Complementarity had won an opinion poll by one vote was no evidence of any real victory. In fact, the people I most wanted to influence – Joe Polchinski, Gary Horowitz, Andy Strominger, and above all Stephen – had voted with the opposition.

In the weeks that followed, Lárus Thorlacius and I put two and two together and figured out the answer to John Preskill's question. It took us a while, but I'm sure that if my conversation with Preskill and Page had gone on for another half hour, we would have solved it right there. In fact, I think John had given half the answer. Simply put, it takes some time for a bit of information to be radiated back out of a black hole. John had speculated that by the time an outside observer could recover the bit and jump into the black hole, the original bit had long since hit the singularity. The only question was how long it takes to recover a bit from the evaporating Hawking radiation.

Amusingly, the answer had already been given in an extraordinary paper that had come out just one month before the Santa Barbara conference. What the paper implied – though without saying so explicitly – was that to recover even a single bit of information, you would have to wait until half of the Hawking photons were radiated. Given the very slow rate at which black holes radiate photons, it would take 10^{68} years – a time vastly longer than the age of the universe – to radiate half of the photons from a stellar-mass black hole. But it would take only a fraction of a second before the original bit would be annihilated at the singularity. Obviously, there was no possibility of capturing the bit in the Hawking radiation, then jumping in and comparing it with the first bit. Black Hole Complementarity was safe. The author of the brilliant paper? Don Page.

16

WAIT! REVERSE THE REWIRING

Once, during the 1960s, I went to a play in a small avant-garde theater in Greenwich Village. An important component – slapstick humor as it turned out – involved the audience participating between acts, replacing the stage crew and moving scenery.

One woman was told to move a chair to the rear of the stage, but the moment she touched it, it collapsed into a bunch of kindling. Someone else grabbed a small suitcase by the handle, but it wouldn't move. My job was to heave a six-foot boulder up to someone on a low balcony. Just to keep in the spirit of things, I put my arms around it and pretended to lift with all my might. A moment of real cognitive dissonance occurred when the boulder sprang into the air as easily as if it weighed just a few ounces. It was a thin, hollow shell of painted balsa wood.

The connection our brains make between the size of an object and its weight must be one of those hardwired instincts – part of our automatic physics groker. Consistently getting it wrong would probably indicate serious brain damage – unless the person just happened to be a quantum physicist.

One of the great rewiring jobs that followed Einstein's 1905 discoveries required undoing the *big is heavy, small is light* instinct and replacing it with exactly the opposite: *big is light, small is heavy.* As with so much else, it was Einstein who had the first inkling of this Alice-in-Wonderland inversion of logic. What was he smoking at the time? Most likely only his pipe. As always, Einstein's most far-reaching conclusions flowed out of the simplest imaginary experiments that he did inside his head.

The Incredible Shrinking Box of Photons

This particular thought experiment begins with an adjustable box — empty except for some photons — that can be made bigger or smaller at will. The interior walls are perfectly reflecting mirrors, so that the photons trapped in the box bounce back and forth between the mirrored surfaces and can't leak out.

A wave confined in an enclosed region of space cannot have a wavelength longer than the size of the region. Try picturing a ten-meter wave fitting in a one-meter box.

It doesn't make sense. A one-centimeter wave, however, would fit comfortably in the box.

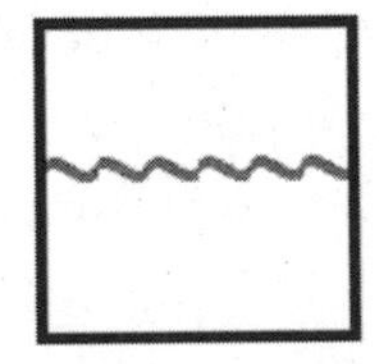

Einstein imagined making the box smaller and smaller, while the photons remain trapped inside. As the box shrinks, the wavelength of the photons cannot remain unchanged. The only possibility is that the wavelength of each photon must shrink along with the box. Eventually, the box will become microscopically small and be filled with very high-energy photons — high energy because their wavelength is so short. Further shrinking the box will increase the energy even more.

But recall Einstein's most famous equation, $E = mc^2$. If the energy in the box increases, so does its mass. Thus, the *smaller* it becomes, the more its mass *increases*. Once again, naive intuition had it upside down. Physicists would have to relearn the rules: small is heavy, and big is light.

The relation between size and mass shows up in another way. Nature seems to be built hierarchically, each level of structure comprising smaller objects. Thus, molecules are made of atoms; atoms of electrons, protons, and neutrons; protons and neutrons of quarks. These layers of structure were discovered by scientists colliding target atoms with particles and seeing what came out. In a sense, it is not so different from ordinary observation, where light (photons) is bounced off objects and then focused on either film or a retina of an eye. But as we've seen, to probe very small sizes, we must use very high-energy photons (or other particles). Obviously, during the time that an atom is being probed by a very energetic photon, a great deal of mass — at least by the standards of elementary particle physics — must be concentrated in a small space.

Let's make a graph to show the reciprocal relation between size and mass/energy. On the vertical axis, we plot the size scale we are trying to probe. On the horizontal axis, we plot the mass/energy of the photon needed to resolve the object.

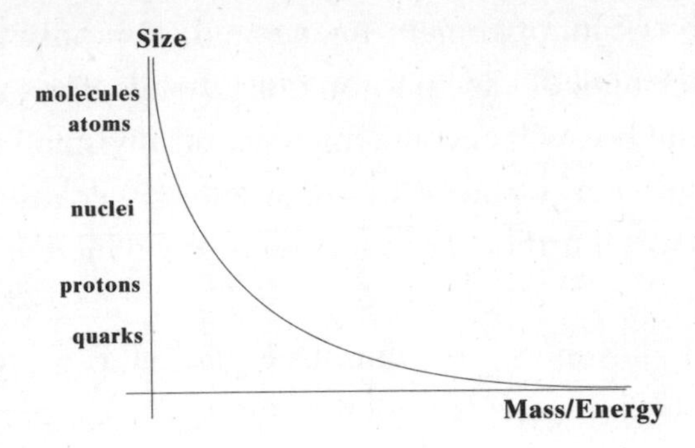

The pattern is clear: the smaller the object, the bigger the mass/energy needed to see it. Wiring ourselves for this inverse relation between size and mass/energy was something that every physics student had to do for most of the twentieth century.

Einstein's box of photons was not an anomaly. The idea that smaller means more massive pervades modern elementary particle physics. But ironically, the twenty-first century promises to undo that rewiring.

To see why, imagine that we want to determine what, if anything, exists on a scale one million times smaller than the Planck length. Perhaps the hierarchical structure of nature continues that deep. The standard twentieth-century strategy would be to probe some target with a photon of one million times the Planck energy. But this strategy would backfire.

Why do I say that? Although we will probably never be able to accelerate particles to the Planck energy, let alone one million times larger, we already know what would happen if we could. When that much mass becomes stuffed into a tiny space, a black hole will form. We will be frustrated by the horizon of the black hole, which will hide everything that we are trying to detect in its interior. As we try to see smaller and smaller distances by increasing the photon energy, the horizon will get bigger and bigger, and hide more and more: another Catch-22.

So what results from the collision? Hawking radiation — that's all. But as the black hole gets bigger, the wavelength of the Hawking photons grows. The sharp image of tiny sub-Planckian objects is replaced by an increasingly fuzzy image made by long-wavelength photons. So at best we can expect that as the energy of the collision increases, we will only rediscover nature on a larger scale. Thus, the true size-versus-energy graph looks like this:

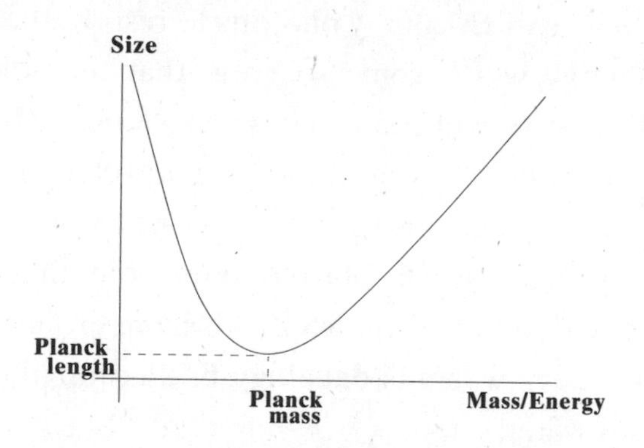

The Infrared-Ultraviolet Connection

Things bottom out at about the Planck scale — we can detect nothing smaller — and beyond that, the new wiring is the same as the preindustrial big = heavy. Thus, the onward march of reductionism — the idea that things are made of smaller things — must end at the Planck scale.

The terms *ultraviolet (UV)* and *infrared (IR)* have taken on a meaning in physics beyond their original implications, which had to do with short- and long-wavelength light. Because of the twentieth-century connection between size and energy, physicists often use the words to denote high (UV) and low (IR) energies. But the new wiring mixes it all up: beyond the Planck mass, higher energy means larger size, and lower energy means smaller size. The confusion is reflected in the terminology: the new trend that equates large size with large energy has confusingly become known as the *Infrared-Ultraviolet connection.*[1]

1. The awful terminology is my own fault. The expression *Infrared-Ultraviolet connection* was first used in a 1998 paper that I wrote with Edward Witten.

In part, it was the lack of understanding of the Infrared-Ultraviolet connection that misled physicists about the nature of information falling on a horizon. In chapter 15, we imagined using Heisenberg's microscope to watch an atom falling toward a black hole. As time progresses and the atom gets closer and closer to the horizon, it requires increasingly high-energy photons to resolve the atom. Eventually, the energy will become so great that the collision of the photon and the atom will create a large black hole. Then the image will have to be assembled out of the long-wavelength Hawking radiation. The result is that instead of becoming sharper, the image of the atom will get increasingly blurred to the point that the atom will appear to spread out over the whole horizon. From the outside, it will look, to use a now familiar analogy, like a drop of ink dissolving in a tub of hot water.

Black Hole Complementarity, even if outrageous, seemed internally consistent. By 1994, I wanted to shake Hawking and say, "Look, Stephen, you are missing the whole point of your own work!" I soon tried, but to no avail. The monthlong attempt had both humor and pathos. Let's take a break from physics as I describe my frustrations at the time.

17

AHAB IN CAMBRIDGE

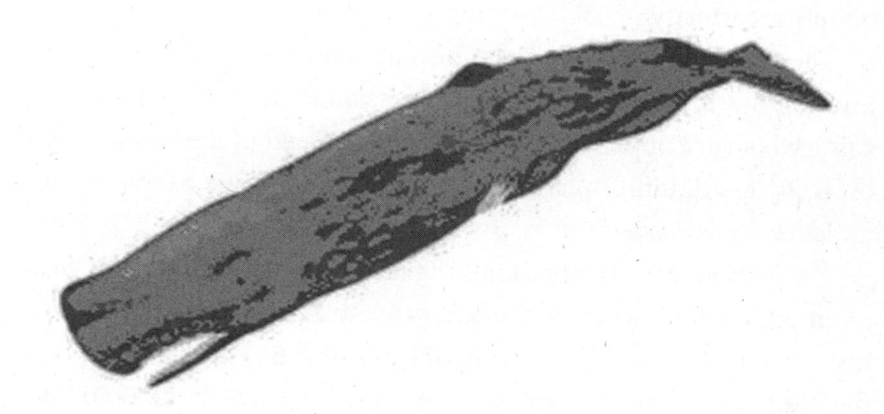

A tiny white dot had grown to fill my entire visual field. But unlike Ahab's obsession, mine was not a hundred-ton whale; it was a hundred-pound theoretical physicist in a motorized wheelchair. My thoughts rarely wandered from Stephen Hawking and his misguided ideas about the destruction of information inside black holes. To my mind, there was no longer any doubt about the truth, but I was consumed with the need to make Stephen see it. I had no desire to harpoon him or even humiliate him; I wanted only to make him see the facts as I saw them. I wanted him to see the profound implications of his own paradox.

What bothered me most was how easily so many experts — basically all or almost all general relativists — accepted Stephen's conclusions. I found it hard to understand how he and the others could be so complacent. Stephen's claim that there was a paradox and that it might portend a revolution was correct. But why did he and the others just roll over?

Even worse, I felt that Hawking and the relativists were blithely throwing away a pillar of science without providing anything to take its place. Stephen had tried and failed with his Dollar-matrix — when pushed, it led to disastrous violations of energy conservation — but all of his followers were content to say, "Ho hum, information is lost in black hole evaporation," and leave it at that. I was irritated by what seemed to me to be intellectual laziness and an abdication of scientific curiosity.

The only relief for my obsession was running, sometimes fifteen miles or more, through the hills behind Palo Alto. Focusing on pursuing whoever happened to be a few yards ahead of me would often clear my mind, until I passed him or her. Then it was Stephen who appeared up ahead.

He invaded my dreams. One night in Texas, I dreamt that Stephen and I were both stuffed into a mechanized wheelchair. With all my strength, I struggled to push him out of the chair. But Stephen the Hulk was incredibly strong. He grabbed me in a choke hold that cut off my air supply. We wrestled until I woke up in a sweat.

The cure for my obsession? Like Ahab, I would have to go to the enemy and hunt him where he lurked. So early in 1994, I accepted an invitation to visit Cambridge University's brand-new Newton Institute. In June, Stephen would be holding court among a group of physicists, most of whom I knew well but didn't count among my allies: Gary Horowitz, Gary Gibbons, Andy Strominger, Jeff Harvey, Steve Giddings, Roger Penrose, Shing-Tung Yau, and other heavyweights. My only ally, Gerard 't Hooft, would not be there.

I was not eager to revisit Cambridge. Twenty-three years earlier, two experiences had left me feeling bruised and annoyed. I was young, unknown, and still suffering the insecurities of being an academic from a working-class background. An invitation to dine at the High Table of Cambridge's Trinity College did little to assuage them.

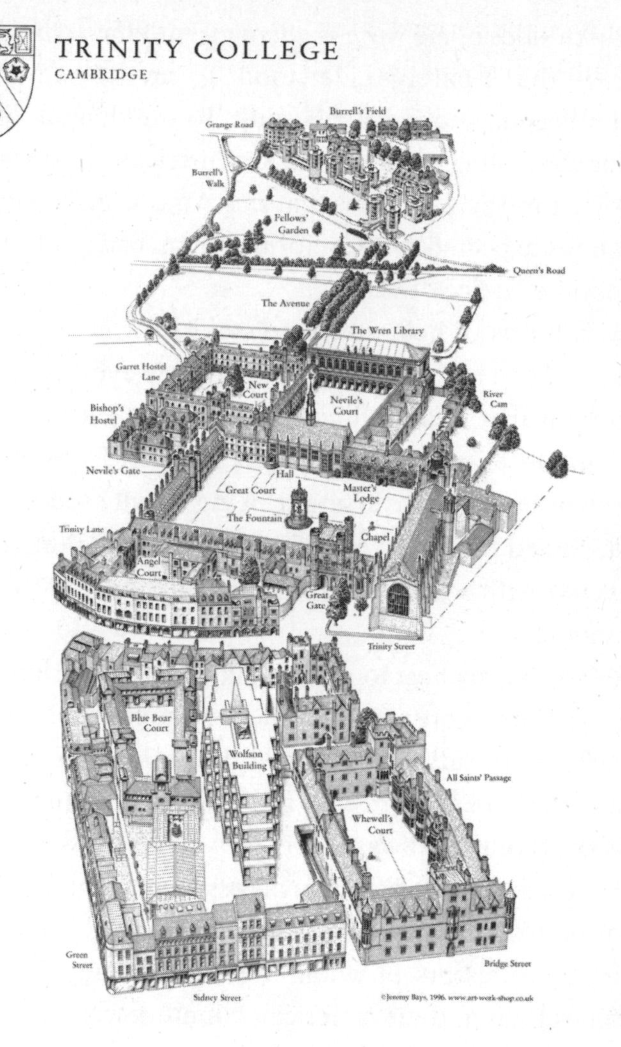

I still don't know what it means to be invited to the High Table. I don't know whether it was an honor and if it was, who or what was being honored. Or was it just a place to have lunch? In any case, my host, a gracious and kindly professor named John Polkinghorne, conducted me into a medieval hall hung with portraits of Isaac Newton and other giants. The undergraduates were seated at the lowest level, dressed in academic gowns. The faculty of science proceeded to the High Table, an elevated stage at one end of the hall. The meal was served by waiters who were far better dressed than I, and on both sides of me were academic gentlemen who mumbled in a language that I barely understood. To my left, an ancient British don was soon snoring in his soup. To my right, a distinguished academic was telling a story about an American guest who had been there some time ago. It seems that this American, lacking in the sophistications expected of a Cambridge man, had made a hilariously inappropriate choice of wine.

Now in terms of being a connoisseur of wine, I am reasonably certain that I can tell red from white with my eyes closed. I am even more certain that I can tell wine from beer. But beyond that, my palate fails me. I could hardly help feeling that I was the butt of the story. The rest of the conversation, being of only special Cambridge interest, passed me by. I was left to enjoy a tasteless meal (boiled fish covered with white flour paste), completely cut off from any conversation.

Another day, my host took me walking around Trinity College. A large greensward, very well-groomed, occupied a place of honor in front of the main entrance to one of the buildings. I noticed that no one walked across the grass. A walkway around the lawn was the only allowed route. So I was surprised when Professor Polkinghorne took my arm and started straight across — diagonally. What was the meaning of this? Were we trespassing on sacred ground? The answer was simple: professors, of whom there were many fewer in British universities than in their American counterparts, were granted the ancient privilege of walking across the grass. No one else, or at least no one of lower rank, was permitted to tread on it.

The next day, I left the college without an escort to go back to my hotel. At the age of thirty-one, I was young to be a professor, but I was one. I naturally assumed that this gave me the right to cross the lawn. But when I was about halfway across, a short stubby gentleman, wearing what looked like a tuxedo and a bowler hat, bolted out from one of the buildings and irately demanded that I leave the lawn immediately. I protested that I was an American professor. My protest had no effect.

Twenty-three years later — bearded, older, and maybe a bit more intimidating in appearance — I again attempted the feat. This time I made it across without any problem. Had Cambridge changed? I really don't know. Had I changed? Yes. Things that a couple of decades earlier had offended me as classist snobbery — the High Table, the special lawn privileges — now struck me as nothing more than pleasant hospitality and perhaps a bit of British eccentricity. My return to Cambridge surprised me in a number of ways. Besides the fact that my dislike of the university's idiosyncrasies had changed to something more like amusement, the notoriously bad British food had definitely improved. I found that I positively liked Cambridge.

My first day there, I woke up very early. I decided to meander across town to my eventual destination — the Newton Institute. I left my wife, Anne, in our apartment on Chesterton Road and walked past the river Cam, past the boathouses where the boats for rowing competitions were stored, and across Jesus Green. (During my first visit, I had been puzzled, even irritated, that so much of Cambridge culture has religious roots).

I walked to Bridge Street and crossed the river Cam. Cam? Bridge? Cambridge? Was it possible that I was on the site of the original bridge — the one that had given the great university its name? Probably not, but it was fun to speculate.

A nearby park bench was occupied by an elderly but elegant "scientific"-looking gentleman with a long handlebar mustache. By God, the man looked like Ernest Rutherford, the discoverer of the atomic nucleus. I sat down and struck up a conversation. It clearly wasn't Rutherford, unless he had risen from the grave: Rutherford had been dead for nearly sixty years. Perhaps a son of Rutherford?

My bench mate was quite familiar with the name Ernest Rutherford — he knew him as the New Zealander who had discovered atomic energy. But although the resemblance was strong, he was not a Rutherford. More likely, he was a relative of mine: a retired Jewish postman with an amateur interest in science. His name was Goodfriend, probably anglicized from Gutefreund a generation earlier.

My early-morning stroll took me past Silver Street, where an ancient building once housed the Department of Applied Mathematics and Theoretical Physics — the building where John Polkinghorne had been my host. But even in Cambridge, things change. The mathematical sciences ("maths" in British academic terminology) were now housed in a new location near the Newton Institute.

Then I saw a towering structure in the distance. It hovered. It loomed. It soared. King's College Chapel is God's house in Cambridge. It physically dominates Cambridge's many houses of science.

How many generations of science students had prayed, or at least pretended to pray, in that cathedral? Out of curiosity, I entered its hallowed interior. In that environment, even I — a scientist with not a religious bone in my body — found a certain hollowness in my belief that nothing exists but electrons, protons, and neutrons, that the evolution of life is no more than a computer-game competition between the most selfish genes. "Cathedralitis," the awe inspired by a cleverly assembled pile of stones and colored glass windows: I am almost, but not quite, immune to it.

All this brought to mind something about the British academic world that had long puzzled me: the incongruous mix of religious and scientific traditions. Cambridge and Oxford, both founded by clerics in the twelfth century, have embraced with equal fervor what we in the United States euphemistically call the faith-based and the reality-based communities. Even odder, they seem to do it with a unique intellectual tolerance that mystifies me. Take, for example, the names of nine of Cambridge's most famous colleges: Jesus College, Christ's College, Corpus Christi College, Magdalene College, Peterhouse, St. Catharine's College, St. Edmund's College, St. John's College, and Trinity College. But then again, there is Wolfson College, named for Isaac Wolfson, a secular Jew. Even more striking is Darwin College, the same Darwin whose masterstroke ejected God from the science of life.

The history is long and colorful. Isaac Newton did more to cast out supernatural beliefs than anyone before him. Inertia (mass), acceleration, and a universal law of gravitation replaced the hand of God, which was no longer needed to guide the planets. But as historians of seventeenth-century science never tire of reminding us, Newton was a Christian, and a passionate religious believer at that. He spent more time, energy, and ink on Christian theology than on physics.

For Newton and his peers, the existence of an intelligent Creator must have been an intellectual necessity: how else to explain the existence of man? Nothing in Newton's vision of the world could explain the creation, from inanimate material, of so complex an object as a sentient human being. Newton had more than enough reason to believe in a divine origin.

But where Newton failed, two centuries later the ultimate (and unwilling) subversive, Charles Darwin (also a Cambridge man), succeeded. Darwin's idea of natural selection, combined with Watson and Crick's double helix (discovered in Cambridge) replaced the magic of creation with the laws of probability and chemistry.

Was Darwin an enemy of religion? Not at all. Although he had lost faith in Christian dogma and considered himself an agnostic, he was a strong supporter of his local parish church, as well as a close friend of the vicar, the Reverend John Innes.

Of course, all was not always entirely friendly. The story of Thomas Huxley's debate (over evolution) with Bishop Samuel "Soapy Sam" Wilberforce had its rough edge. The bishop asked Huxley whether it had been his grandmother or grandfather who had been the ape. Huxley returned the compliment by calling Wilberforce a prostitute of the truth. Still, no one was shot, stabbed, or even punched. It was all done in the civil tradition of British academic intercourse.

And what about today? Even now, the genteel coexistence of religion and science persists. John Polkinghorne, who had escorted me across the lawn, is no longer a professor of physics. In 1979 he resigned his professorship to study for the Anglican priesthood. Polkinghorne is one of the leading proponents of the popular view that science and religion are entering into a period of perfect convergence, that God's plan is expressed through the extraordinary design of the *laws of nature.* These laws are not only utterly improbable but also exactly right to guarantee the existence of intelligent life — life that, incidentally, can appreciate God and his laws.[1] Today Polkinghorne is one of the most eminent churchmen in Great Britain. However, I don't know whether he is still permitted to cross the lawn.

Meanwhile, Oxford's renowned evolutionist, Richard Dawkins, leads the charge against any imagined convergence of science and religion. According to Dawkins, life, love, and morality are a mere playing out of a deadly competition not between people, but between selfish genes. The British intellectual world seems to be big enough for both Dawkins and Polkinghorne.

But back to King's College Chapel. It was hard to think in purely optical terms of the morning light as it filtered through the stained glass. So, with a slight case of cathedralitis, I sat down on a bench with a good view of the impressive interior.

1. For my own views on this subject, see my book *The Cosmic Landscape: String Theory and the Illusion of Intelligent Design* (2005).

Before long I was joined by a serious-looking man – tall, big boned but not fat, with what seemed to me a particularly un-British demeanor. His shirt was a rough blue cotton type that in my youth I would have called a work shirt, his pants brown corduroy, held up by a pair of wide suspenders, giving him the look of a nineteenth-century denizen of the American West. In fact, I was not so far off base. His accent was western Montana, not East Anglican.

After we established our common American identity, the conversation turned to religion. No, I explained, I was not there to pray. In fact, I was not a Christian, but a son of Abraham who had come to admire the architecture. He was a building contractor and had wandered into King's College Chapel to look at the stonework. Though a man of deep religious conviction, he wasn't sure whether it was proper to pray in this church. His own religious allegiance was to the Church of Jesus Christ of Latter-day Saints. The Church of England was a source of suspicion to him. As for myself, I saw no reason to disappoint him with my profound skepticism – my complete rejection of religious faith, which I consider to be a strong belief in supernatural powers.

I was familiar with next to nothing about Mormonism. My only experience with the religion was that I had once lived next door to a very nice Mormon family. All I knew was that the Mormons had rigid rules against drinking coffee, tea, and Coca-Cola. My presumption was that the Mormon faith was a typical offshoot of northern European Protestantism. So I was surprised when my acquaintance told me that the Mormons were like the Jews. With no land to call home, they had followed their own Moses through the desert, braving every conceivable danger and deprivation, until they had at last arrived in their land of milk and honey — the region of the Great Salt Lake in Utah.

My acquaintance sat hunched over, forearms resting on spread knees, big hands clasped between them. The tale he told was not of misty antiquity, but an American story that began around 1820. I suppose that it should have been familiar to me, but it wasn't. Here are the rough details as well as I can recall them, supplemented with the historical record that I later looked up.

Joseph Smith was born in 1805 to a mother who suffered from epilepsy and powerful visions of religiosity. One day the angel Moroni came to him and whispered the secret of hidden, ancient, pure gold plates inscribed with the words of God. These words were intended to be revealed only to Smith, but there was one catch: the inscriptions were written in a language that no living person could decipher.

But Moroni told Joseph not to worry. He would provide Joseph with a pair of magic transparent stones — a pair of supernatural spectacles. The stones had names, Urim and Thummim. Moroni instructed Joseph to place Urim and Thummim in his hat, and then by peering into the hat, Joseph could see the magic script revealed in plain English.

My reaction to this story was to sit quietly as if I were in deep thought. I suppose that either one is a man of faith or one is not, and if one is not, a story of gold plates, viewed through magical spectacles placed in a man's hat, is very funny. But funny or not, several thousand believers followed Joseph Smith, and later, after Smith's violent death at age thirty-eight, they followed his successor, Brigham Young, through harrowing dangers and tribulations. Today the religious descendants of these believers number in the tens of millions.

Incidentally, you may ask what happened to the gold plates that Urim and Thummim helped Joseph to decipher. The answer is that after translating them into English, he lost them.

Now, Joseph Smith was a very charismatic man who held a good deal of attraction for and to the opposite sex. This must have been part of a divine plan. God ordered Joseph to marry and impregnate as many young girls as possible. He also told Joseph to gather a multitude of followers and lead them to the first version of the promised land, a place called Nauvoo, Illinois. When he and his followers got to Nauvoo, he quickly announced that he would run for the U.S. presidency. But the fine people of Nauvoo were good Christians — conventional Christians — and didn't much like his ideas about polygamy. So they shot him dead.

Just as the mantle of Moses passed to Joshua, Smith's authority passed to Brigham Young, another man of multiple loves and many children. The Mormon exodus began with a very quick exit out of Nauvoo. Eventually, Young saw them through a long, arduous, and dangerous journey to Utah.

I was fascinated, and still am, by this story. I believe that at the time, it affected my feelings — no doubt completely unfairly — about Stephen and his powerful charismatic influence over many physicists. Obsessed with my own frustration, I imagined him a pied piper leading a false crusade against Quantum Mechanics.

But that morning, neither Stephen nor black holes were on my mind. King's College Chapel had left me with an entirely new scientific paradox to obsess over. It had nothing to do with physics, except in an indirect way. It was a paradox having to do with Darwinian evolution. How is it possible that human beings have evolved so powerful an impulse to create irrational belief systems and hold on to them with such tenacity? One might have thought that Darwinian selection would reinforce a tendency toward rationality and cull any genetic disposition toward superstitious, faith-based belief systems. After all, an irrational belief can get one killed, as it did Joseph Smith. Undoubtedly, it has killed many millions. One might expect that evolution would eliminate tendencies toward following reckless leaders on the grounds of faith. But it seems that exactly the opposite is true. This scientific paradox provoked my curiosity for the first time in Cambridge. Ever since, I've been fascinated by it and have spent a good deal of time trying to unravel it.

During the few weeks that I spent in Cambridge, I seemingly strayed very far from the subject that had brought me there — the quantum behavior of black holes. But that was not entirely the case. What was nagging at the back of my mind was the question of whether scientists such as Hawking, 't Hooft, myself, and all the other participants in the Black Hole War might be victims of our own faith-based illusions.

Those weeks in Cambridge were troubling and full of melodramatic thoughts. The story of Ahab and the whale is an ambiguous one: was it the maddened whale that led Ahab to the bottom of the sea, or was it crazed Ahab who led the weak Starbuck to his doom? More to the point, was I, like Ahab, following a foolish obsession, or was Stephen tempting others with a false idea?

I have to admit that today I find the idea of Stephen the Pied Piper or Stephen the Hermit (after the French crusader Peter the Hermit) leading his enthralled followers to intellectual destruction quite hilarious. Evidently, obsession is a powerful hallucinogenic drug.

Now, I don't mean to give you the impression that I spent several weeks aimlessly wandering the streets of Cambridge, a prisoner of my dark thoughts. I was scheduled to give a series of talks about Black Hole Complementarity at the Newton Institute. I spent a lot of time at the institute preparing those lectures and arguing the various points with my skeptical colleagues.

To the Newton

It must have been 10:00 a.m. by the time I left King's College Chapel and walked out into the sunlit June day. The Darwinian mystery of irrational faith had wormed its way into my brain, but a much more pressing technical problem required an immediate solution: I still had to find the Newton Institute.

My all-but-useless map directed me out of the center of old Cambridge to a modern-looking residential area of somewhat less character. I hoped that this was a mistake; it was a disappointment to my romantic sentimentality. I saw a sign for Wilberforce Road. Could this be the same Wilberforce, the one known as "Soapy Sam," who had asked Huxley whether it had been his grandfather or grandmother who had been the ape? Perhaps the romance of history was not entirely absent.

In fact, the truth was even better. Wilberforce Road was named for Samuel's father, the Reverend William Wilberforce. William had played an admirable role in British history, being one of the leaders of the abolitionist movement to eliminate slavery in the British Empire.

Finally, I turned the corner from Wilberforce to Clarkson Road. My immediate impression on seeing the Newton Institute was again disappointment. It was a contemporary building — not ugly, but built, in the ordinary modern manner, of glass, brick, and steel.

The Newton Institute

But dismay turned to admiration as soon as I entered the building. It had the perfect architecture for its purpose: arguing and exchanging ideas – old, new, and untried – in vigorous debate; skewering wrong theories; and, I hoped, encountering and defeating the enemy. There was a large, very well-lit area with many comfortable chairs, tables to write on, and blackboards on most of the walls. Several small knots of people were seated around coffee tables, each table covered with the scraps of paper that physicists are forever scribbling on.

I intended to join Gary Horowitz, Jeff Harvey, and a couple of other friends at their table, but before I could do so, something captured my attention. A conversation of a different sort was taking place, and I succumbed to the temptation to eavesdrop. In a separate corner of the room, the king was holding court: Stephen, seated at the center, slightly elevated on his mechanical throne, was regaling a group of British journalists. The interview was obviously not about physics but about Stephen. When I arrived, he was talking about his own personal history and his debilitating disease. His story must have been prerecorded, but as always some ineffable essence of his distinctive personality had overwhelmed the flatness of the robotic voice.

The journalists were spellbound — each one studying Stephen's face for subtle signs as he told of his early life before being diagnosed with Lou Gehrig's disease. According to his testimony, those early years were dominated by a sense of boredom — the boredom of a young man who seemed to be going nowhere. He was twenty-four years old, an unexceptional graduate student of physics, not making much progress: a bit of a layabout with little ambition. Then came the early stroke of midnight, the terrifying diagnosis, a certain death sentence. We all live under a death sentence, but in Stephen's case it appeared to be immediate: a year, maybe two, probably not even enough time to finish his Ph.D.

Initially, Stephen was gripped by terror and depression. He had nightmares of being summarily executed. But then something unexpected happened. Somehow the idea of imminent death was replaced by the prospect of a few years of reprieve. The result was a sudden, powerful zest for life. Boredom was replaced by the fierce need to make his mark on physics, to marry, to have children, and to experience the world and all it offers in whatever time remained. Stephen said something amazing and unforgettable to the reporters, something I would have dismissed as pure BS if it had come from anyone else. He said that getting ill — cripplingly ill — was the best thing that could have happened to him.

I am not prone to hero worship. I have admired certain scientists and literary figures for their clarity and depth, but I would not call them personal heroes. Until that day, the only giant in my pantheon of heroes was the great Nelson Mandela. But while eavesdropping at the Newton, I suddenly came to see Stephen as a truly heroic figure: a man big enough to fill the shoes of Moby Dick (if whales wore shoes).

But I could also see — or imagined I could see — how easy it would be for a man like Stephen to become a pied piper. I remind you of the awesome cathedral-like silences that fill large lecture halls while Stephen composes an answer to a question.

It wasn't just in academic settings that Stephen would get such treatment. On one occasion, I was having dinner with Stephen; his wife, Elaine; and one of his eminently successful former students, Raphael Bousso. We were in central Texas at a generic roadside restaurant — the kind you can find along any U.S. highway. We were already eating — Elaine, Raphael, and I conversing, Stephen mostly listening — when a worshipful waiter recognized Stephen. He approached with the awe, reverence, fear, and humility of a devout Catholic unexpectedly encountering the pope in a diner. He practically threw himself at Stephen's feet, begging for a blessing, as he revealed the deep personal affinity he had always felt for the great physicist.

Stephen certainly enjoys being a super-celebrity; it is one of the few outlets he has for communicating with the world. But does he enjoy or encourage the almost religious veneration? It's not easy to tell what he is thinking, but I have spent enough time with him to be able to read his facial expressions, at least to some extent. The subtle signal that emerged in the Texas restaurant suggested annoyance, not pleasure.

Let me return to my original purpose in coming to England: convincing Stephen that his belief in information loss was wrong. Unfortunately, direct discussions with Stephen are almost impossible for me. I don't have the patience to wait several minutes for a response of just a few words. But there were others — such as Don Page, Gary Horowitz, and Andy Strominger — who had spent a good deal of time interacting and collaborating with him. They had learned to communicate with him far more efficiently than I.

My strategy depended on two things. The first was that physicists like to talk, and I am very good at getting conversations going. In fact, I'm so good at it that physicists, even though they might disagree with me, tend to flock to the discussions I initiate. Whenever I visit a physics department, mini-seminars blossom, even in the quietest places. So I knew that it would be easy to gather a few of Stephen's and my mutual friends (they *were* friends, even though I saw them as the enemy in the Black Hole War) and start an argument. I was also sure that Stephen would be drawn in – keeping him away from a physics argument is like keeping a cat away from catnip – and before long he and I would go at it, hammer and tongs, until one side or the other admitted defeat.

My strategy also depended on the strength of my arguments and the weakness of those on the other side. I had no doubt that I would eventually prevail.

It all worked splendidly – except for one detail: Stephen never joined in. It turned out that this was a period when he was particularly unwell, and we saw little of him. As a result, the battles were exactly the same ones that I had been having in the United States for several years. The whale was slipping away without my getting a shot at him.

A day or two before I left Cambridge, I was scheduled to give a formal seminar to the entire institute about Black Hole Complementarity. This was the last chance for a confrontation with Stephen. The lecture room was full to capacity. Stephen arrived just as I was starting and sat in the back. Normally, he sits up front near the blackboard, but this time he was not alone; his nurse and another assistant were in attendance, just in case he needed medical attention. It was obvious that he was having trouble, and about halfway through the seminar, he left. That was it. Ahab had lost his opportunity.

The seminar ended at around 5:00 p.m., by which time I had had enough of the Newton Institute. I wanted to get out of Cambridge. Anne was visiting with a friend, and she had left me the rental car. Instead of driving back to our apartment, I drove out past the neighboring village, Milton, and stopped in a pub. I am not a drinking man, and drinking alone is definitely not a habit of mine, but in this case I really did want to sit and have a beer by myself. It wasn't solitude that I wanted; it was just the absence of physicists.

It was a typical country pub, with a middle-aged barmaid and a few local customers standing at the bar. One of the customers was a man I would guess to be in his eighties, dressed in a brown suit and a bow tie and leaning on a cane. I don't believe he was Irish, but he had a strong resemblance to the actor Barry Fitzgerald, who played opposite Bing Crosby in *Going My Way.* (Fitzgerald portrayed a crusty but good-hearted Irish priest.) The customer was engaged in a good-natured argument with the barmaid, who called him Lou.

Being pretty sure he wasn't a physicist, I bellied up to the bar next to Lou and ordered my beer. I can't remember exactly how the conversation got started, but he told me that he had had a short military career that had ended with the loss of a leg during the war, which I took to mean World War II. The missing leg didn't seem to hamper his ability to stand at the bar.

The conversation inevitably turned to who I was and what I was doing in Milton. I wasn't in the mood to explain physics, but I didn't want to lie to the old gentleman. I told him that I was in Cambridge for a black hole conference. Whereupon he told me that he was quite an expert on the subject and could tell me many things that I might not know. The conversation began to veer off in a bizarre direction. He claimed that according to family legend, one of his ancestors had been in the black hole but had gotten out at the last moment.

What black hole was he talking about? Crackpots with theories about black holes are a dime a dozen and usually very boring, but this man didn't seem like the usual nutcase. Taking a sip of beer, he went on to say that the Black Hole of Calcutta was a damned ugly place, as nasty as it gets.

The Black Hole of Calcutta! He evidently thought I was in Cambridge for some kind of historical meeting about Anglo-Indian history. I had heard of the Black Hole of Calcutta, but I had no idea what it was. My very vague impression was that it was a whorehouse where unwary British soldiers were robbed and murdered.

I chose not to clarify the situation but instead to learn as much as I could about the original Black Hole. The story is controversial but it seems that it was a cellar, or possibly a dungeon, in a British fort that was overrun by enemy forces in 1756. A large number of British soldiers were trapped in the cellar overnight and, probably by accident, smothered to death. According to the family legend, seven generations back, one of Lou's ancestors had just barely escaped being among the dead.

I had thus found a case of information escaping a black hole. If only Stephen had been there to listen.

18

THE WORLD AS A HOLOGRAM

Subvert the dominant paradigm.

—SEEN ON A BUMPER STICKER

By the time I left Cambridge, I realized that the fault didn't lie with Stephen or the relativists. Hours of discussion, especially with Gary Horowitz (H of CGHS), a card-carrying relativist, had convinced me otherwise. Besides being a technical wizard at the equations of General Relativity, Gary is a deep thinker who likes to get to the bottom of things. He had given many hours of thought to Stephen's paradox, and although he had a clear understanding of the dangers of information loss, he had concluded that Stephen had to be right — he could see no way around the conclusion that information must be lost when black holes evaporate. When I explained Black Hole Complementarity to Gary (not for the first time), he understood the point very well but felt that it was too radical a step. Postulating quantum mechanical uncertainties that operate on a scale as large as a huge black hole seemed far-fetched. It was very clearly *not* a matter of intellectual laziness. It all came down to one question: which principles do you trust?

On my flight out of Cambridge, I realized that the real problem was the absence of a firm mathematical foundation for Black Hole Complementarity. Even Einstein had been unable to convince most other physicists that his particle theory of light was correct. It took about twenty years, a critical experiment, and the abstract mathematical theories of Heisenberg and Dirac before the case was closed. Obviously, I assumed, there would never be an experiment to test Black Hole Complementarity. (I was wrong about that.) But perhaps a more rigorous theoretical foundation was possible.

On the way out of England, I still had no idea that in less than five years, mathematical physics would come to embrace one of the most philosophically disturbing ideas of all time: in a certain sense, the solid three-dimensional world of experience is a mere illusion. And I had no idea how this radical breakthrough would change the course of the Black Hole War.

Holland

Good-bye, merry old England. Hello, windmills and towering Dutchmen. Before returning home, I would cross the North Sea to visit my friend Gerard 't Hooft. After a brief flight to Amsterdam, Anne and I drove to Utrecht, another city of canals and narrow houses, where Gerard is a physics professor — some would say *the* physics professor. In 1994 he had not yet been awarded the Nobel Prize, but no one doubted that it would happen soon.

Among physicists, the name 't Hooft is synonymous with scientific greatness, and in Holland — a country that, I suspect, has had more great physicists per capita than any other — he is a national treasure. So when I arrived at the University of Utrecht, I was surprised by the modest office that Gerard occupied. That summer Europe was a humid hothouse, and Holland, despite its reputation as a cold, wet place, was unbearable. 'T Hooft's cramped office was like all the others: not even an air conditioner. As I remember, he was on the sunbaked side of the building, and I wondered what miracle protected his large, green, exotic plant from dying of the heat. As a guest, I was put around the corner in a shadier office, but it was still too hot to work or even to discuss our common passion: black holes.

On the weekends, Gerard, Anne, and I would get in Gerard's car and tour the smaller villages near Utrecht, where the air was a little cooler. Like many great scientists, 't Hooft has a tremendous curiosity about the natural world — not just physics but all of nature. A curiosity about how animals might evolve in a world dominated by urban pollution has led him to design a whole menagerie of futuristic creatures. Here is one of his creations. You can find more on his home page, www.phy.uu.nl/~thooft.

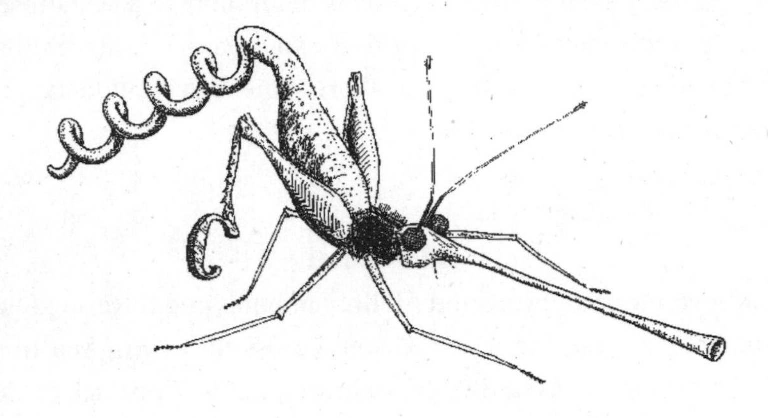

Het Wijndiefje (Wine theif), Bacchus deliriosus. This parasite can be found near pubs. Fully equipped to open bottles and cans of all kinds, it can be quite a nuisance if your wine cellar happens to be infected by it.

'T Hooft is also an amateur painter and a musician. Anne, too, is a painter and piano player, so in the car and at lunch in the local villages — over Dutch pancakes, cold mineral water, and lots of ice cream — we talked about everything from the shape of seashells and the future evolution of life on a polluted planet to Dutch painters and piano technique. But not black holes.

During the workweek, we discussed physics a little. Gerard is a contrarian who loves to argue, and our conversations often went like this: "Gerard," I would say, "I completely agree with you." "Yes," he would respond, "but I completely disagree with you."

There was one particular point that I wanted to talk about. It was something I had been thinking about for almost twenty-five years, and it had to do with String Theory. But Gerard didn't like String Theory, and getting him to delve into it was a chore. The point that I wanted to make concerned the location of individual bits of information. There is something crazy about String Theory that I first came across in 1969, but it is so crazy that string theorists don't even want to think about it.

String Theory says that everything in the world is made of microscopic, one-dimensional elastic strings. Elementary particles such as photons and electrons are extremely small loops of string, each not much bigger than the Planck scale. (Don't worry if you don't get the details. In the next chapter, I'll walk you through the main ideas. For now just accept the premise.)

The Uncertainty Principle makes these strings vibrate and fluctuate with zero point motion (see chapter 4), even when they have no extra energy. Different parts of the same string are in constant motion relative to each other, stretching and spreading the tiny parts over some distance. In itself, this spreading is not a problem; electrons in an atom are distributed over a much larger volume than the nucleus, the reason also being zero point motion. All physicists take it for granted that elementary particles are not infinitely small points of space. We all expect that electrons, photons, and other elementary particles are at least as big as the Planck length, and possibly bigger. The problem is that the mathematics of String Theory implies an absurdly violent case of quantum jitters, with fluctuations so ferocious that the pieces of an electron would spread out to the very ends of the universe. To most physicists, including string theorists, that seems so crazy that it is unthinkable.

How can it be that an electron is as big as the universe and we don't notice it? You might wonder what keeps the strings in your body from hitting or getting tangled up with the strings in my body, even if we happen to be separated by hundreds of miles. The answer is not simple. First of all, the fluctuations are exceedingly rapid, even on the infinitesimal time scale set by the Planck time. But they are also very delicately tuned, so that the fluctuations of one string are subtly matched to those of a second string in just the right way to make bad effects cancel. Nevertheless, if you could watch the most rapid internal zero point motions of an elementary particle, you would discover that its parts fluctuate out to the edges of the universe. At least that's what String Theory says.

This wildly bizarre behavior reminded me of my joke to Lárus Thorlacius (see page 241), that the world inside a black hole might be like a hologram, with the real information being far away on the two-dimensional horizon. String Theory, if you take it seriously, goes even further. It places every bit of information, whether in black holes or black newsprint, at the outer edges of the universe, or at "infinity" if the universe has no end.

Every time I started to discuss this idea with 't Hooft, we got stuck at the beginning. But shortly before I left Utrecht for home, Gerard said something that startled me. He said that if we could look at the microscopic Planck-sized details on the walls of his office, in principle they would contain every bit of information about the interior of the room. I don't recall him using the word *hologram,* but he was clearly thinking the same thing I was: in some way that we don't understand, every bit of information in the world is stored far away at the most distant boundaries of space. In fact, he had the jump on me: he mentioned a paper, a few months old, in which he had speculated about this idea.

The conversation ended on that note, and we didn't talk much about black holes during my last two days in Holland. But when I went back to my hotel that evening, I worked out a detailed argument that proved the point: the maximum amount of information that can possibly be contained in any region of space cannot be greater than what can be stored on the boundary of the region, using no more than one-quarter of a bit per Planck area.

Now let me make a comment about the ubiquitous *one-quarter* that keeps recurring. Why *one-quarter of a bit per Planck area* instead of *one bit per Planck area*? The answer is trivial. Historically, the Planck unit was poorly defined. Indeed, physicists should go back and redefine the Planck unit so that four Planck areas become one Planck area. I will lead the way; from now on, the rule will be as follows:

> *The maximum entropy in a region of space is one bit per Planck area.*

Let's return to Ptolemy, whom we met in chapter 7. There we imagined that he was so fearful of a conspiracy that the only information allowed in his library had to be visible from the outside. Therefore, it had to be written only on the outer walls. With one bit per Planck area, Ptolemy's library could hold a maximum of 10^{74} bits. That's a huge amount of information, far more than any real library could hold, but nevertheless much smaller than the 10^{109} Planck-sized bits that could have been stuffed into the interior of his library. What 't Hooft had guessed and what I proved in my hotel room was that Ptolemy's imaginary law corresponds to a genuine physical limitation on the amount of information that a region of space can hold.

Pixels and Voxels

A modern digital camera doesn't need film. It has a two-dimensional "retina" filled with microscopic, light-sensitive area-cells called *pixels.* All pictures, whether they are modern digital photographs or ancient cave paintings, are deceptions; they fool us into seeing what's not there, depicting three-dimensional images even though they contain only two-dimensional information. In *The Anatomy Lesson,* Rembrandt tricks us into seeing substance, layers, and depth, when there is really only a thin stratum of paint on a two-dimensional canvas.

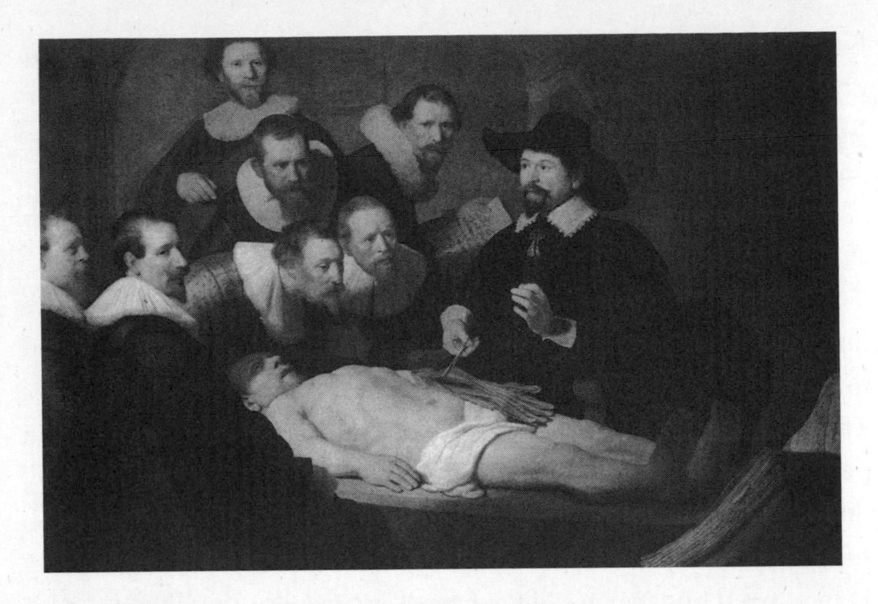

Why does the trick work? It all happens in the brain, where specialized circuits create an illusion based on previous experience: you see what your brain has been trained to see. In reality, there isn't enough information on the canvas to tell you whether the dead man's feet are really closer to you or are just too big for the rest of his body. Is his body foreshortened, or is he just very short? The organs, blood, and guts under his skin are all in your head. For all you know, the man is not a man at all, but a plaster dummy — or even a two-dimensional painting. Do you want to see what's written on the scroll behind the tallest man's head? Try walking around the painting to get a better look. Sorry, the information is just not there. The image on the pixel-filled screen of your camera also doesn't store real three-dimensional information; it's just as much of a deception.

Is it possible to build an electronic system to store genuine three-dimensional information? Of course it is. Instead of filling a surface with two-dimensional pixels, imagine filling a volume of space with microscopic three-dimensional cells, or, as they are sometimes called, *voxels.* Since the array of voxels is truly three-dimensional, it's easy to imagine how the coded information can faithfully represent a solid chunk of the three-dimensional world. It's tempting to conjecture a principle: two-dimensional information can be stored in a two-dimensional array of pixels, but three-dimensional information can be stored only in a three-dimensional array of voxels. We could give it a fancy name, something like the *Invariance of Dimensionality.*

The apparent rightness of this principle is what makes holograms so surprising. A hologram is a two-dimensional sheet of film or a two-dimensional array of pixels that can store the full details of a three-dimensional scene. It's not a fake created in your brain. The information is really there on the film.

The principle of the ordinary hologram was first discovered in 1947 by the Hungarian physicist Dennis Gabor. Holograms are unusual photographs consisting of crisscrossed, zebra-striped interference patterns similar to the ones that light makes when it passes through two slits. In a hologram, the pattern is made not by slits, but by light scattering off different parts of the objects being depicted. The photographic film is filled with information in the form of microscopic dark and light patches. It looks nothing like the real three-dimensional object; under a microscope, all you can see is random optical noise,[1] something like this:

1. The term *noise* in this context does not refer to sound. It indicates random, unstructured information, such as the white noise on the screen of a defective TV set.

The three-dimensional objects have been taken apart and put back together in what appears to be a completely scrambled two-dimensional form. It's only by means of that scrambling that a piece of the three-dimensional world can be faithfully represented on a two-dimensional surface.

The scrambling can be undone, but only if you know the trick. The information is there on the film, and it can be reconstructed. Light shined on the scrambled pattern will scatter off, reconstituting itself as a free-floating, realistic three-dimensional image.

The ghostly reality of a holographic image can be seen from any angle and looks solid. Given the right technology, Ptolemy might have coated the walls of his library with pixels containing a scrambled holographic image of thousands of scrolls. In the right lighting conditions, those scrolls would have appeared as three-dimensional images in the interior of his library.

You can probably see that I am taking you to very strange territory, but it's all part of the intellectual rewiring process that physics is once again undergoing. Here, then, is the conclusion that 't Hooft and I had reached: the three-dimensional world of ordinary experience — the universe filled with galaxies, stars, planets, houses, boulders, and people — is a hologram, an image of reality coded on a distant two-dimensional surface. This new law of physics, known as the Holographic Principle, asserts that everything inside a region of space can be described by bits of information restricted to the boundary.

To put it in concrete terms, consider the room I am working in. I in my chair, the computer in front of me, my messy desk piled high with papers I'm afraid to throw out — all that information — is precisely coded in Planckian bits, far too small to see but densely covering the walls of the room. Or instead, think of everything within a million light-years of the Sun. That region also has a boundary — not physical walls, but an imaginary mathematical shell — that contains everything within it: interstellar gas, stars, planets, people, and all the rest. As before, everything inside that giant shell is an image of microscopic bits spread over the shell. Moreover, the required number of bits is at most one per Planck area. It is as if the boundary — office walls or mathematical shell — were made of tiny pixels, each occupying one square Planck length, and everything taking place in the interior of the region is a holographic image of the pixelated boundary. But as in the case of an ordinary hologram, the information encoded in the distant boundary is a very scrambled representation of the three-dimensional original.

The Holographic Principle is a shocking departure from what we have been accustomed to in the past. That information is distributed throughout the *volume* of space seems so intuitive that it's hard to believe that it could be wrong. But the world is not voxelated; it is pixelated, and all information is stored on the boundary of space. But what boundary and what space?

In chapter 7, I raised this question: where is the information that Grant is buried in Grant's Tomb? After rejecting a few false answers, I concluded that the information is in Grant's Tomb. But is that really right? Begin with the region of space enclosed in Grant's coffin. According to the Holographic Principle, Grant's remains are a holographic illusion — an image reconstructed from information stored on the walls of the coffin. In addition, the remains, and the coffin itself, are contained within the walls of the large monument called Grant's Tomb.

So Grant's remains, his wife Julia's remains, the coffins, and the tourists who come to see them are all images of information stored on the walls of the tomb.

But why stop there? Imagine a huge sphere enclosing the entire solar system. Grant, Julia, coffins, tourists, tomb, the Earth, the Sun, and the nine planets (Pluto *is so* a planet!) are all coded by information stored on the great sphere. And thus it goes, until we come to either the boundary of the universe or infinity.

It is evident that the question of where a particular bit of information is located does not have a unique answer. Ordinary Quantum Mechanics introduces a degree of uncertainty into such questions. Until one looks at a particle, or for that matter any other object, there is quantum uncertainty in its location. But once the object is observed, everyone will agree on where it is. If the object happens to be an atom of Grant's body, ordinary Quantum Mechanics will make its location slightly uncertain, but it won't put it out at the edges of space, or even on the walls of his coffin. So if asking where a bit of information is located is not the right question, what is?

As we try to be more and more exact, especially when we account for both gravity and Quantum Mechanics, we are driven to a mathematical representation involving patterns of pixels dancing across a distant two-dimensional screen and a secret code for translating the scrambled patterns into coherent three-dimensional images. But, of course, there is no screen covered with pixels surrounding every region of space. Grant's coffin is part of Grant's Tomb, which is part of the Solar System, which is contained in a galactic-sized sphere surrounding the Milky Way . . . until the entire universe is surrounded. At each level, everything enclosed may be described as a holographic image, but when we go looking for the hologram, it's always out at the next level.[2]

As weird as the Holographic Principle is – and it is very weird – it has become part of the mainstream of theoretical physics. It is no longer just a speculation about quantum gravity; it has become an everyday working tool, answering questions not only about quantum gravity but also about such prosaic things as the nuclei of atoms (see chapter 23).

Although the Holographic Principle is a violent restructuring of the laws of physics, the proof requires no fancy mathematics. Start with a spherical region of space delineated by an imaginary mathematical boundary. The region contains some "stuff," anything at all – hydrogen gas, photons, cheese, wine, whatever – as long as it doesn't overflow the boundary. I'll just call it stuff.

2. The Holographic Principle raises strange questions – questions of the kind that one might have read about in *Amazing Stories* or some other pulp science fiction magazine in the 1950s. "Is our world a three-dimensional illusion of some two-dimensional pixel world, perhaps programmed into some cosmic quantum computer?" Even more thrilling, "Will future hobbyists be able to simulate reality on a screen of quantum pixels and become masters of their own universes?" The answer to both questions is yes – but . . .

Sure, the world could all be in some futuristic quantum computer, but I don't see that the Holographic Principle adds much to the idea, except that the number of circuit elements might be somewhat smaller than expected. It would take 10^{180} of them to fill the universe. Future world builders might be relieved that, thanks to the Holographic Principle, they will need only 10^{120} pixels. (For comparison's sake, a digital camera has a few million.)

The most massive thing that can be squeezed into the region is a black hole, whose horizon coincides with the boundary. The stuff must have no more mass than that, or it will overflow the boundary, but is there any limit on the number of bits of information in the stuff? That's what we are interested in: determining the maximum number of bits that can be stuffed into the sphere.

Next imagine a spherical shell of material – not an imaginary shell, but one made of real matter – surrounding the whole setup. The shell, being made of real material, has its own mass. Whatever the shell is made of, it can be squeezed, by either external pressure or the gravitational attraction of the stuff on the inside, until it perfectly fits in the region.

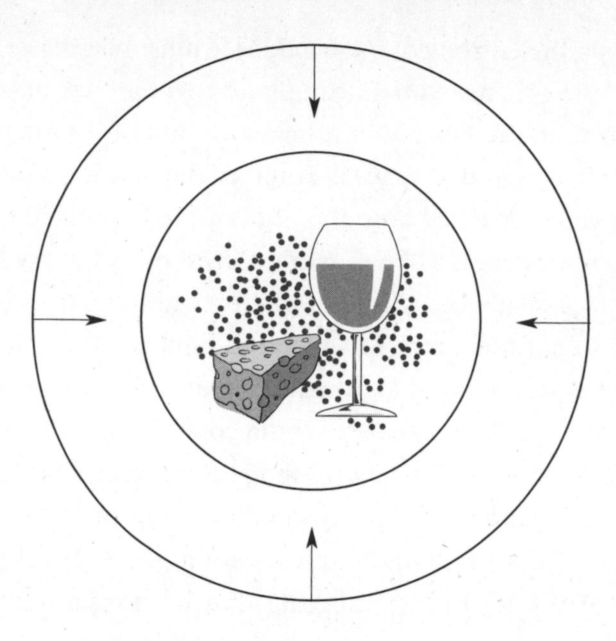

By tuning the mass of the shell, we can create a horizon that coincides with the boundary of the region.

The original stuff that we started with had some amount of entropy — hidden information — whose value we left unspecified. But there is no doubt about the *final* entropy: it's the entropy of the black hole — its area in Planck units.

To complete the argument, we need only remember that the Second Law of Thermodynamics demands that entropy always increases. Thus, the entropy of the black hole must be greater than the entropy of the original stuff. Putting it all together, we have proved an amazing fact: the maximum number of bits of information that can ever fit in a region of space is equal to the number of Planckian pixels that can be packed onto the area of the boundary. Implicitly, this means that there is a "boundary description" of everything that can take place inside the region of space; the boundary surface is a two-dimensional hologram of the three-dimensional interior. For me, this is the best kind of argument: a couple of basic principles, a thought experiment, and a far-reaching conclusion.

There is another way to picture the Holographic Principle. If the boundary sphere is very big, any small portion of it will approximately look like a plane. In the past, people were deceived by the large size of the Earth into thinking that it was flat. For an even more extreme case, suppose the boundary happens to be a sphere a billion light-years in diameter. From the perspective of a point inside the sphere but only a few light-years from the boundary, the

spherical surface seems to be flat. This means that everything taking place within a few light-years of the boundary can be thought of as a hologram on a flat sheet of pixels.

Of course, you shouldn't get the idea that I am talking about an ordinary hologram. Needless to say, the graininess of an ordinary sheet of photographic film is far coarser than a sheet of Planck-sized pixels. Moreover, this new kind of hologram can change as time goes on; it's a cinematic hologram.

But the biggest difference is that the hologram is quantum mechanical. It flickers and shimmers with the uncertainty of a quantum system, in order that the three-dimensional image have the quantum jitters. We are all made of bits moving in complicated quantum motions, but when we look closely at those bits, we find that they are located out at the farthest boundaries of space. I don't know anything less intuitive about the world than this. Getting our collective head around the Holographic Principle is probably the biggest challenge that we physicists have had since the discovery of Quantum Mechanics.

Somehow 't Hooft's paper, which preceded my own by several months, went largely unnoticed. Partly that was due to its title, "Dimensional Reduction in Quantum Gravity." The term "Dimensional Reduction" happens to be a technical physics term that has an entirely different meaning than what 't Hooft intended. I made sure that my paper would not suffer the same fate. I titled it "The World as a Hologram."

On the way home from Holland, I started to write it all down. I was very excited about the Holographic Principle, but I also knew that it was going to be very hard to convince anyone else. The world as a hologram? I could almost hear the skeptical reaction: "He used to be a good physicist, but he's gone completely bonkers."

Black Hole Complementarity and the Holographic Principle might have been the kind of ideas — the existence of atoms is another example — that physicists and philosophers argued about for hundreds of years. Making and studying a black hole in the laboratory is at least as hard as it was for the ancient Greeks to see atoms. But in fact it took less than five years for a consensus to form. How did the paradigm shift happen? The weapon that brought the war to a close was largely the rigorous mathematics of String Theory.

PART IV

Closing the Ring

19

WEAPON OF MASS DEDUCTION

Actually, I would not even be prepared to call String Theory a "theory" rather a "model" or not even that: just a hunch. After all, a theory should come together with instructions on how to deal with it to identify the things one wishes to describe, in our case the elementary particles, and one should, at least in principle, be able to formulate the rules for calculating the properties of these particles, and how to make new predictions for them. Imagine that I give you a chair, while explaining that the legs are still missing, and that the seat, back and armrest will perhaps be delivered soon; whatever I did give you, can I still call it a chair?

—GERARD 'T HOOFT

By itself, the Holographic Principle was not enough to win the Black Hole War. It was too imprecise, and it lacked a firm mathematical foundation. The reaction to it was skepticism: The universe a hologram? Sounds like science fiction. The fictitious future physicist Steve passing to the "other side" while the emperor and the count watch him being immolated? Sounds like spiritualism.

What is it that takes a fringe idea, something that may have lain dormant for years, and abruptly tips the scale in its favor? In physics it often happens without warning. A crucial, dramatic event suddenly catches the attention of a critical mass of physicists, and within a short time, the bizarre, the fantastic, the unthinkable, becomes the ordinary.

Sometimes it's an experimental result. Einstein's particle theory of light was slow to catch on, with most physicists believing that some new twist would eventually rescue the wave theory. But in 1923 Arthur Compton scattered X-rays from carbon atoms and showed that the pattern of angles and energies was unmistakably that of colliding particles. Eighteen years had passed between Einstein's original claim and Compton's experiment, but then, within months, resistance to the particle theory of light evaporated.

A mathematical result, especially if it is unexpected, can be the catalyst. The basic elements of the Standard Model (of elementary particle physics) date back to the mid-1960s, but there were arguments — some of them by the originators of the theory — that the mathematical foundations were inconsistent. Then, in 1971, a young, unknown student carried out an extremely intricate and subtle calculation and announced that the experts were wrong. Within a very short time, the Standard Model truly became standard, and the unknown student — Gerard 't Hooft — burst into the physics universe as its most luminous star.

Another example of how mathematics can tip the scales in favor of a "screwball" idea is Stephen Hawking's calculation of the temperature of a black hole. The early response to Bekenstein's claim that black holes have entropy was skepticism and even derision, not least by Hawking. In retrospect, Bekenstein's arguments were brilliant, but at the time they were too fuzzy and approximate to be convincing, and they led to an absurd conclusion: black holes evaporate. It was Hawking's difficult technical calculation that shifted the black hole paradigm from cold, dead star to object aglow with its own internal heat.

The critical events that I've described share some features. First, they were surprising. A totally unanticipated result, whether experimental or mathematical, is a powerful attention getter. Second, for a mathematical result, the more technical, precise, unintuitive, and difficult it is, the more it shocks people into recognizing the value of a new way of thinking. Part of the reason is that intricate calculations can go wrong at many points. Surviving potential dangers makes them difficult to ignore. Both 't Hooft's and Hawking's calculations had this quality.

Third, paradigms change when new ideas provide lots of more straightforward work for others to do. Physicists are always looking for new ideas to work on and will jump on anything that creates opportunities for their own research.

Black Hole Complementarity and the Holographic Principle were certainly surprising, even shocking, but by themselves they didn't have the other two qualities, at least not yet. In 1994, an experimental confirmation of the Holographic Principle seemed totally out of the question, and so did a convincing mathematical demonstration. In fact, both may have been closer than anyone realized. Within two years, a precise mathematical theory began to take shape, and a decade later we may now be on the brink of some fascinating experimental confirmation.[1] It was String Theory that made both possible.

Before I tell you about the details of String Theory, let me give you an overall perspective. No one knows for certain whether String Theory is the right theory of our world, and we may not be sure for many years. But for our purpose, that's not the most important point. We do have impressive evidence that String Theory is a mathematically consistent theory of *some* world. String Theory is based on the principles of Quantum Mechanics; it describes a system of elementary particles similar to those in our own universe; and unlike other theories (Quantum Field Theory being a case in point),

1. See chapter 23.

all material objects interact through gravitational forces. Most important, String Theory contains black holes.

How do we use String Theory to prove something about nature if we don't know that it's the right theory? For some purposes, it doesn't matter. We take String Theory to be a model of some world and then calculate, or prove mathematically, whether or not information is lost in black holes *in that world.*

Let's suppose we discover that information is not lost in our mathematical model. Once we find that out, we can look more closely and discover just where Hawking went wrong. We can try to see whether Black Hole Complementarity and the Holographic Principle are correct in String Theory. If so, it doesn't prove that String Theory is right, but it does prove that Hawking was wrong, since he claimed to prove that black holes *must* destroy information in *any* consistent world.

I am going to keep the explanation of String Theory to the bare-bones minimum. If you want more detail, you can find it in a number of books, including my earlier book *The Cosmic Landscape,* Brian Greene's *The Elegant Universe,* and Lisa Randall's *Warped Passages.* String Theory was almost an accidental discovery. Originally, it had nothing to do with black holes or the remote Planckian world of quantum gravity. It was about the more pedestrian subject of hadrons. The word *hadron* is not an everyday household term, but hadrons are among the most common, and most widely studied, particles in nature. They include protons and neutrons — the particles that make up the atomic nucleus — as well as some close relatives called mesons and the flippantly named glueballs. In their heyday, hadrons were at the cutting edge of elementary particle physics, but today they are often relegated to the somewhat old-fashioned subject of nuclear physics. However, in chapter 23, we will see that a closing circle of ideas is making hadrons the "comeback kids" of physics.

Elementary, My Dear Watson

There is an old story about two Jewish ladies who meet on a street corner in Brooklyn. One says to the other, "You must have heard by now that my son is a doctor. By the way, whatever became of your son – the one who had all that trouble learning arithmetic?" The other lady answers, "Ah, my boy became a Harvard professor of elementary particle physics." The first lady sympathetically replies, "Oh, dear, I'm terribly sorry to hear that he never graduated to advanced particle physics."

What exactly do we mean by an elementary particle, and what would the opposite be? The simplest answer is that a particle is elementary if it is so small and simple that it can't be taken apart into smaller pieces. The opposite is not an advanced particle, but a composite particle – one made out of smaller, simpler pieces.

Reductionism is the scientific philosophy that equates understanding with taking things apart into components. So far it's worked very well. Molecules are explained as composites of atoms; in turn, atoms are collections of negatively charged electrons orbiting a central positive nucleus; the nucleus is revealed to be a glob of nucleons; finally, each nucleon is made of three quarks. Today all physicists agree that molecules, atoms, nuclei, and nucleons are composite.

But at some time in the past, each of these objects was thought to be elementary. Indeed, the word *atom* comes from the Greek word for indivisible and has been in use for about 2,500 years. More recently, when Ernest Rutherford discovered the atomic nucleus, it seemed so small that it might as well have been a simple point. Evidently, what one generation calls elementary, their descendants may call composite.

All of this raises the question of how we decide – at least for the moment – whether a certain particle is elementary or composite. Here's a possible answer: Bang two of them together, really hard, and see if anything comes out. If something comes out, it must have been inside one of the original particles. Indeed, when two very fast electrons collide with a great deal of energy, all sorts of junk spews out. Photons, electrons, and positrons[2] will be especially numerous. If the collision is very energetic, protons and neutrons, as well as their antiparticles,[3] will emerge. And to top it all off, every so often a whole atom may appear. Does that mean that electrons are made of atoms? Obviously not. Smashing things together with lots of energy may be helpful in figuring out the properties of particles, but what comes out is not always a good guide to what the particles are made of.

Here is a better way to tell whether something is made of parts. Begin with an object that is obviously composite — a rock, a basketball, or a lump of pizza dough. There are many things that you can do to such an object — squeeze it into a smaller volume, deform it into a new shape, or start it spinning about an axis. Squeezing, bending, or spinning an object requires energy. For example, a spinning basketball has kinetic energy; the faster it spins, the greater the energy. And since energy is mass, the rapidly rotating ball has greater mass. A good measure of the rate of rotation — a combination of how fast the ball spins, its size, and its mass — is called *angular momentum.* As the ball is revved up with more and more angular momentum, it gains energy. The following graph illustrates the way the energy of a spinning basketball increases.

2. Positrons are the antimatter twins of electrons. They have exactly the same mass as electrons but the opposite electric charge. Electrons have a negative charge, and positrons have a positive charge.

3. All particles have antimatter twins with the opposite value of the electric charge and other similar properties. Thus, there are antiprotons, antineutrons, and the antiparticles of the electron called the positron. Quarks are no exception. The antiparticle of a quark is called an antiquark.

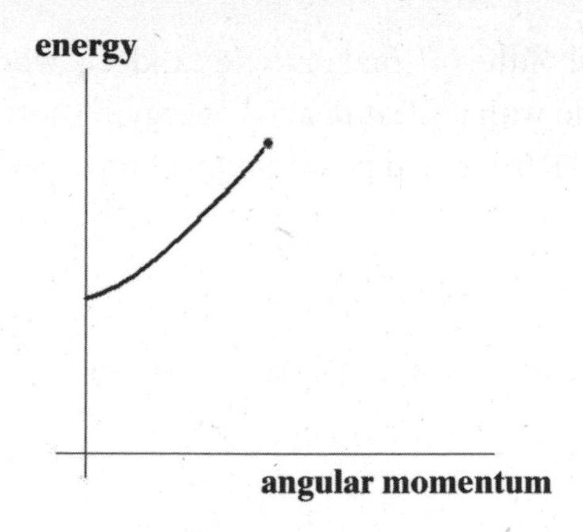

A Spinning Basketball

But why does the curve come to an abrupt end? The answer is easy to understand. The material that the ball is made of (leather or rubber) can withstand only so much stress. At some point, the ball will be torn apart by centrifugal force.

Now imagine a particle no bigger than a point of space. How do you get a mathematical point to rotate about an axis? What would it even mean to do so? Or for that matter, what would it mean to change its shape? The ability to set an object into rotation, or to start its shape oscillating, is a sure sign that it is made of smaller parts — parts that can move relative to one another.

Molecules, atoms, and nuclei also can be spun up, but in the case of these microscopic balls of matter, Quantum Mechanics plays a central role. As with all other oscillating systems, energy and angular momentum can only be added in discrete steps. Spinning a nucleus is not a process of gradually ramping up its energy. It's more like bumping it up a staircase. The graph of energy and angular momentum is a sequence of separate points.[4]

4. The Italian mathematical physicist Tullio Regge was the first to study the properties of such graphs. The sequence of points is called a Regge trajectory.

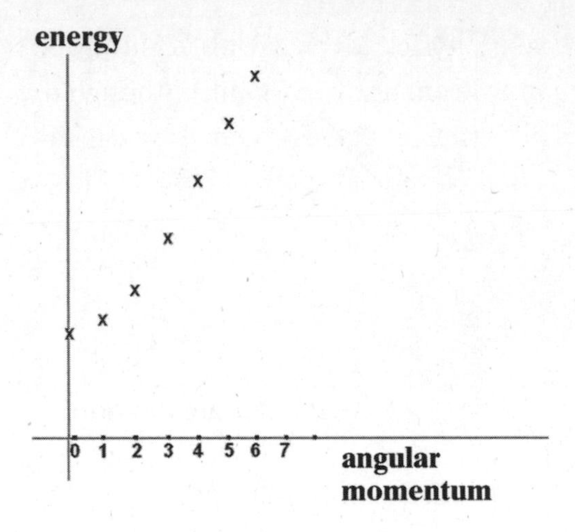

A Spinning Nucleus

Apart from the fact that the steps are discrete, the graph looks a great deal like the one for the basketball, including the fact that it comes to a sudden end. Like the basketball, the nucleus can withstand only so much centrifugal force before it flies apart.

What about electrons? Can we rotate them? With all our efforts, and they have been considerable over the years, no one has ever succeeded in giving an electron any additional angular momentum. We will come back to electrons, but let's first turn to hadrons: protons, neutrons, mesons, and glueballs.

Protons and neutrons are very similar to each other. They have almost exactly the same mass, and the forces that act to bind them into nuclei are practically identical. The only important difference is that the proton has a small positive electric charge and the neutron, as its name implies, is electrically neutral. It's almost as if a neutron is a proton that has somehow shed its electric charge. It's this similarity that led physicists to combine them, linguistically, into a single object: the nucleon. The proton is the positive nucleon, and the neutron is the neutral nucleon.

In the early days of nuclear physics, the nucleon, though almost 2,000 times heavier than the electron, was also believed to be an elementary particle. But the nucleon is nowhere near as simple as the electron. As nuclear physics advanced, objects 100,000 times smaller than atoms began to seem not very small. Although the electron has remained a point of space — at least as far as we can tell at the present time — the nucleon has been shown to possess a rich, complex inner machinery. It turns out that the nucleons are a lot less like electrons and a lot more like nuclei, atoms, and molecules. Protons and neutrons are conglomerates of many smaller objects. We know that because we can set them rotating and vibrating, and we can change their shape.

Just as for a basketball or an atomic nucleus, we can draw a graph showing the spinning of a nucleon, angular momentum on the horizontal axis and energy on the vertical axis. When this was first done more than forty years ago, the pattern that showed up was surprising in its simplicity: the sequence of points turned out to be almost exactly a *straight line.* Even more surprising, it apparently went on without end.

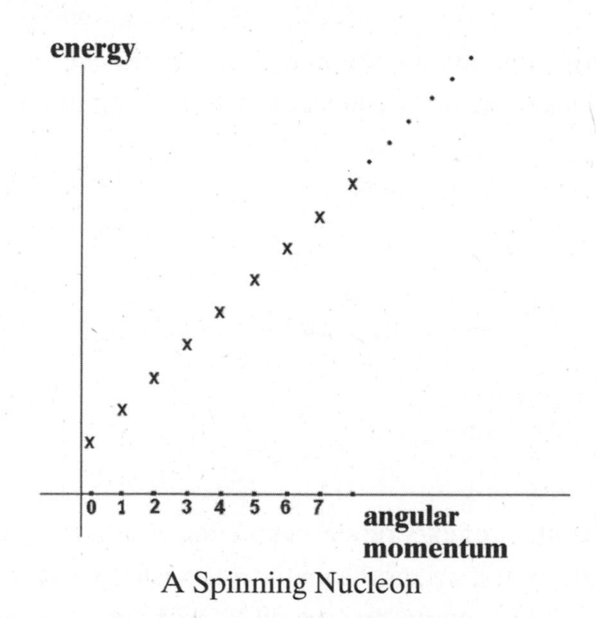

A Spinning Nucleon

There are clues in this kind of diagram about the internal construction of the nucleon. Two remarkable features have great meaning to those who know how to read the hidden message. Just the fact that the nucleon can be spun around an axis indicates that it is not a point particle; it is made of parts than can move in relation to one another. But there is a lot more. Instead of terminating abruptly, the sequence appears to continue on indefinitely, implying that nucleons don't fly apart when spun too fast. Whatever holds the parts together is much more tenacious than the forces holding a nucleus together.

Not surprisingly, the nucleon stretches out as it rotates, but not like a spinning pizza dough, which forms a two-dimensional disk.

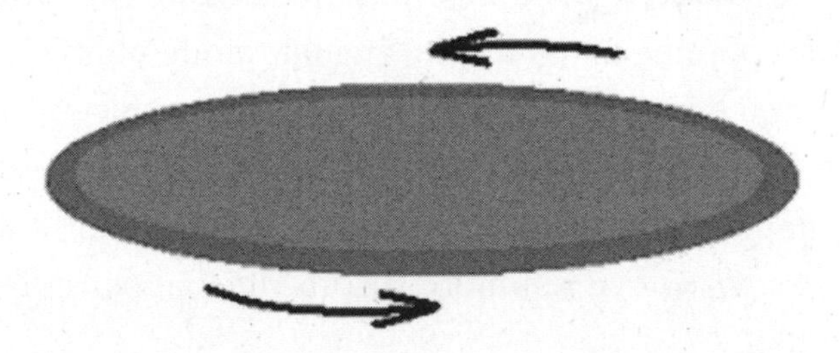

The pattern of points for the nucleon is a straight line, indicating that the nucleon stretches out into a long, thin, elastic, stringlike object.

A half century of experimenting on nucleons has made it certain that they are elastic strings that can stretch, rotate, and vibrate when excited by adding energy. In fact, all hadrons can be spun out into long, stringlike objects. Evidently, they are all made of the same sticky, stringy, stretchable stuff – something like maddeningly stubborn bubble gum that just won't let go. Richard Feynman used the term *partons* to indicate the parts of a nucleon, but it was Murray Gell-

Mann's terms – *quarks* and *gluons* – that stuck. *Gluon* refers to the sticky material that forms long strings and keeps quarks from flying apart.

Mesons are the simplest hadrons. Many different kinds of mesons have been discovered, but they all share the same structure: one quark and one antiquark, joined by a sticky string.

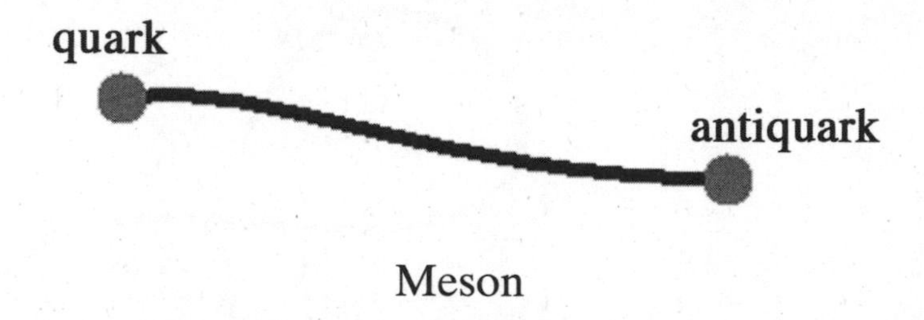

Meson

A meson can vibrate like a spring, twirl around an axis like a cheerleader's baton, or bend and flop around in a variety of ways. Mesons are examples of *open strings,* meaning that they have ends. In that respect, they are unlike rubber bands, which we would call *closed strings.*

Nucleons contain three quarks, each attached to a string, and the three strings are joined at the center like a gaucho's bolo. They can also twirl and vibrate.

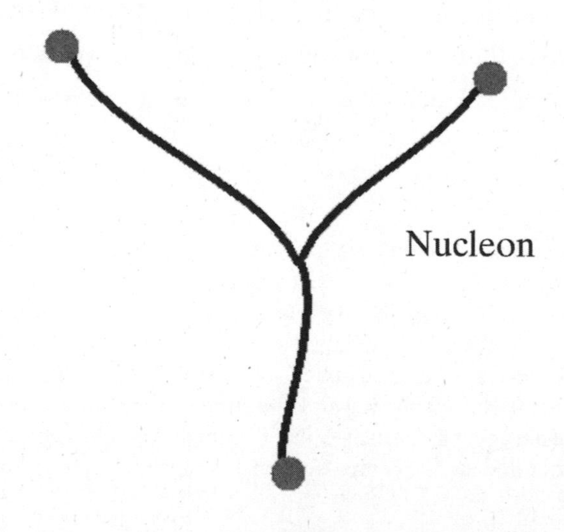

Nucleon

The rapid rotation or vibration of a hadron adds energy to the string, stretches it out, and increases its mass.[5]

One more kind of hadron exists: a family of "quarkless" particles made only of string, closing on themselves and forming a loop. Hadron physicists call them *glueballs,* but to a string theorist, they are just *closed strings.*

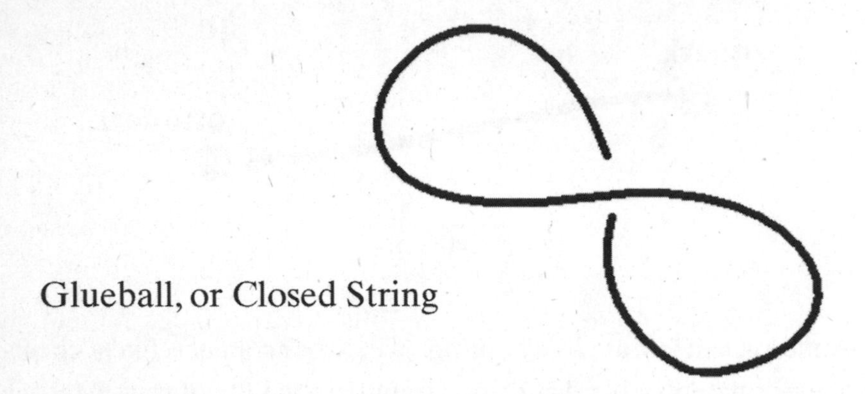

Quarks don't appear to be made of smaller particles. Like electrons, they are so small that their size is undetectable. But the strings that bind quarks together are definitely made of other objects, and those objects aren't quarks. The sticky particles that combine to form strings are called gluons.

In a sense, gluons are very tiny pieces of string. Although they are extremely small, they nonetheless seem to have two "ends" – one positive and one negative – almost as if they were tiny bar magnets.[6]

5. At first, particle physicists didn't realize that many hadrons were spinning or vibrating versions of nucleons and mesons; it was thought that they were completely new and distinct particles. The published tables of elementary particles from the 1960s include long lists that exhausted the entire Greek and Latin alphabets several times over. But in time the "excited states" of hadrons became familiar and were recognized for what they are: spinning and vibrating mesons and nucleons.

6. The two ends of a magnet are usually called the north and south poles. I don't want to imply that gluons line up like compass needles, so I will call the gluon poles positive and negative.

+ — –

Gluon

The mathematical theory of quarks and gluons is called Quantum Chromodynamics (QCD), a name that sounds as if it has more to do with color photography than with elementary particles. The terminology will become clear shortly.

According to the mathematical rules of QCD, a gluon cannot exist by itself. The positive and negative ends are required, by mathematical law, to attach themselves either to other gluons or to quarks: every positive end must be attached to the negative end of another gluon or to a quark; every negative end must be attached to a positive end of another gluon or to an antiquark. Finally, three positive or three negative ends can join together. With these rules, nucleons, mesons, and glueballs can be readily assembled.

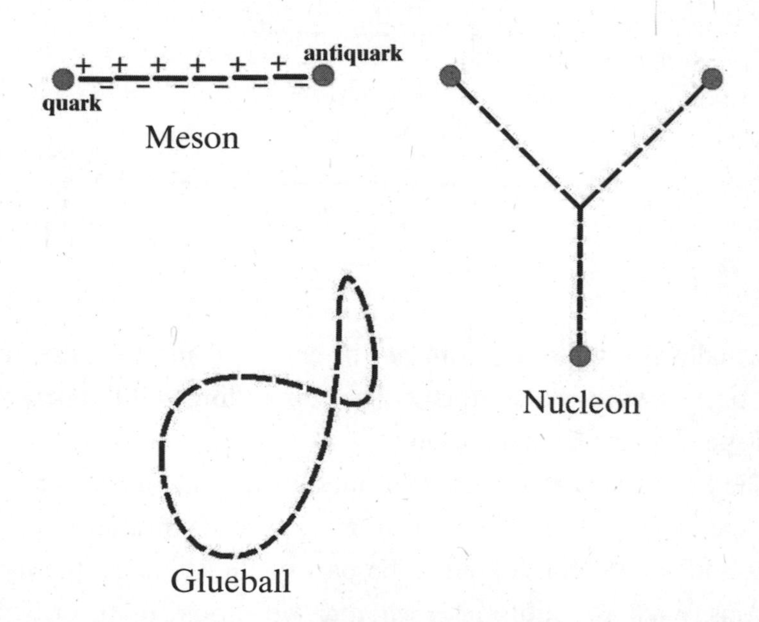

Now consider what happens if the quark in a meson is struck with a very large force. The quark will start to move rapidly away from the antiquark. If it were similar to an electron inside an atom, it would fly off and escape, but that's not at all what happens here. As it separates from its partner, gaps form between the gluons, just as they do between the molecules of a rubber band as it becomes overstretched. However, instead of snapping, the gluons clone themselves, producing more gluons to fill the gaps. In this way, a string forms between the quark and the antiquark, which frustrates the quark's escape. The following figure shows a time sequence of a high-speed quark attempting to escape its antiquark partner in a meson.

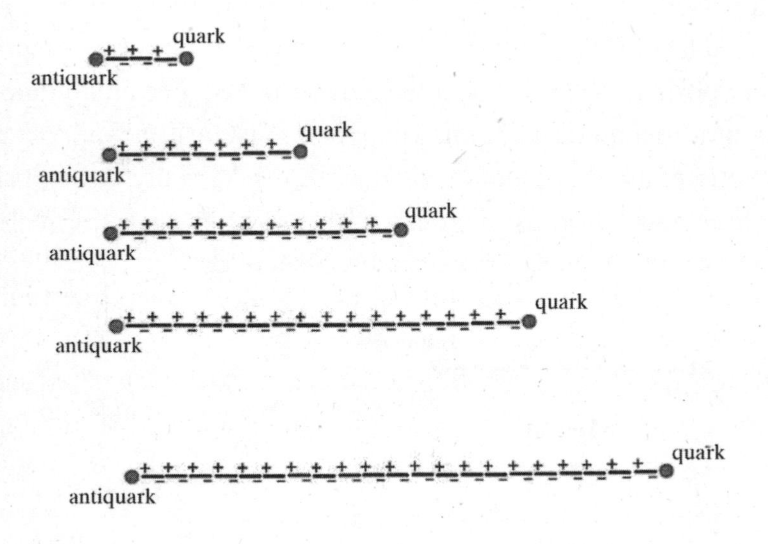

Eventually, the quark will run out of energy, come to a stop, and turn back toward the antiquark. The same thing will happen to a high-speed quark from a nucleon.

The String Theory of nucleons, mesons, and glueballs is not an idle speculation. It has been extremely well confirmed over the years and is now considered to be part of the standard theory of hadrons. What is confusing is whether we should think of String Theory as a being a consequence of Quantum Chromodynamics — in other words, should strings be thought of as long chains of the more fundamental gluons — or whether it is the other way around — that is, gluons are nothing but short segments of string. Probably both are right.

Quarks seem to be as small and elementary as electrons. They cannot be spun, squeezed, or deformed. Despite the fact that they seem to have no internal parts, they have a degree of complexity that seems paradoxical. There are many types of quarks with different electric charges and masses. What gives rise to these distinctions is a mystery; the internal machinery that underlies the differences is much too small to detect. So we call them elementary, at least for the moment, and like botanists give them different names.

Before World War II, when physics was primarily a European enterprise, physicists used the Greek language to name particles. *Photon, electron, meson, baryon, lepton,* and even *hadron* originated from the Greek. But later brash, irreverent, and sometimes silly Americans took over, and the names lightened up. *Quark* is a nonsense word from James Joyce's *Finnegan's Wake,* but from that literary high point, things went downhill. The distinctions between the different quark types are referred to by the singularly inappropriate term *flavor.* We might have spoken of chocolate, strawberry, vanilla, pistachio, cherry, and mint chocolate chip quarks but we don't. The six flavors of quarks are up, down, strange, charmed, bottom, and top. At one point, bottom and top were considered too risqué, so for a brief time they became truth and beauty.

My main purpose in telling you about flavor is only to illustrate how little we know about the building blocks of matter and how tentative our assignment of the term *elementary particle* may be. But there is another distinction that *is* very important to the way QCD works. Each quark — up, down, strange, charmed, top, bottom — comes in three *colors:* red, blue, and green. That's the origin of the "Chromo" in Quantum Chromodynamics.

Now wait a minute. Surely quarks are much too small to reflect light in the usual sense. Colored quarks are only marginally less silly than chocolate, strawberry, and vanilla quarks. But people need names for things; calling quarks red, green, and blue is no more ridiculous than calling liberals blue and conservatives red. And although we may not understand the origin of quark color much better than the origin of quark flavor, color plays a far more important role in QCD.

Gluons, according to QCD, don't have flavor, but individually they are even more colorful than quarks. Each gluon has a positive and a negative pole, and each pole has a color: red, green, or blue. It is a slight oversimplification, but essentially correct, to say that there are nine types of gluons.[7]

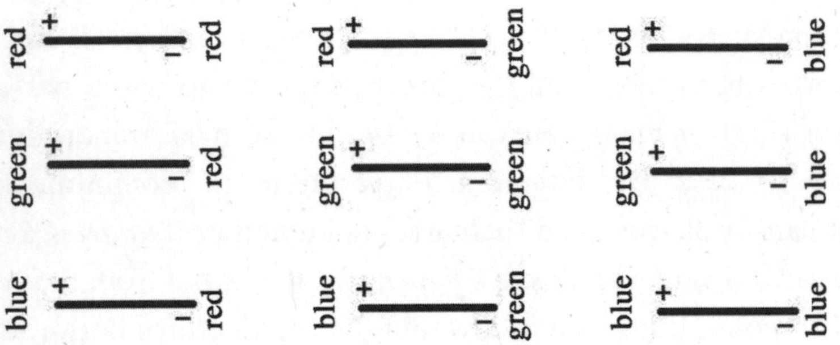

red + — − red
red + — − green
red + — − blue
green + — − red
green + — − green
green + — − blue
blue + — − red
blue + — − green
blue + — − blue

The Nine Kinds of Gluons

Why are there three colors and not two or four or some other number? It has nothing to do with the fact that color vision relies on three primary colors. As I mentioned before, the color labels are arbitrary and have nothing to do with the colors you and I see. In fact, no one knows for sure why there are three; it is one of those mysteries that tell us that we are still far from a complete understanding of elementary particles. But from the way they combine into nucleons and mesons, we know that there are three and only three quark colors.

7. The experts reading this will note that there are only eight *distinct* types of gluons. One quantum mechanical combination — the gluon with equal probability to be red-red, blue-blue, and green-green — is redundant.

I have a confession to make. Despite the fact that I have been an elementary particle physicist for more than forty years, I really don't like particle physics very much. The whole thing is too messy: six flavors, three colors, dozens of arbitrary numerical constants — that's hardly the stuff of simplicity and elegance. Why keep doing it then? The reason (and I'm sure it's not only mine) is that the very messiness must be telling us something about nature. It seems hard to believe that infinitesimal point particles could be capable of so many properties and so much structure. At some yet undiscovered level, there must be a lot of machinery underlying these so-called elementary particles. It's curiosity about that hidden machinery, as well as its implications for the basic principles of nature, that pushes me on through the miserable swamp of particle physics.

As particles go, quarks are well known to the general public. But if I had to guess which of the particles holds the best hints about the hidden machinery, I would put my money on gluons. What is the sticky pair of positive and negative ends trying to tell us?

In chapter 4, I explained that there is more to Quantum Field Theory than a list of particles. The two other ingredients are propagators — world lines showing the motion of a particle from one space-time point to another — and vertices. First let's look at the propagators. Because gluons have two poles, each labeled by a color, physicists often draw world lines as double lines. To indicate a particular type of gluon, we could write the colors next to the individual lines.[8]

8. For some of my colleagues, the so-called double-line propagator is just a trick to keep track of the mathematical possibilities. For others, including myself, it is a deep hint of some microscopic structure that at present is just too small to detect.

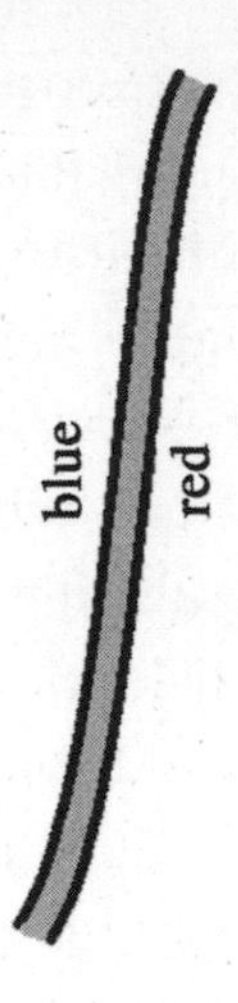

The last ingredient in Quantum Field Theory is the list of vertices. Most important for us is the vertex that describes a single gluon splitting in two.[9] The pattern is fairly simple: when a gluon with two ends splits, two new ends must materialize. According to the mathematical rules of QCD, the new ends must both have the same color. Following are two examples. Reading from bottom to top, the first shows a blue-red gluon splitting to blue-blue and blue red; the second shows a blue-red gluon splitting to blue-green and green-red.

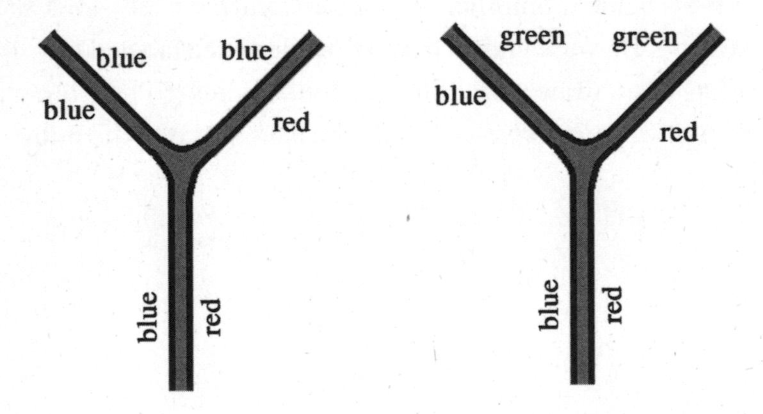

9. You may wonder how we know that gluons can split into pairs of gluons. The answer is buried deep in the mathematics of QCD. According to the mathematical rules of Quantum Field Theory, gluons can do only one of two things — split in two or emit a pair of quarks. In fact, they can do both.

The vertices could be turned upside down to show how two gluons can coalesce into a single gluon.

Although it is not obvious, and it took a while to fully understand, gluons have a strong propensity to stick together and form long chains: positive end to negative end, red to red, blue to blue, and green to green. Those chains are the strings that bind quarks and give hadrons their stringy properties.

Strings in the Basement

The idea of elastic strings turns up again in the study of quantum gravity, except that everything is smaller and faster by about twenty orders of magnitude. These miniature, swift, and terribly powerful threads of energy are called *fundamental strings.*[10]

Let me say it again so that there will be no confusion later: String Theory has two very distinct applications in modern physics. The application to hadrons takes place on size scales that seem minute by ordinary human standards but are gigantic from the viewpoint of modern physics. That the three types of hadrons — nucleons, mesons, and glueballs — are stringlike objects described by the mathematics of String Theory is an accepted fact. The laboratory experiments that underpin hadronic String Theory date back almost half a century. Strings that bind hadrons, and are themselves made of gluons, are called *QCD strings.* Fundamental strings — the ones associated with gravity and physics somewhere near the Planck scale — are the ones that have created all the excitement, controversy, blog rants, and polemical books of late.

Fundamental strings may be as much smaller than a proton as a proton is smaller than the state of New Jersey. Among them, the graviton ranks first in importance.

Gravitational forces are very similar to electrical forces in many

10. Whether fundamental strings are the ultimate explanation of elementary particles or just another stage in the reductionist march to smaller things is a matter of debate. Whatever the term's origin, *fundamental strings* is now used for convenience.

ways. The force law between electrically charged particles is called Coulomb's law; the law of gravitational forces is called Newton's law. Both electrical and gravitational forces are *inverse square laws.* This means that the strength of the force decreases according to the square of the distance. Doubling the distance between particles has the effect of dividing the force by a factor of four; tripling the distance diminishes the force by a factor of nine; quadrupling the distance decreases the force by a factor of sixteen, and so on. The Coulomb force between two particles is proportional to the product of their electric charges; the Newton force is proportional to the product of their masses. These are the similarities, but there are also differences: electrical force can be repulsive (between like charges) or attractive (between opposite charges), but gravity is always attractive.

One important similarity is that both types of forces can create waves. Imagine what happens to the force between two distant charged particles if one of them is suddenly moved — let's say away from the other charge. One might think that the force on the second particle will instantly change when the first one is displaced. But there is something wrong with this picture. If the force on a distant charge were really to change suddenly, with no delay, we could use this effect to send instant messages to distant regions of space. But instant messages violate one of the deepest principles of nature. According to the Special Theory of Relativity, no signal can travel faster than the speed of light. You cannot send a message in less time than it takes for light to travel.

In fact, the force on a distant particle does not change instantly when a nearby particle is suddenly moved. Instead, a disturbance spreads out (at the speed of light) from the displaced particle. The force on the distant particle changes only when the disturbance reaches it. The disturbance that spreads out resembles an oscillating wave. When the wave finally arrives, it shakes the second particle, causing it to behave like a cork bobbing on a ripple in a pond.

When it comes to gravity, the situation is analogous. Imagine a giant hand shaking the Sun. The Sun's movement will not be felt on the Earth for eight minutes, the time it takes light to travel the distance between the two. The "message" spreads out — again with the speed of light — in the form of a *ripple of curvature* or a *gravitational wave.* Gravitational waves are to mass what electromagnetic waves are to electric charge.

Now let's add some quantum theory. As we know, the energy of oscillating electromagnetic waves comes in indivisible quanta called photons. Planck and Einstein had very good reasons to believe that oscillating energy comes in discrete units, and unless we are very mistaken, those same arguments apply to gravitational waves. The quanta of the gravitational field are called gravitons.

I should say here that the existence of gravitons, unlike photons, is an experimentally untested conjecture — one that most physicists think is based on solid principles, but nonetheless a conjecture. Even so, the logic behind the existence of gravitons is compelling to most physicists who have thought about it.

The similarity between photons and gravitons raises interesting questions. Electromagnetic radiation is explained (in Quantum Field Theory) by a vertex diagram in which a charged particle — an electron, for example — emits a photon.

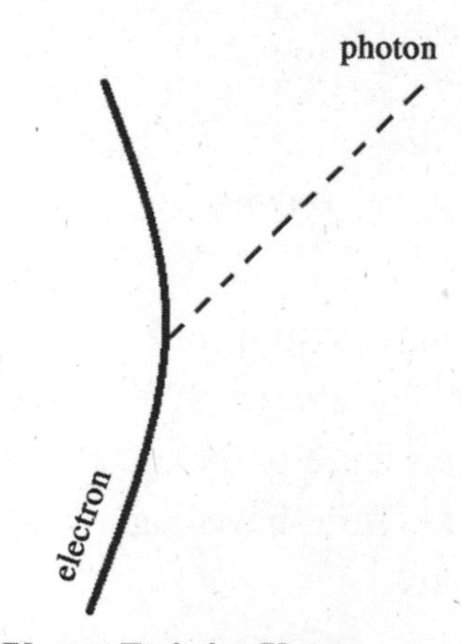

Photon Emission Vertex

It is natural to expect that gravitational waves are created when particles emit gravitons. Since everything gravitates, all particles must be able to emit gravitons.

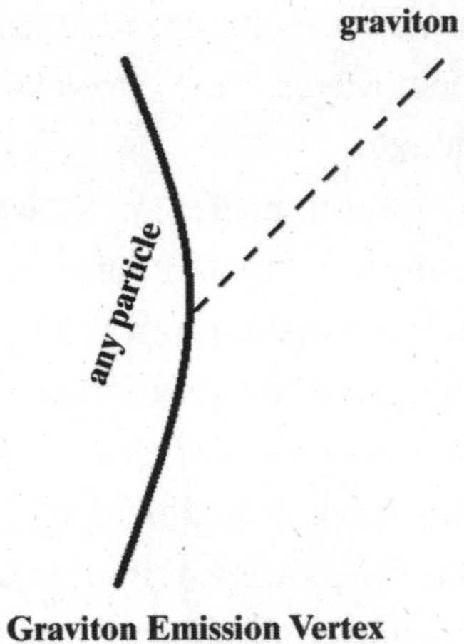

Graviton Emission Vertex

Even a graviton can emit a graviton.

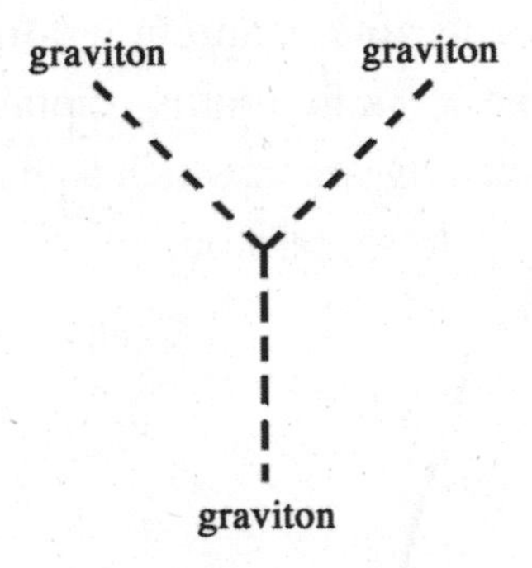

Unfortunately, including gravitons in Feynman diagrams leads to a mathematical debacle. For almost half a century, theoretical physicists have tried to make sense of a Quantum Field Theory of gravitons, and the repeated failures have convinced many of us that it's a fool's errand.

The Trouble with Quantum Field Theory

One of the brighter incidents during my 1994 trip to Cambridge was lunch with my old friend Sir Roger Penrose. Sir Roger had just become Sir Roger, and Anne and I made a visit to Oxford to congratulate him.

All four of us – Roger, I, and our wives – were sitting on the bank of the river Cherwell, in a pleasant outdoor restaurant, watching the punters go by. Punting, in case you are unfamiliar with the sport, is a form of genteel boating using a long pole to push the boat at a leisurely pace. It is a bucolic activity that always makes me think of Renoir's *Luncheon of the Boating Party,* but it does have its dangers. As one boat went by with a party of singing undergraduates, the pretty girl operating the pole got it stuck in the mud. Reluctant to let it go, she provided our lunch entertainment, clinging frantically to the pole as the boat glided away.

Meanwhile, we four were concentrating on a single chocolate mousse that we were sharing for dessert. The ladies had finished their portion, and Roger and I, while laughing at the stranded punter (who was also laughing), were working on the remaining chunk of dark, delicious chocolate. I began to notice with some fascination that as Roger and I alternated forkfuls of chocolate, we would each cut the remaining piece in half. Roger also noticed, and so began a competition to see who would be able to split the last remaining bit.

Roger remarked that the Greeks had wondered whether matter was infinitely divisible, or whether there was a smallest indivisible bit of every substance – what they called an atom. "Do you think there are chocolate atoms?" I asked. Roger claimed not to remember whether chocolate was one of the elements on the periodic table. In any case, we eventually split the mousse to what seemed like the smallest atom of chocolate, and if I remember correctly, Roger got it. The punting incident also ended happily when the next boat came by.

The problem with Quantum Field Theory is that it is based on the idea that space (and space-time) is like an infinitely divisible chocolate mousse. No matter how finely you slice it, you can always subdivide it further. All the great puzzles of mathematics are about infinity: How can the numbers go on forever, but how can they not? How can space be infinitely divisible, but how can it not be? I suspect that infinity has been a prime cause of insanity among mathematicians.

Insane or not, an infinitely divisible space is what mathematicians call a *continuum.* The problem with a continuum is that an awful lot can go on at the smallest distances. In fact, a continuum has no smallest distance – you can disappear into an infinite regression of smaller and smaller cells, and things can take place at every level. To put it another way, a continuum can hold an infinite number of bits of information in every tiny volume of space, no matter how small.

The problem of the infinitely small is especially troublesome in Quantum Mechanics, where anything that can jitter does jitter, and "everything not forbidden is compulsory." Even in empty space at absolute zero, fields, such as the electric and magnetic fields, fluctuate. These fluctuations take place on every scale, from the largest wavelengths of billions of light-years all the way down to dimensions no bigger than a mathematical point. This jitter of quantum fields can store an infinity of information in every tiny volume. This is a recipe for mathematical disaster.

The potentially infinite number of bits in every small volume shows up in Feynman diagrams as an infinite regress of smaller and smaller subdiagrams. Start with the simple idea of a propagator showing an electron moving from one space-time point to another. It begins and ends with a single electron.

b

a

There are other ways for an electron to go from a to b — for example, by juggling photons along the way.

Obviously, there is no end to the possibilities, and according to Feynman's rules, they all have to be added together to find the actual probability. Every diagram can be decorated with more structure. Each propagator and vertex can be replaced by a more complicated history involving diagrams within diagrams within diagrams, until they are just too small to see. But with the aid of a high-powered magnifying glass, even finer structure can be added — endlessly.

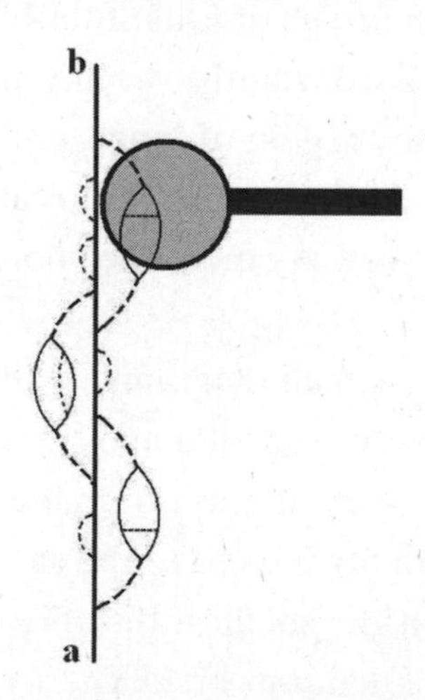

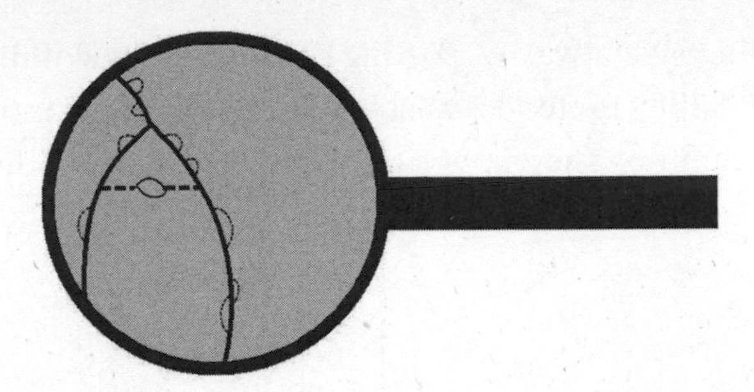

The infinite potential to add ever smaller structure to a Feynman diagram is one of the disquieting consequences of the space-time continuum of Quantum Field Theory: chocolate mousse, all the way down.

With all of this going on, it's hardly surprising that Quantum Field Theory is a mathematically dangerous subject. Making all the fluctuations in the infinitely many infinitesimal cells of space assemble themselves into a coherent universe is not easy. In fact, most versions of Quantum Field Theory go haywire and produce nonsense. Even the Standard Model of elementary particles may not be mathematically consistent in the final analysis.

But nothing compares with the difficulties of trying to build a Quantum Field Theory of gravity. Remember, gravity is geometry. In trying to combine General Relativity with Quantum Mechanics, at least according to the rules of Quantum Field Theory, one finds that space-time itself is constantly varying its shape. If you were able to zoom in on a tiny region of space, you would see space violently shaking, twisting itself into tiny lumps and knots of curvature. Moreover, the deeper you zoomed, the more violent the fluctuations would become.

The hypothetical Feynman diagrams involving gravitons reflect this perversity. The infinity of smaller and smaller diagrams becomes wildly out of control. Every attempt to make sense out of a Quantum Field Theory of gravity has led to the same conclusion: there is too much happening at the smallest distance scales. Applying conventional methods of Quantum Field Theory to gravity leads to a mathematical fiasco.

Physicists have a way of putting off the disaster of infinitely divisible space: they pretend that space, like a chocolate mousse, is not a true continuum. The supposition is that if you subdivide space past a certain point, you will discover an indivisible nugget that can no longer be divided. To put it another way, stop drawing Feynman diagrams when the substructure gets too small. A limitation on the smallness of things is called a *cutoff.* Basically, a cutoff is nothing more than dividing space into indivisible voxels and never allowing more than one bit per voxel.

A cutoff sounds like a cop-out, but there is an excuse. Physicists have long speculated that the Planck length is the ultimate atom of space. Feynman diagrams, even those involving gravitons, make perfect sense as long as you cease adding structures smaller than the Planck length — or so the argument goes. This was the almost universal expectation about space-time — that it would have an indivisible, granular, voxelated structure at the Planck scale.

But that was before the discovery of the Holographic Principle. As we saw in chapter 18, replacing continuous space by an array of finite Planck-sized voxels is the wrong idea. Voxelating space vastly overestimates the amount of variation that can go on in a region. It would have led Ptolemy to the wrong conclusion about the number of bits his library could hold, and it would lead theoretical physicists to the wrong conclusion about the amount of information a region of space can store.

It was appreciated almost from the beginning that String Theory solves the puzzle of infinitely small Feynman diagrams. It does this in part by eliminating the idea of an infinitely small particle. But until the advent of the Holographic Principle, it was not appreciated just how radically different String Theory is from a cutoff or voxelated version of Quantum Field Theory. The remarkable fact is that String Theory is quintessentially a holographic theory describing a pixelated universe.

Modern String Theory, just like its older incarnation, has both open and closed strings. In most, but not all, versions of the theory, the photon is an open string similar to a meson, except much smaller. In all versions, the graviton is a closed string most closely resembling a miniature glueball. Could it be that there is some unexpected deep sense in which these two types of strings — fundamental strings and QCD strings — are somehow the same objects? From the discrepancy in their sizes, it would seem unlikely, but string theorists are beginning to suspect that the huge difference in scales is misleading. In chapter 23, we will see that there is a unity to String Theory, but for now we will think of the two versions of String Theory as distinct phenomena.

A string is any flexible object that is much longer than it is thick: shoelaces and fishing lines are strings. In physics, the word *string* also implies elasticity: strings are stretchable as well as bendable, like bungee cords and rubber bands. QCD strings are strong — you could lift a good-sized truck on the end of a meson — but fundamental strings are even stronger. Indeed, despite the extreme thinness of fundamental strings, they are incredibly strong — vastly stronger than anything made of normal matter. The number of trucks that could be suspended from a fundamental string is about 10^{40}. That enormous tensile strength makes it extremely difficult to stretch a fundamental string to any appreciable length. As a result, the typical size of a fundamental string may be almost as small as the Planck length.

Quantum Mechanics plays no important role in the strings that we encounter in everyday life — the bungee cords, rubber bands, and stretched wads of gum — but both QCD strings and fundamental strings are highly quantum mechanical. Among other things, this means that energy can be added only in discrete, indivisible units. Passing from one value of energy to another can be done only in "quantum jumps" up the staircase of energy levels.

The bottom of the energy staircase is called the *ground state.* Adding a single unit of energy leads to the *first excited state.* Another step in energy leads to the *second excited state,* and so on up the steps. The ordinary elementary particles, such as electrons and photons, are at the bottom of the staircase. If they vibrate at all, it is only with quantum zero point motion. But if String Theory is right, they can be made to rotate and vibrate with increased energy (and therefore increased mass).

A guitar string can be excited by plucking it with a pick, but as you may imagine, a guitar pick is much too big to pluck an electron. The simplest way is to hit the electron with another particle. In effect, we use one particle as a "pick" to pluck the other. If the collision is violent enough, it will leave both strings vibrating in excited states. The obvious next question is, "Why don't experimental physicists excite electrons or photons in accelerator laboratories and settle, once and for all, the question of whether particles are vibrating fundamental strings?" The problem is the step size: it's just too big. The energy needed to rotate or vibrate a hadron is modest by the standards of modern particle physics, but the energy needed to excite a fundamental string is outrageously large. Adding one unit of energy to an electron would increase its mass to almost the Planck mass. Even worse, that energy must be concentrated in an incredibly small space. Roughly speaking, we would have to squeeze the mass of a billion billion protons into a diameter a billion billion times smaller than a proton. No accelerator ever built comes anywhere near being able to do that. It has never been done, and it probably never will be.[11]

Strings that are highly excited are bigger on average than their ground state counterparts; the additional energy whips them around and stretches them to a longer length. If you could bombard a string with enough energy, it would spread out and become as big as a violently jittering, tangled ball of yarn. And there is no limit; with even more energy, the string could be excited to any size.

11. This is why some physicists claim that String Theory remains an experimentally unproved theory. There is merit to this claim, but the fault lies no more with theorists than with experimental physicists. Those lazy dogs need to go out and build a galactic-sized accelerator. Oh, and also collect the trillions of barrels of oil that would be needed every second to fuel it.

There is one way that immensely excited strings are created in nature, if not in the laboratory. As we will see in chapter 21, black holes — even those giants at the centers of galaxies — are enormously large, tangled "monster strings."

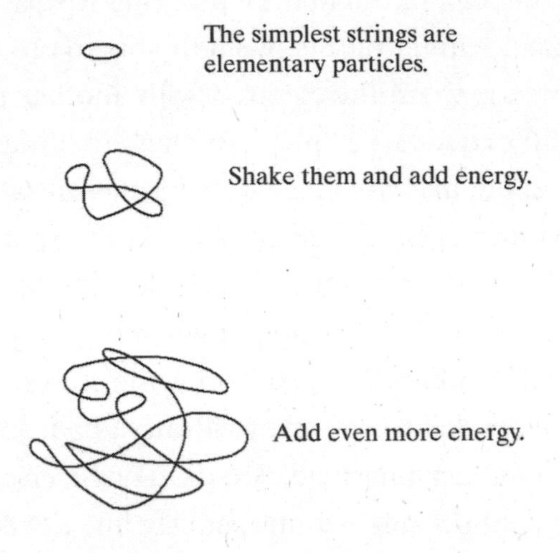

There is yet another important and fascinating consequence of Quantum Mechanics, which is very subtle and far too technical to explain in these pages. Space as we ordinarily perceive it is three-dimensional. There are many terms for the three dimensions: for example, longitude, latitude, and altitude; or length, width, and height. Mathematicians and physicists often describe the dimensions using three axes labeled *x, y,* and *z.*

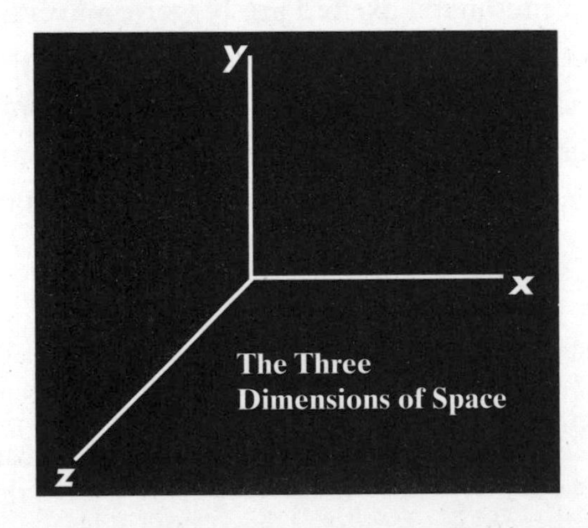

But fundamental strings are not happy with only three dimensions to move in. By that I mean that the subtle mathematics of String Theory goes haywire unless more dimensions of space are added. String theorists discovered many years ago that the mathematical consistency of their equations breaks down unless *six more dimensions of space* are added. I have always felt that if a thing is understood well enough, it should be possible to explain it in nontechnical terms. But String Theory's need for six extra dimensions has eluded a simple explanation, even after more than thirty-five years. I fear that I will have to take the scoundrel's way out and say, "It can be shown that . . ."

I would be quite surprised to meet anyone who can visualize four or five dimensions, let alone nine.[12] I can't do it any better than you can, but I can add six more letters of the alphabet — *r, s, t, u, v, w* — to the usual *x, y,* and *z,* then push the symbols around using algebra and calculus. With nine directions to move in, "it can be shown that" String Theory is mathematically consistent.

Now, you may well ask: if String Theory requires nine dimensions and space is observed to have only three, isn't that prima facie evidence that String Theory is wrong? But it's not so simple. Many very famous physicists — including Einstein, Wolfgang Pauli, Felix Klein, Steven Weinberg, Murray Gell-Mann, and Stephen Hawking (none of them string theorists) — have seriously contemplated the idea that space has more than three dimensions. They were obviously not hallucinating, so there must be some way of hiding the existence of the extra dimensions. The buzzwords for hiding extra dimensions are *compact, compactify,* and *compactification:* string theorists make the six extra dimensions of space compact, thus compactifying them by the process of compactification. The idea is that the extra dimensions of space can be wrapped up in very small knots, so that we enormous creatures are far too big to move around in them, or to even notice them.

12. One often hears that String Theory is ten-dimensional. The additional dimension is nothing but time. In other words, String Theory is (9 + 1)-dimensional.

The notion that one or more dimensions of space may be curled up into tiny geometries, and are therefore too small to detect, is a common theme of much modern high-energy physics. Some people think that *extra dimensions* is too speculative an idea – "science fiction with equations," as one wit put it. But that's a misunderstanding based on ignorance. All modern theories of elementary particles make use of some form of extra dimensions to provide the missing machinery that makes particles complicated.

String theorists did not invent the concept of extra dimensions, but they have used it in particularly creative ways. Although String Theory requires six extra dimensions, we can get the general idea by adding just a single new dimension to space. Let's explore the concept of an extra dimension in its simplest context. Starting with a world with only one space dimension – let's call it Lineland – we will eventually add one extra compact dimension. Locating a point in Lineland requires only one coordinate; the inhabitants call it X.

To make Lineland interesting, we need to add some objects, so let's create particles to move along the line.

Think of them as tiny beads that can stick together to form one-dimensional atoms, molecules, and maybe even living creatures. (I rather doubt that life can exist in a world with only one dimension, but let's suspend disbelief on that score.) Think of both the line and the beads as being infinitely thin so that they don't stick out into the other dimensions. Or even better, try to visualize the line and the beads without the other dimensions.[13]

A clever person could design many alternative versions of Lineland. The beads could be all alike, or for a more interesting world, several different kinds of beads might exist. To keep track of the different types, we could label them by colors: red, blue, green, and

13. The CGHS model that I explained in chapter 15 is Lineland, but with a massive (and no doubt dangerous) black hole at the end of the Linelanders' space.

so forth. I can imagine endless possibilities: Red beads attract blue beads but repel green beads. Black beads are very heavy, but white beads are massless and move through Lineland with the speed of light. We could even allow the beads to be quantum mechanical, the color of any given bead being uncertain.

Life in only one dimension is very constricted. With freedom to move only along a line, the Linelanders invariably bump into one another. Can they communicate? Easily: they can launch their end beads at one another to send messages. But their social life is very dull; each creature has only two acquaintances – one to the right and one to the left. You need at least two dimensions to have a social circle.

But appearances are deceptive. When the Linelanders look through a very high-powered microscope, they are startled to discover that their world is really two-dimensional. What they see is not an ideal mathematical line of zero thickness, but rather the surface of a cylinder. Under ordinary circumstances, the circumference of the cylinder is far too small to be detected by the Linelanders, but under the microscope, much smaller objects, smaller even than the Lineland atoms, are discovered – objects so small that they can move about in *two* directions.

Like their Brobdingnagian brothers, these Lineland Lilliputians can move along the length of the cylinder, but they are small enough that they also can move around its circumference. They can even move in both directions simultaneously, spiraling around the cylinder. Oh joy, they can even pass each other without colliding. Justifiably, they claim to live in two-dimensional space, but with one peculiarity: if they move in a straight line around the extra dimension, they soon come back to the same place.

The Linelanders need a name for the new direction, so they call it Y. But unlike X, they can't move very far along Y without returning to the starting point. The Linelander mathematicians say that the Y direction is *compact.*

The cylinder shown on page 341 is what you get by adding one extra compact direction to an original one-dimensional world. Adding six extra dimensions to a world that already has three is far beyond the capacity of the human brain to visualize. What separates physicists and mathematicians from other people is not that they are mutants who can visualize any number of dimensions, but rather that they have undergone an arduous mathematical retraining — again, that rewiring of the mind — to "see" the extra dimensions.

A single extra dimension doesn't provide much opportunity for variety. Moving in the compact direction would be like going around in a circle without realizing it. But just two extra dimensions allows an endless variety of new opportunities. The two extra dimensions could form a sphere,

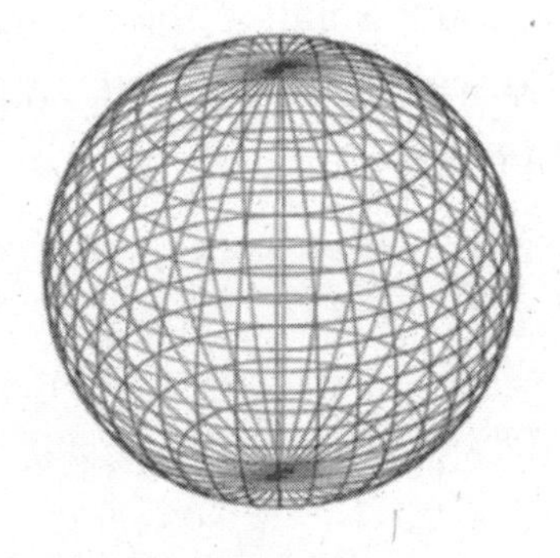

a torus (the surface of a donut),

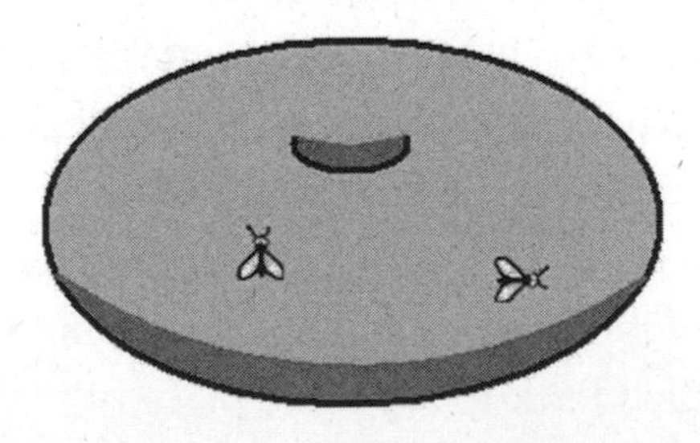

a donut with two or three holes,

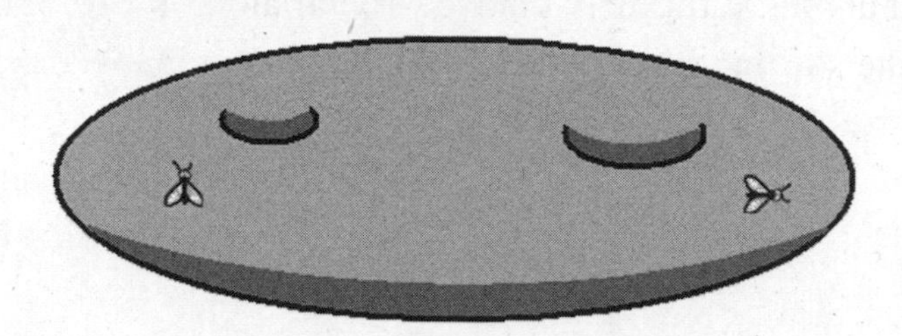

or even the bizarre space called the Klein bottle.

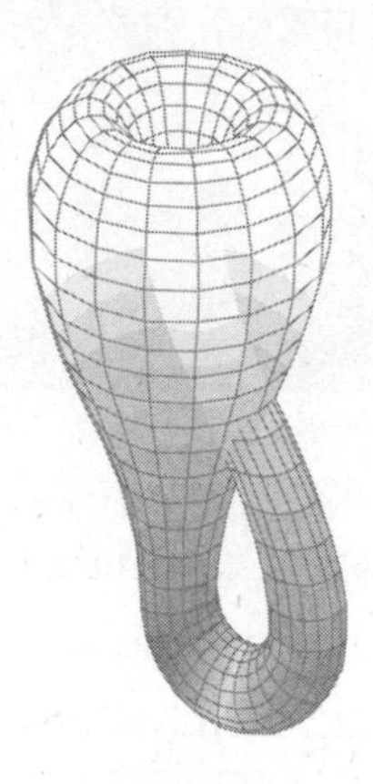

Picturing two extra dimensions is not so difficult — we just did it — but as the number grows, they become harder and harder to visualize. By the time you get to the six extra dimensions required by String Theory, visualizing without mathematics becomes hopeless. The special geometric spaces that string theorists use to compactify the six extra dimensions are called *Calabi Yau manifolds,* and there are millions of them, no two of which are the same. Calabi Yau manifolds are extremely complex, with hundreds of six-dimensional donut holes and unimaginable pretzel twists. Nevertheless, mathematicians make pictures of them by slicing them up into lower-dimensional drawings, similar to embedding diagrams. Here is a picture of a two-dimensional slice of a typical Calabi Yau space.

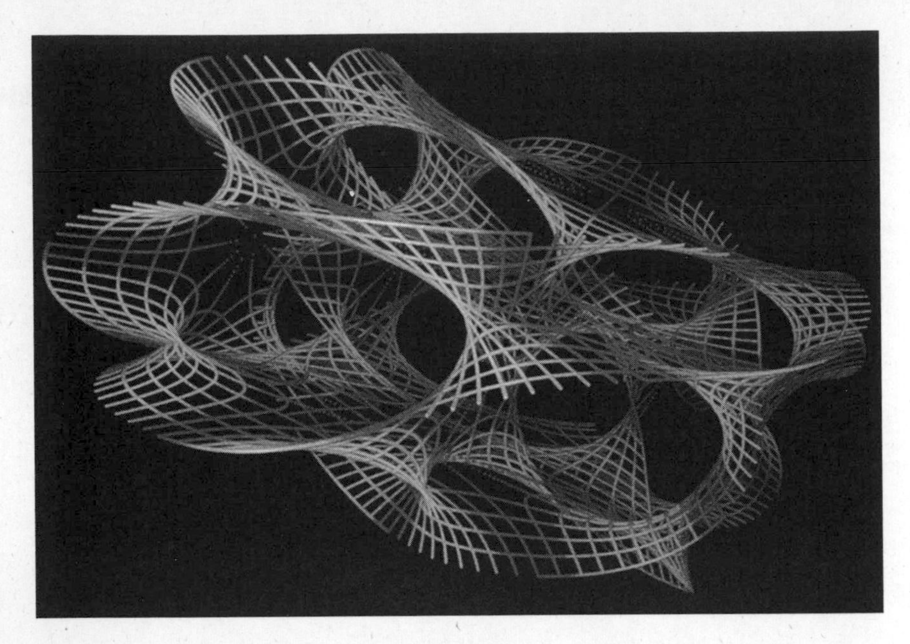

I will try to give you some idea of what ordinary space looks like when a six-dimensional Calabi Yau manifold is added at every point. First, let's look at the usual dimensions, in which large objects such as humans can move about. (I have drawn it as two-dimensional, but by now you should be able to add the third dimension in your imagination.)

At every point of three-dimensional space, there are also six other compact dimensions in which very small objects can move. By necessity, I have drawn the Calabi Yau spaces separated from one another, but you should visualize them at every point of ordinary space.

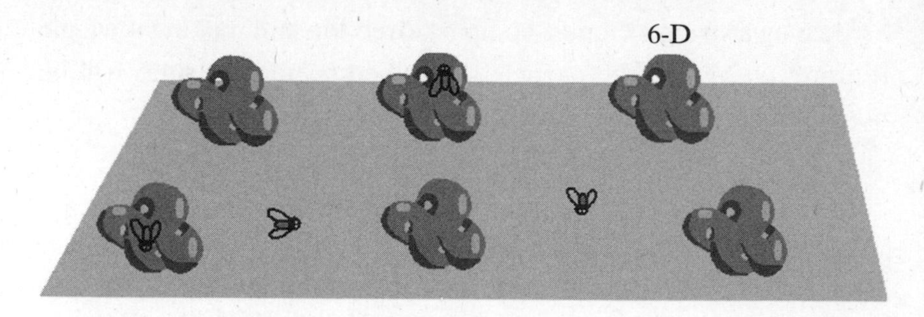

Now let's return to strings. An ordinary bungee cord can be stretched in many directions – for example, along the east-west axis, the north-south axis, or the up-down axis. It can be stretched at various angles, such as north by northwest with a 10-degree tilt from the horizontal. But if there are extra dimensions, the possibilities multiply. In particular, strings can be stretched around a compact direction. A closed string could loop around the Calabi Yau space one or more times, while not being stretched at all along the ordinary directions of space.

Let me make it even more complicated. The string could wrap around the compact space and wiggle at the same time, with the wiggles propagating around the string like a snake.

Stretching a string around a compact direction and making it wiggle requires energy, so the particles described by these strings will be heavier than ordinary particles.

Forces

Our universe is a world not just of space, time, and particles, but also of forces. Electrical forces acting between charged particles can move bits of paper and dust (think of static electricity), but more important, these same forces keep atomic electrons in their orbits around nuclei. Gravitational forces acting between the Earth and the Sun keep the Earth in orbit.

All forces ultimately originate from microscopic forces between individual particles. But where do these interparticle forces come from? To Newton, the universal gravitational force between masses was just a fact of nature — a fact that he could describe but not explain. During the nineteenth and twentieth centuries, however, physicists such as Michael Faraday, James Clerk Maxwell, Albert Einstein, and Richard Feynman had brilliant insights that explained force in terms of more basic underlying concepts.

According to Faraday and Maxwell, electric charges don't directly push and pull each other; there is an intermediary in the space between the charges that transmits the force. Imagine a Slinky — one of those lazy toy springs — stretched between two distant balls.

Each ball exerts a force only on the adjacent piece of the Slinky. Then each piece of the Slinky exerts a force on its neighbors. The force is transmitted down the Slinky till it tugs at the object at the end. It may seem as though the two objects are pulling at each other, but it's an illusion created by the intermediate Slinky.

When it comes to electrically charged particles, the intermediate agents are the electric and magnetic fields that fill the space between them. Though invisible, these fields are quite real: they are smooth, invisible disturbances of space that transmit the forces between the charges.

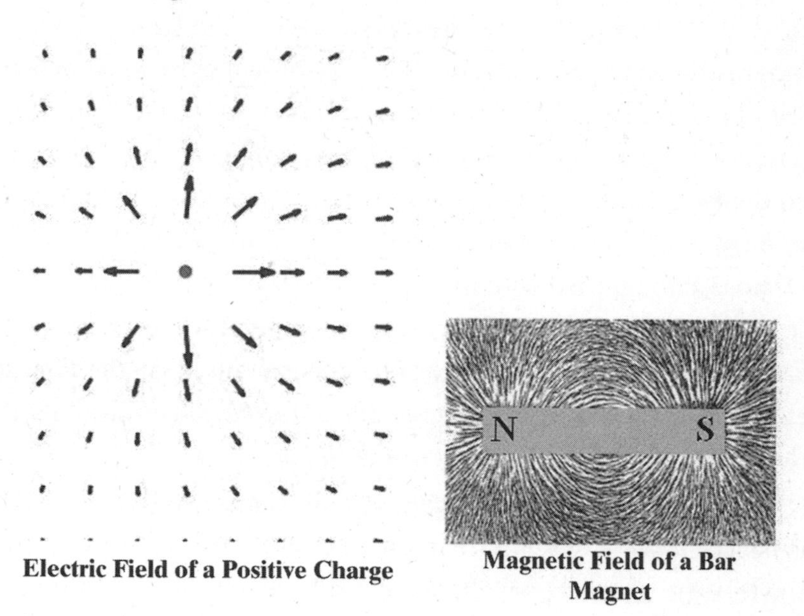

Electric Field of a Positive Charge

Magnetic Field of a Bar Magnet

Einstein went even deeper in his theory of gravity. Masses warp the geometry of space-time in their vicinity, and by doing so, they distort the trajectories of other masses. The distortions of geometry can also be thought of as fields.

One might have thought that was the end of it. It was, until Richard Feynman showed up with a quantum theory of force, which at first sight seems totally different from the Faraday-Maxwell-Einstein theory of fields. His theory begins with the notion that electrically charged particles can emit (throw) and absorb (catch) photons. There was nothing controversial about that idea; it had long been understood that X-rays are emitted when electrons are abruptly stopped by an obstruction in an X-ray tube. The reverse process of absorption was already part of Einstein's paper in which he first introduced the idea of light quanta.

Feynman pictured a charged particle as a juggler of photons, constantly emitting, absorbing, and creating numerous photons in the space surrounding the charge. An electron, standing still, is a perfect juggler, never missing a catch. But just like a human juggler in a railroad car, sudden acceleration can upset things. The charge can be pulled away from its position, causing it to be in the wrong place to absorb the photon. The missed photon flies off and becomes a bit of radiated light.

Back in the railroad car, the juggler's partner gets on the train, and the two decide to practice some coordinated team juggling. For the most part, each juggler catches his own throws, but when they get close enough, from time to time each one may catch a ball thrown by the other. The same thing happens when two electric charges get close. The clouds of photons surrounding the charges get mixed up, and one charge may absorb a photon emitted by the other. This process is called *photon exchange.*

As a result of photon exchange, the charges exert forces on each other. The hard question of whether the force is attractive (a pull) or repulsive (a push) can be answered only by the subtleties of Quantum Mechanics. Suffice it to say that when Feynman did the calculations, he found the same thing that Faraday and Maxwell had predicted: like charges repel, and opposite charges attract.

It's interesting to compare the juggling skills of electrons with human jugglers. A human can probably make a few throws and catches per second, but an electron emits and absorbs about 10^{19} photons every second.

According to Feynman's theory, all matter juggles, not just electric charges. Every form of matter emits and absorbs gravitons — the quanta of the gravitational field. The Earth and the Sun are surrounded by clouds of gravitons that intermingle and get exchanged. The result is the gravitational force that holds the Earth in orbit.

So how frequently does a single electron emit a graviton? The answer is surprising: not very often at all. On average, it takes more time than the entire age of the universe for an electron to emit a single graviton. That's why, according to Feynman's theory, the gravitational force between elementary particles is so weak compared to the electrical force.

So which theory is right: the Faraday-Maxwell-Einstein field theory or Feynman's particle-juggling theory? They sound much too different to both be true.

But they are. The key is the quantum complementarity between waves and particles that I explained in chapter 4. Waves are a field concept: light waves are nothing but a rapid undulation of electromagnetic fields. But light is particles — photons. So Feynman's particle picture of force and Maxwell's field picture are one more example of quantum complementarity. The quantum field created by the cloud of juggled particles is called a *condensate.*

A String Joke

Let me tell you the latest joke that's been going around among string theorists.

Two strings walk into a bar and order a couple of beers. The bartender says to one of them, "Hey, long time no see. How are ya?" Then he turns to the other string and says, "You're new around here, ain't ya? Are you a closed string like your friend?" The second string answers, "No, I'm a frayed knot."

Well, what did you expect from a string theorist?

The joke ends there, but the story goes on. The bartender feels a little woozy. Maybe it's the result of too many clandestine drinks behind the bar, or maybe the shimmering quantum fluctuations of the two customers are making him dizzy. But no, it is more than the standard jitters; the strings seem to be moving very strangely, as though some hidden force is tugging them and binding them together. Whenever one string makes a sudden move, an instant later the other gets pulled from his barstool, and vice versa. Yet there doesn't seem to be anything connecting them.

Fascinated by the bizarre behavior, the bartender peers into the space between them, looking for a clue. At first all he can see is a faint shimmering, a dizzying distortion of geometry, but after staring for about a minute, he notices that little bits of string are constantly breaking off the two customers' bodies, forming a condensate between them. It is the condensate that is pulling and jerking them around.

Strings do emit and absorb other strings. Let's take the case of closed strings. In addition to just jittering with zero point motion, a quantum string can split into two strings. I will describe this process in chapter 21, but for now a simple picture should give you the idea. Here is a picture of a closed string.

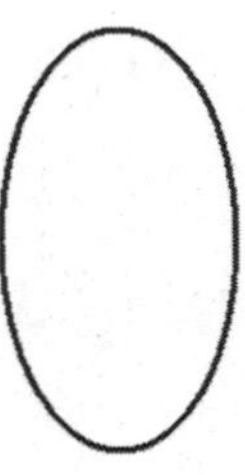

The string wiggles in a kind of pinching motion until an earlike appendage appears.

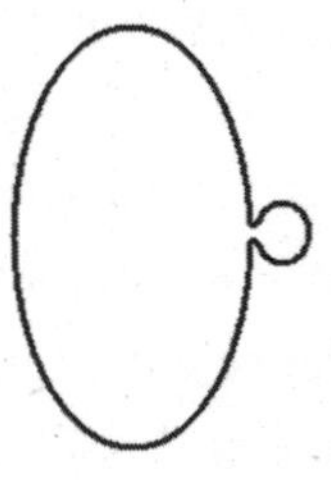

The string is now ready to split, emitting a small piece of itself.

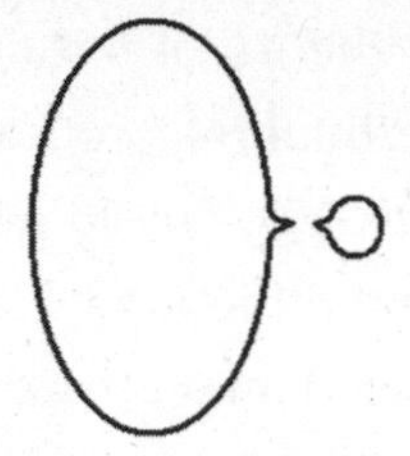

The opposite is also possible: a small string, encountering a second, larger string, can be absorbed by the reverse process.

What the bartender saw was a condensate of small strings — like a cloud of quantum flies — surrounding his customers. But when he looked less closely, the fuzzy condensate merely appeared to warp his vision — exactly like a region of curved space-time would.

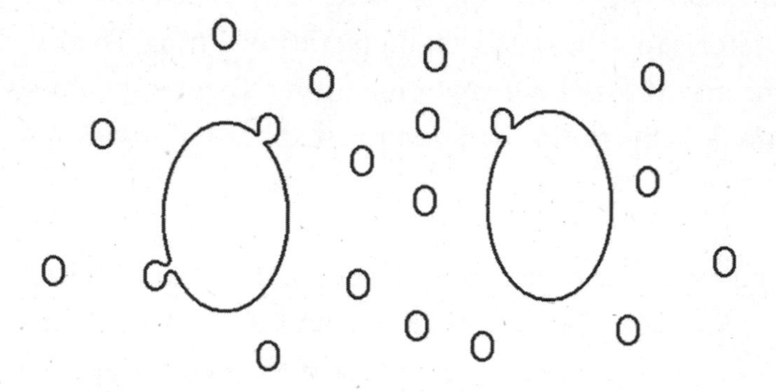

The small, closed loops of string are gravitons, swarming about the larger strings and forming a condensate that closely mimics the effects of a gravitational field. Gravitons — the quanta of the gravitational field — are similar in structure to the glueballs of nuclear physics, but 10^{19} times smaller. One wonders what, if anything, all this might mean for nuclear physics.

Some physicists in other fields of physics have found the string theorists' enthusiasm irritating. The string theorists argue that "the beautiful, elegant, consistent, robust mathematics of String Theory leads to the amazing, unbelievable, fantastic fact of gravitational forces, so it must be right." But to skeptical outsiders, no amount of superlatives, even if justified, adds up to a convincing argument. If

String Theory is the right theory of nature, the way to confirm it is through compelling experimental predictions and empirical tests, not superlatives. They are right, but so are the string theorists. The real problem is the extreme difficulty of experimenting on objects a billion billion times smaller than a proton. But whether String Theory is eventually confirmed by experimental data or not, in the meantime it is a consistent mathematical laboratory in which we can test various ideas about how gravity fits together with Quantum Mechanics.

Given gravity's emergence in String Theory, we can assume that when enough massive strings are brought together, a black hole will form. So String Theory is a framework in which Hawking's paradox can be examined. If Hawking is correct that black holes inevitably cause information loss, the mathematics of String Theory should confirm this. If Hawking is wrong, String Theory should show us how it is possible for information to escape from a black hole.

During the period in the early 1990s when Gerard 't Hooft and I visited each other twice in Stanford and once in Utrecht (if I remember correctly), 't Hooft generally distrusted String Theory, despite the fact that he wrote one of the seminal papers explaining the relation between String Theory and Quantum Field Theory. I have never been certain what the source of his dislike was, but I can guess that part of it had to do with the fact that since 1985, the American theoretical physics establishment had become overwhelmingly homogeneous, dominated by string theorists. 'T Hooft, ever the contrarian, believes (as I do) that there is strength in diversity. The more different ways in which you come at a question, and the more different styles of thought that can be brought to bear, the better the chances of solving the really hard problems of science.

There was more to Gerard's skepticism than just dyspeptic grumpiness about the takeover of physics by too narrow a group, however. As far as I can tell, he accepts that there is value to String

Theory, but he rebels against the claim that String Theory is the "final theory." String Theory was discovered by accident, and its development proceeded in bits and snatches. At no point did we ever have a comprehensive set of principles or a small set of defining equations. Even today it consists of an interrelated web of mathematical facts that have held together with remarkable consistency, but those facts do not add up to the kind of compact set of principles that characterize Newton's theory of gravity, the General Theory of Relativity, or Quantum Mechanics. Instead, there is a network of pieces that fit together like a very complicated jigsaw puzzle whose overall picture we only dimly perceive. Remember 't Hooft's quote from the beginning of this chapter: "Imagine that I give you a chair, while explaining that the legs are still missing, and that the seat, back and armrest will perhaps be delivered soon; whatever I did give you, can I still call it a chair?"

It is true that String Theory is not yet a full-blown theory, but at the moment it is far and away our best mathematical guide to the ultimate principles of quantum gravity. And, I might add, it has been the most powerful weapon in the Black Hole War, particularly in confirming Gerard's own beliefs.

In the next three chapters, we will see how String Theory helps explain and confirm Black Hole Complementarity, the origin of black hole entropy, and the Holographic Principle.

20

ALICE'S AIRPLANE, OR THE LAST VISIBLE PROPELLER

To most physicists, especially those who specialized in the General Theory of Relativity, Black Hole Complementarity seemed too crazy to be true. It was not that they were uncomfortable with quantum ambiguity; ambiguity at the Planck scale was entirely acceptable. But Black Hole Complementarity was proposing something far more radical. Depending on the state of motion of the observer, an atom might remain a tiny microscopic object, or it might spread out over the entire horizon of an enormous black hole. This degree of ambiguity was too much to swallow. It seemed strange even to me.

As I thought about it during the weeks that followed the Santa Barbara conference in 1993, the peculiar behavior began to remind me of something I had seen before. Twenty-four years earlier, during the infancy of String Theory, I had become bothered by an unsettling property of the tiny stringlike objects — I called them "rubber bands" at the time — that represented elementary particles.

According to String Theory, everything in the world is made out of one-dimensional elastic strings of energy that can be stretched, plucked, and whirled around. Start by thinking of a particle as a miniature rubber band not much bigger than the Planck length. A rubber band, if plucked, will start to jiggle and vibrate, and if there were no friction between the pieces of rubber, the jiggling and vibrating would go on forever.

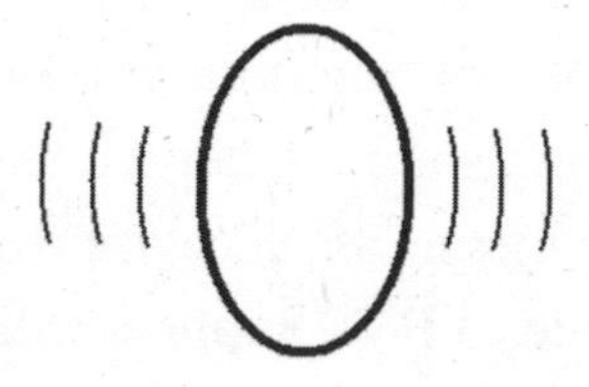

Adding energy to a string makes it oscillate even more violently, sometimes to the point that it resembles a huge, wildly fluctuating tangle of yarn. These oscillations are *thermal fluctuations,* which add real energy to the string.

But let's not forget the quantum jitters. Even if all the energy is removed from a system, leaving it in its ground state, the jitters never completely go away. This complicated motion of an elementary particle is subtle, but with the help of an analogy, I can give you some insight into it. First, however, I want to tell you about dog whistles and airplane propellers.

For whatever reason, dogs are sensitive to high-frequency sounds that go undetected by humans. Perhaps a dog's eardrum is lighter and capable of higher-frequency vibrations. Thus, if you need to call your dog but you don't want to bother the neighbors, you use a dog whistle. A dog whistle makes a sound of such high frequency that the human auditory system doesn't respond.

Now imagine Alice diving into a black hole and blowing her dog whistle to signal Rex, whom she has left in Bob's care.[1] At first Bob hears nothing; the frequency is too high for his ears. But remember what happens to signals that originate close to the horizon. According to Bob, Alice and all her functions seem to slow down. That includes the high-frequency sound of her whistle. Although the sound is initially out of Bob's range, as Alice approaches the horizon, the whistle becomes audible to Bob. Suppose Alice's dog whistle has a whole variety of high frequencies, some even out of Rex's range. What will Bob hear? At first nothing, but soon he will begin to hear the lowest frequency emitted by the whistle. As time goes on, the next-higher pitch will become audible. In time Bob will hear the whole symphony of sounds made by Alice's whistle. Keep this story in mind as I tell you about airplane propellers.

Most likely you have had an opportunity to watch an airplane propeller as it slows to a stop. At first the blades are invisible, and all you see is the central hub.

But as the propeller slows and the frequency goes below about thirty revolutions per second, the blades come into view, and the whole assembly gets bigger.

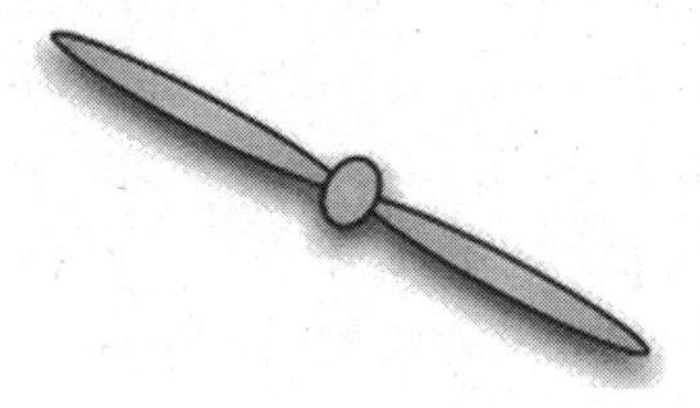

1. Strictly speaking, sound cannot propagate in empty space. Either you can go back to the drain hole analogy, or you can substitute an ultraviolet flashlight for Alice's whistle.

Now imagine an airplane with a new kind of "compound" propeller. Let's call it Alice's Airplane. At the tip of each propeller blade, there is another hub with additional "second-level" blades attached to it. The second-level blades whirl much faster than the original blades — let's say ten times faster.

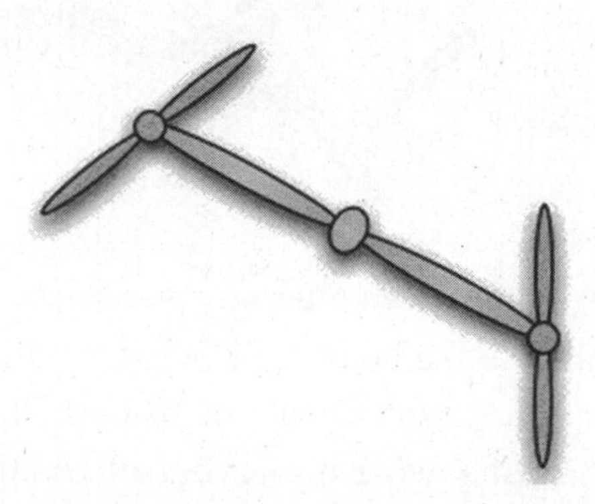

When the first-level blades come into view, the second-level blades are still invisible. As the propeller slows down even more, the second-level blades come into view. Again, the structure seems to grow. A third level of blades is attached to the ends of the second-level blades. Those blades rotate ten times faster than the second-level blades. It will take even more slowing down, but in time the compound propeller will seem to spread out over an increasing area.

Alice's Airplane does not stop at three levels. Its propeller goes on endlessly, and as it slows, more and more of it becomes visible. Bigger and bigger, it eventually grows to enormous proportions. But unless the propeller comes to a complete stop, all you can see is a finite number of levels.

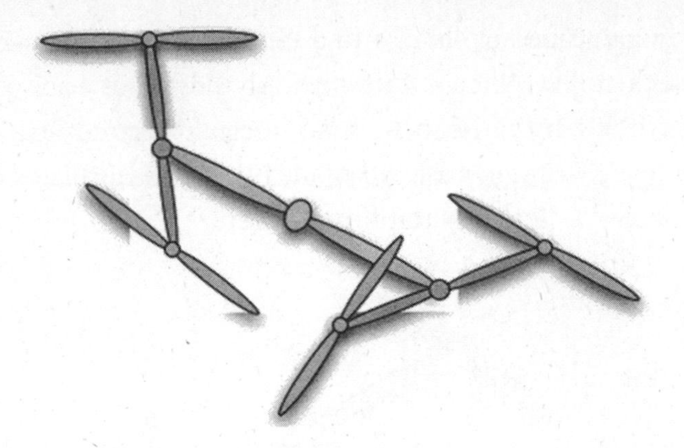

The next step, if you haven't already guessed it, is to have Alice fly her plane straight into a black hole. What would Bob see? From all that I've told you, especially about black holes and time machines, you can probably work it out yourself. As time unfolds, the propeller would appear to slow down. Eventually, the first blade would appear, and then more and more of the assembly would become visible, sprouting more levels and eventually growing to cover the entire horizon.

That's what Bob would see. But what would Alice, moving beside the propeller, see? Nothing very unusual. If she were blowing her dog whistle, the sound would remain inaudible to her. If she looked toward the propeller, it would continue to whirl too fast for her eyes or her camera to detect. She would see what you and I see while looking at a high-speed propeller — the hub and nothing more.

You may think there is something wrong with this picture. Alice might not be able to see the rapidly spinning propellers, but to say that they are undetectable seems too strong. After all, they could easily chop her to bits. Indeed, that's true for real propellers, but the motion that I am describing is more subtle. Recall that in chapters 4 and 9, I explained that there are two kinds of jitters in nature: quantum jitters and thermal jitters. Thermal jitters are dangerous; they can painfully transfer energy to your nerve endings or cook a steak. They can tear apart molecules or atoms if the temperature is high enough. But no matter how long you leave your steak out in the cold, empty vacuum of space, the quantum fluctuations of the electromagnetic field will leave it completely raw.

In the 1970s, black hole theorists such as Bekenstein, Hawking, and especially William Unruh showed that near a black hole horizon, the thermal and quantum jitters get mixed up in an odd way. Jitters that appear to be innocent quantum fluctuations to someone falling through the horizon become exceedingly dangerous thermal fluctuations to anything that remains suspended outside the black hole. It is as if the invisible motion of Alice's propellers (invisible to Alice) are quantum jitters, but as they slow down in Bob's frame of reference, they become thermal jitters. The benign quantum motion that Alice fails to perceive would be extremely dangerous to Bob if he were to hover just above the horizon.

By now you have probably made the connection with Black Hole Complementarity. Indeed, the similarity with the things I explained in chapter 15 about atoms falling into a black hole are striking. Since that was five chapters ago, here's a quick refresher.

Imagine that while falling toward the horizon, Alice has her eye on an atom falling next to her. The atom looks perfectly ordinary, even as it passes the horizon. Its electrons continue to revolve around the nucleus at the usual rate, and it looks no bigger than any other atom — about one-billionth the size of this page.

As for Bob, he sees the atom slow down as it approaches the horizon, and at the same time thermal motions smash it apart and spread it over an ever widening area. The atom seems to resemble a miniature Alice's Airplane.

Do I mean that atoms have propellers that have propellers that have propellers ad infinitum? Surprisingly, that's almost exactly what I mean. Elementary particles are usually imagined to be very small objects. The central hub of Alice's compound propeller also seems to be small, but the whole assemblage, including all the levels of structure, is huge or even infinite. Could we be mistaken about particles when we say they are small? What does experimentation say about it?

In thinking about experimental observations of particles, it is useful to imagine each experiment as a process similar to photographing a moving object. The ability to capture rapid motions depends on how fast the camera can act to record the image. Shutter speed is the important measure of time resolution. Obviously, shutter speed would play a central role in photographing Alice's compound propeller. A slow camera would capture only the main hub. A faster camera would capture additional high-frequency structure. But even the fastest camera could capture only so much of the compound structure of the propeller — unless it happened to catch the plane as it fell into the black hole.

The shutter speed in a particle physics experiment is related to the energy of the colliding particles: the higher the energy, the faster the shutter. Unfortunately for us, shutter speed is severely limited by the capacity to accelerate particles to very high energy. Ideally, we would like to resolve motions taking place over time periods smaller than the Planck time. This would require accelerating particles to energies above the Planck mass — in principle easy, but in practice impossible.

This is a good time to pause and consider the extraordinary difficulties facing modern physics. To observe the smallest objects and the most rapid motions, physicists throughout the twentieth century have relied on larger and larger accelerators. The first accelerators were simple tabletop arrangements that could probe the structure of atoms. Nuclei required larger machines, some as big as buildings. Quarks were discovered only when accelerators grew to be miles in length. Today's biggest accelerator, the Large Hadron Collider in Geneva, Switzerland, is nearly twenty miles in circumference, but still it is far too small to accelerate particles to the Planck mass. How much bigger would an accelerator have to be to resolve motions of Planckian frequency? The answer is discouraging, to say the least: in order to accelerate a particle to the Planck mass, an accelerator would have to be at least as big as our galaxy.

To put it in simple terms, looking at Planckian motions with modern technology is comparable to photographing a rotating airplane propeller with a camera whose shutter stays open for about ten million years. Not surprisingly, elementary particles appear to be very small because all we can see is the hub.

If experiments cannot tell us whether particles have outlying, high-frequency, vibrating structures, we have to appeal to our best theories. For the second half of the twentieth century, the most powerful mathematical framework for the study of elementary particles was Quantum Field Theory. Quantum Field Theory is a fascinating subject that begins by postulating particles that are so small they can be regarded as mere points of space. But that picture soon breaks down. Particles quickly surround themselves with more particles that come and go at a tremendous pace. These new comers-and-goers are themselves surrounded by even more rapidly appearing and disappearing particles. Photographing with ever faster shutter speeds would reveal more and more structure inside particles — more and more rapidly oscillating particles coming into and out of existence. A slow camera sees a molecule as an unresolved blur. It reveals itself only as a collection of atoms if the shutter speed is fast enough to catch the atomic motions. The story repeats itself at the atomic level. The blur of electric charge around a nucleus requires an even faster experiment to resolve into electrons. Nuclei resolve into protons and neutrons, which become quarks, and so it goes.

But these progressively faster photos would not show the main feature that we are looking for: an expanding structure that fills more and more space. Instead, it would show smaller and smaller particles forming a kind of Russian matryoshka (nesting doll) hierarchy. This is not what we need to explain how particles behave near horizons.

String Theory is much more promising. What it says is so counterintuitive that physicists for many years did not know what to make of it. The elementary particles described by String Theory — the supposedly tiny loops of string — are just like compound propellers. Start with a slow shutter. An elementary particle looks almost like a point; think of it as the hub. Now speed up the shutter to where it stays open for a bit longer than the Planck time. The image begins to show that the particle is a string.

Speed up the shutter even more. What you see is that every piece of the string is fluctuating and vibrating, so that the new picture looks more tangled and spread out.

But it doesn't end there; the process repeats itself. Every little loop, every bend of the string, resolves itself into more rapidly fluctuating loops and squiggles.

What does Bob see as he watches a stringlike particle fall toward the horizon? At first the oscillating motion is much too fast to resolve, and all he sees is the tiny hublike center. But soon the peculiar nature of time near the horizon begins to assert itself, and the motion of the string appears to slow down. He gradually sees more and more of the oscillating structure in the same way that he saw Alice's compound propeller. As time goes on, even more rapid oscillations come into view, and the string seems to grow and spread over the entire horizon of the black hole.

But what if we fall alongside the particle? Then time behaves normally. The high-frequency fluctuations remain high frequency, far out of range of our slow camera. Being near the horizon gives us no advantage. As in the case of Alice's Airplane, we see only the tiny hub.

String Theory and Quantum Field Theory share the property that things appear to change as the shutter speed increases. But in Quantum Field Theory, the objects do not grow. Instead, they appear to break down into progressively smaller objects — ever smaller Russian dolls. But as the constituents become as small as the Planck length, a totally new pattern emerges: the Alice's Airplane pattern.

In Russell Hoban's allegory *The Mouse and His Child,* there is an amusing (unintended) metaphor for how Quantum Field Theory works. Sometime during their nightmarish adventure, the toy mechanical mice — father and son — discover an endlessly fascinating can of Bonzo Dog Food. On the label of the can is a picture of a dog holding a can of Bonzo Dog Food, whose label shows a dog holding a can of Bonzo Dog Food, whose label . . . Deeper and deeper the mice peer, in a frustrating quest to see "The Last Visible Dog," but they are never quite sure that they have seen it.

Things inside things inside things — that's the story of Quantum Field Theory. Unlike the Bonzo label, however, the things move, and the smaller they are, the faster they move. So to see them, you need both a more powerful microscope and an ever faster camera. But notice one thing: neither the resolved molecule nor the can of Bonzo Dog Food appears to grow larger as more and more structure is uncovered.

String Theory is different and works more like Alice's Airplane. As things slow down, more and more stringy "propellers" come into view. They occupy an increasing amount of space so that the entire complex structure grows. Of course, Alice's Airplane is an analogy, but it does capture a lot of the mathematical properties of String Theory. Strings, like anything else, have the quantum jitters, but in a special way. Just like Alice's Airplane, or the symphonic version of her dog whistle, strings vibrate at many different frequencies. Most of the vibrations are too fast to detect, even at the rapid shutter speeds provided by powerful particle accelerators.

As I began to understand these things in 1993, I also began to understand Hawking's blind spot. To most physicists who had been brought up on Quantum Field Theory, the notion of growing particles with unbounded, jittering structures was extremely foreign. Ironically, the only other person who had hinted at such a possibility was the world's greatest quantum field theorist, my comrade in arms Gerard 't Hooft. Although he presented this idea in his own way – not in the language of String Theory – his work also expressed a sense that things grow as they are examined with increasing time resolution. By contrast, Hawking's bag of tricks included the Bonzo Dog Food label but not Alice's Airplane. For Stephen, Quantum Field Theory, with its point particles, was the be-all and end-all of microscopic physics.

21

COUNTING BLACK HOLES

One morning, when I went down to breakfast, my wife, Anne, remarked that my T-shirt was on backward; the V shape woven into the fabric was in the back. Later in the day, when I came home from a jog, she laughed and said, "Now it's inside out." That set me thinking: how many ways are there to wear a T-shirt? Anne mockingly said, "That's the sort of stupid thing you physicists are always thinking about." Just to prove my superior cleverness, I quickly declared that there are 24 ways to wear a T-shirt. You can stick your head through any of 4 holes. That leaves 3 holes for your torso. Having picked a neck hole and a torso hole, that leaves 2 possibilities for your left arm. Once you decide where your left arm goes, there is only 1 choice for your right arm. So that means $4 \times 3 \times 2 = 12$ ways to choose from. But then you can turn the shirt inside out, giving another 12, so I proudly announced that I had solved the problem: 24 ways to wear a T-shirt. Anne was not impressed. She replied, "No, there are 25. You forgot one." Puzzled, I asked, "What did I miss?" With a look that would freeze hell, she said, "You can roll it in a ball and shove it . . ." You get the idea.[1]

1. Since writing this, Anne has discovered at least 10 more ways to wear a T-shirt.

Physicists (and, even more so, mathematicians) are very good at counting things — in particular, counting possibilities. Counting possibilities is at the heart of understanding entropy, but in the case of black holes, what exactly do we count? It's certainly not the number of ways a black hole can wear a T-shirt.

What made the counting of black hole possibilities so important? After all, Hawking had already given the answer when he calculated that entropy equals the horizon area in Planck units. But there was enormous confusion surrounding black hole entropy. Let me remind you why.

Stephen argued that the whole idea of entropy as hidden information — information that could be counted if you knew the details — must be wrong when black holes were involved. He was hardly the only one to say this. Almost all black hole experts had come to the same conclusion: black hole entropy was something different, having nothing to do with counting quantum states.

Why would Hawking and the relativists have such a radical view? The problem was Stephen's persuasive argument that one could just keep throwing more and more information into a black hole — like squeezing an infinite number of clowns into a clown car — without any information leaking back out. If entropy had its usual meaning — the total number of possible bits that could be hidden in a black hole — the amount of information that could be hidden would have to be limited. But if an indefinite number of bits could be lost in the black hole, that would mean that the calculation of black hole entropy could not be counting all the hidden possibilities — and *that* would mean that a revolutionary new basis would have to be given for one of the oldest and most trusted subjects of physics, thermodynamics. Thus, it became urgent to know whether black hole entropy really counts the possible configurations of a black hole.

In this chapter, I'm going to tell you how string theorists went about this counting and how, in the process, they gave a firm quantum mechanical basis for the Bekenstein-Hawking entropy — a basis that left no room for information loss. This was a major accomplishment, and it went a long way toward undermining Stephen's claim that an indefinite amount of information could be swallowed by a black hole.

But first let me explain a point of view originally suggested by Gerard 't Hooft.

'T Hooft's Conjecture

There are lots of different elementary particles, and I think it is fair to say that physicists don't fully understand what makes one different from another. But without asking the deep questions, we can still take an empirical look at all of the particles that are either known from experiment or expected to exist on theoretical grounds. One way to exhibit them is to plot them on an axis and make a kind of (not to scale) elementary particle spectrum. The horizontal axis represents mass, with the left end corresponding to the lightest objects. The mass increases toward the right. The vertical lines mark specific particles.

At the lower end are all the familiar particles whose existence is certain. Two of them have no mass and move with the speed of light: the photon and the graviton. Then come the various types of neutrinos, the electron, some quarks, the mu-lepton, some more quarks, the W-boson, the Z-boson, the Higgs-boson, and the tau-lepton. The names and details are not important.

At somewhat larger mass, there is a whole collection of particles whose existence is only conjectural, but a lot of physicists (including me) think they may exist.[2] For reasons that are not important to us here, these hypothetical particles are called *superpartners.* Above the superpartners is a big gap that I've indicated by question marks. It's not that we know there is a gap; we just don't have any special reason to postulate particles in this region. Furthermore, no accelerator being built or even contemplated will be powerful enough to create particles of such large mass. So the gap is terra incognita.

Then, with masses far beyond those of the superpartners, there are the *Grand Unification particles.* These are also conjectural, but there are very good reasons to believe that they exist — in my opinion, even better reasons than for the superpartners — but their discovery will at best be indirect.

The most controversial particles in my diagram are the *string excitations.* According to String Theory, these are the very heavy, rotating and vibrating *excited states* of ordinary particles. Then at the very top, we have the *Planck mass.* Before the early 1990s, most physicists would have expected the Planck mass to be the end of the elementary particle spectrum. But Gerard 't Hooft had a different point of view. He argued that there were certainly objects with greater mass. The Planck mass, though huge on the scale of the electron or quark mass, is comparable to the mass of a speck of dust. Obviously, heavier things exist — bowling balls, steam locomotives, and Christmas fruitcakes among them. But special among those heavier things are the ones that are the smallest in size for a given mass.

Take an ordinary brick. Its mass is roughly one kilogram. "Solid as a brick" is what we say. But bricks, solid as they seem, are almost entirely empty space. Put under enough pressure, they can be squeezed to a much smaller size. If the pressure were high enough, a brick could be squeezed to the size of a pinhead or even a virus. And it would still be mostly empty space.

2. We will know within a few years, when the European accelerator called the LHC (Large Hadron Collider) starts operating.

But there is a limit. I don't mean a practical limit based on the limitations of present-day technology. I am talking about laws of nature and fundamental physical principles. What is the smallest diameter that a one-kilogram object can occupy? The Planck size is an obvious guess, but it's not the right answer. An object can be squeezed until it becomes a black hole whose mass is one kilogram,[3] and no further: that's the smallest, most concentrated possible object with a given mass.

Just what is the size of a one-kilogram black hole? The answer is probably smaller than you think. The Schwarzschild radius (radius of the horizon) of such a black hole is about one hundred million Planck lengths. That radius may sound big, but the truth is that it's a trillion times smaller than a single proton. It seems to be as small as an elementary particle, so why not count it as one?

'T Hooft did just that. Or at least he said that there is no important way in which it is fundamentally different from an elementary particle. He then proposed the following bold idea:

The spectrum of particles does not terminate at the Planck mass. It continues on to indefinitely large mass in the form of black holes.

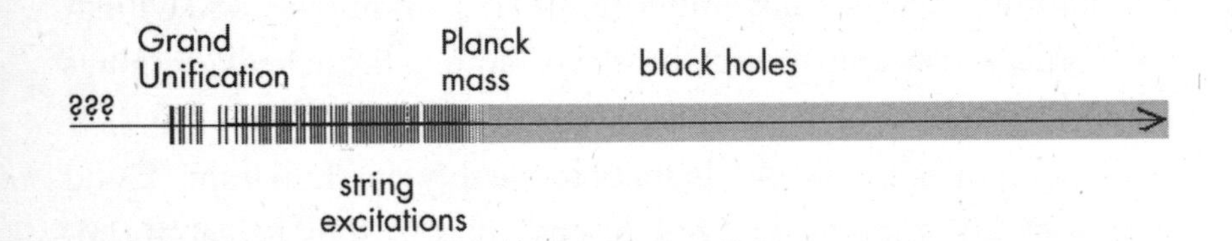

'T Hooft also argued that black holes could not have arbitrary mass, but like the ordinary particles, only certain discrete masses are possible. These allowed values become so dense and closely

3. There is a technical subtlety here. Squeezing a brick or other object increases its energy, and because of $E=mc^2$, it also increases its mass. But we can compensate for that in a variety of ways. What we want is to end up with the smallest possible one-kilogram object.

spaced above the Planck mass, however, that they practically become a blur.[4]

The transition from ordinary particles (or string excitations) to black holes is not as sharp as I depict in the figure. Most likely, the spectrum of string excitations fades into the black hole spectrum with no sharp distinction somewhere near the Planck mass. This was 't Hooft's conjecture, and as we will see, there is very good reason to believe it.

Counting Strings and Counting Black Holes

Alice's Airplane is a metaphor for how appearances are in the eye of the beholder. Alice, from the cockpit, doesn't see anything exceptional at the horizon. But seen from outside the black hole, the airplane seems to have more and more propellers that gradually spread out over the horizon. Alice's Airplane is also a metaphor for how String Theory works. As a string falls toward the horizon, an observer on the outside will detect more and more bits of string materializing and filling the horizon.

The entropy of black holes implies that they have a hidden microscopic substructure similar to the molecules in a tub of warm water. But in itself, the existence of entropy holds no clue to the nature of the "horizon-atoms," although it does give a rough count of their number.

In Alice's world, the horizon-atoms are propellers. Maybe there really is a theory of quantum gravity based on propellers, but I think String Theory has a better claim, at least for now.

The idea that strings have entropy goes back to the earliest days of String Theory. The details are mathematical, but the general idea

4. Why so dense? It's the entropy. As the mass increases, the horizon area also increases; thus the black hole entropy rises as well. But remember: *entropy* means hidden information. When we say that the mass of a black hole is one kilogram, we really mean *approximately* one kilogram. A more accurate statement would be that the mass is one kilogram with a certain margin of error. If there are many possible black hole masses within the margin of error, we have left a lot of information out of our description. That missing information is the black hole entropy. Knowing that black hole entropy grows with mass, 't Hooft reasoned that the spectrum of black hole masses must become a very dense blur.

is easy to understand. Start with the simplest string, one representing an elementary particle of given energy. To be definite, let it be a photon. The presence (or absence) of a photon is one single bit of information.

But now let's do something to the photon, assuming that it really is a tiny string — shake it or hit it with other strings or just put it in a hot frying pan.[5] Just like a small rubber band, it will start to vibrate, rotate, and stretch itself out. If enough energy is added, it will begin to resemble a huge, tangled mess: a ball of yarn that the cat got hold of. This is not the quantum jitters, but the *thermal jitters.*

A tangled ball of yarn soon becomes too complicated to describe in detail, but we still might have some rough information. The total length of yarn might be one hundred yards. The tangle might form a ball approximately six feet in diameter. That sort of description would be useful even if it left out the details. The unspecified details are the hidden information that gives the ball of string its entropy.

Energy and entropy — that sounds like heat. And, indeed, the tangled balls of string that make up very excited elementary particles do have a temperature. This is also something that has been known since the early days of String Theory. In many ways, these tangled, excited strings sound a lot like black holes. By 1993 I was seriously wondering whether black holes might be nothing but huge, randomly tangled balls of string. The idea seemed intriguing, but the details were all wrong.

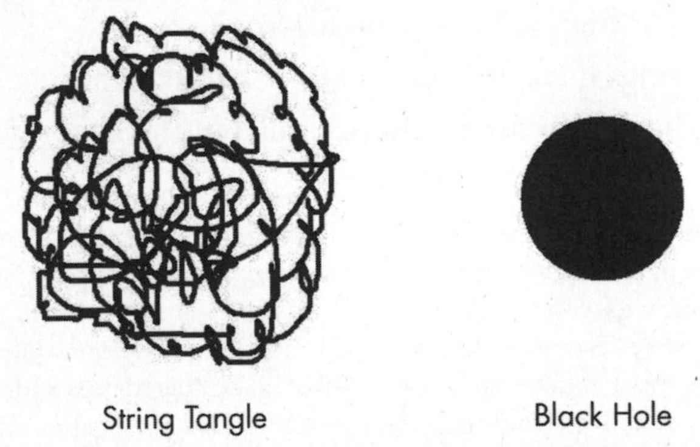

5. And raise the temperature to 10^{33} degrees Kelvin.

For example, the mass (or energy) of a string is proportional to its length. If 1 yard of yarn has a mass of 1 gram, then 100 yards of yarn has a mass of 100 grams, and 1,000 yards of yarn has a mass of 1,000 grams.

But the entropy of a string is also proportional to its length. Imagine moving along the string as it turns and twists. Each turn and twist is a few bits of information. A simplified picture of a string is to pretend that it is a series of rigid links on a lattice. Each link is either horizontal or vertical.

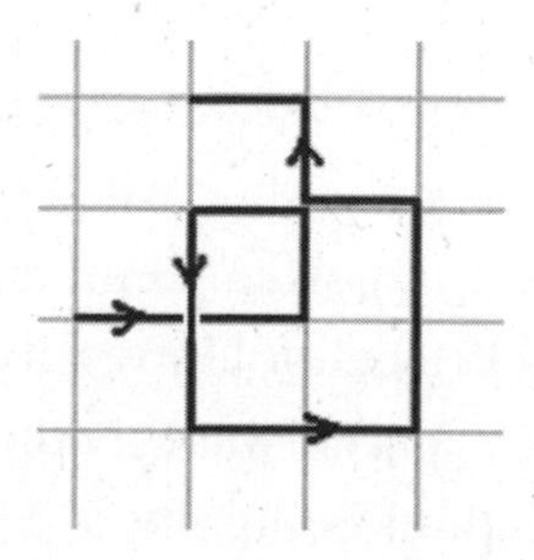

Start out with a single link; it can point up, down, left, or right. There are four possibilities. That is equivalent to two bits of information. Now add a link. It can continue in the same direction, make a right-angle turn (left or right), or make a U-turn. That's two more bits. Every new link adds its two bits. This means that the hidden information is proportional to the total length of the string.

If both the mass and the entropy of a tangled string are proportional to its length, then it doesn't take any sophisticated mathematics to see that entropy must be proportional to the mass:

$$\text{Entropy} \sim \text{Mass}$$

(The mathematical notation for proportionality is ~).

We know that the entropy of an ordinary black hole also grows with its mass. But it turns out that the particular relation Entropy ~ Mass is *not* the right relation for black holes. To see why, just follow the chain of proportionalities: entropy is proportional to area of horizon; area is proportional to square of Schwarzschild radius;

Schwarzschild radius is proportional to mass. Put them together, and you'll see that the entropy is proportional not to the mass, but to the *square of the mass* of a black hole.

$$\text{Entropy} \sim \text{Mass}^2$$

If String Theory is right, everything is made of strings. Everything means *everything,* and that should include black holes. This was a disappointment to me, and a source of frustration, in the summer of 1993.

In fact, I was being stupid. I was missing something obvious, but it didn't occur to me until September, when I visited New Jersey for a month. Two of the most important centers of theoretical physics, Rutgers University and Princeton University, are both in New Jersey, roughly twenty miles apart. I was scheduled to give a lecture at each institution, both lectures being titled "How String Theory Can Explain the Entropy of Black Holes." When I had initially made the arrangements, I had gone out on a limb, hoping that I would figure out what was wrong well before the lectures were to take place.

I don't know whether I am the only physicist who has the same recurrent nightmare. I've had it in various forms since I started out more than forty-five years ago. In the dream, I'm supposed to give an important lecture on some new research, but as the lecture gets closer and closer, I find that I have nothing to say. I have no notes, and sometimes I can't even remember the topic. Pressure and panic set in. Sometimes I even find myself in front of the audience in my underwear, or even worse without my underwear.

This time it was no dream. The first of the two lectures was to take place at Rutgers. As the time approached, I felt more and more pressure to get the story right, but it kept coming out wrong. Then, with about three days to spare, I realized my own stupidity. I had left gravity out of the story.

Gravity acts to pull objects together and concentrate them. Take a huge rock — the Earth, for example. Without gravity, it might merely stick together the way any rock does. But gravity has a powerful effect, pulling the parts of the Earth and squeezing the core, shrinking it to a smaller size. The attractive force of gravity has another effect: it changes the mass of the Earth. The negative potential energy due to gravity subtracts a bit from the Earth's mass. The actual mass is somewhat less than the sum of its parts.

I should stop here and explain this somewhat unintuitive fact. Let's recall for a moment poor Sisyphus, as he endlessly pushed his boulder to the top of the hill, just to watch it roll back down. The Sisyphus cycle of energy conversion is as follows:

chemical → potential → kinetic → thermal

Forget for the moment the chemical energy (the honey that Sisyphus ate) and begin the cycle with the potential energy of the boulder on the hilltop. The water above Niagara Falls also has potential energy, and in both cases, as the mass falls to a lower altitude, the potential energy is diminished. Eventually, it is converted to heat, but imagine that the heat is radiated out into space. The net result is that boulder and water lose potential energy when they lose altitude.

Exactly the same thing happens to the material making up the Earth if it is squeezed (by gravity) closer to the Earth's center: it loses potential energy. The lost potential energy shows up as heat, which eventually gets radiated away into space. The result: the Earth experiences a net loss of energy, and therefore a net loss of mass.

Thus, I began to suspect that the mass of a long, tangled string might also be diminished by gravity and not be proportional to its length, once the effect of gravity was properly included. Here is the thought experiment I imagined. Suppose there was a dial that could be gradually turned to increase and decrease the strength of gravity. Turn the dial one way to diminish gravity, and the Earth would expand a little and become a bit heavier. Turn the dial the other way to increase gravity, and the Earth would shrink and become a bit lighter. Turn it more, and gravity would become even stronger. Eventually, it would become so strong that the Earth would collapse and become a black hole. Most important, the mass of the black hole would be a good deal *less* than the original mass of the Earth.

The giant ball of string that I was imagining would do the same thing. I had forgotten to turn up the gravity dial when I was thinking about the connection between balls of string and black holes. So one evening, with nothing else to do — remember, this was central New Jersey — I imagined turning up the gravity dial. In my imagination, I could see a ball of string pulling itself together into a tight, shrunken sphere. But more important, I realized that the new, smaller ball of string would also have a much smaller mass than it started with.

There was one more point. If the size and mass of the ball of string changed, wouldn't its entropy also change? Luckily, entropy is precisely the thing that *doesn't* change when you slowly turn dials. This is perhaps the most basic fact about entropy: if you change a system slowly, the energy may change (it usually does), but the entropy remains exactly as it was. This foundation of both classical mechanics and Quantum Mechanics is called the *Adiabatic Theorem.*

Let's redo the thought experiment, replacing the Earth with a big tangle of string. Start with the gravity dial set to zero.

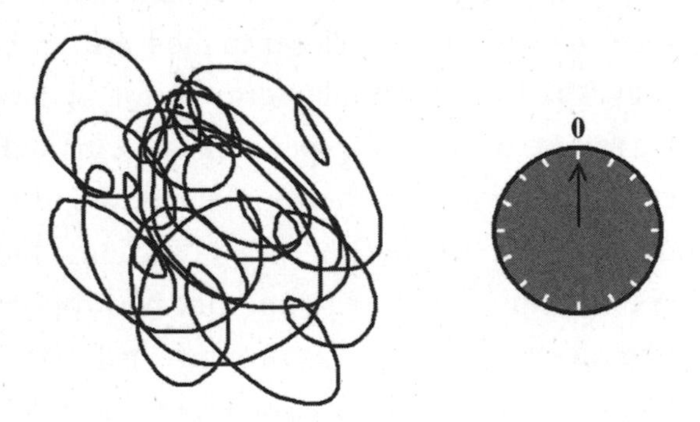

Without gravity, the string doesn't resemble a black hole, but it does have entropy and mass. Next, slowly turn up the gravity dial. The parts of the string begin to pull on one another, and the ball of string becomes compressed.

Continue to turn up the dial until the string becomes so compact that it forms a black hole.

The mass and size have shrunk, but — this is the important point — the entropy remains unchanged. What happens if we turn the dial back toward zero? The black hole begins to puff up and eventually turns back into a big ball of string. If we slowly turn the dial back and forth, the object alternates between a big, loosely tangled ball of string and a tightly compressed black hole. But as long as we move the dial slowly, the entropy remains the same.

In an aha moment, I realized that the problem with the ball-of-string picture of a black hole was not that the entropy came out wrong. It was that the mass needed to be corrected to account for the effects of gravity. When I did the calculation on a single sheet of paper, it all fell into place. As the ball of string shrank and morphed into a black hole, the mass changed in just the right way. In the end, the entropy and mass had the right relation, Entropy ~ Mass^2.

But the calculation was frustratingly incomplete. Remember that the little wiggly ~ sign means "proportional to," not "equal to." Was the entropy exactly equal to the square of the mass? Or was it equal to twice that?

The picture of the black hole horizon that was emerging was a tangle of string flattened out onto the horizon by gravity. But the same quantum fluctuations that Feynman and I had imagined in the West End Café in 1972 would cause some parts of the string to stick out a bit, and these bits would be the mysterious horizon-atoms. Roughly speaking, someone outside the black hole would detect bits of string, each with two ends firmly attached to the horizon. In String Theory language, the horizon-atoms are open strings (strings with ends) attached to a kind of membrane. In fact, these string bits could break loose from the horizon, and that would explain how a black hole radiates and evaporates.

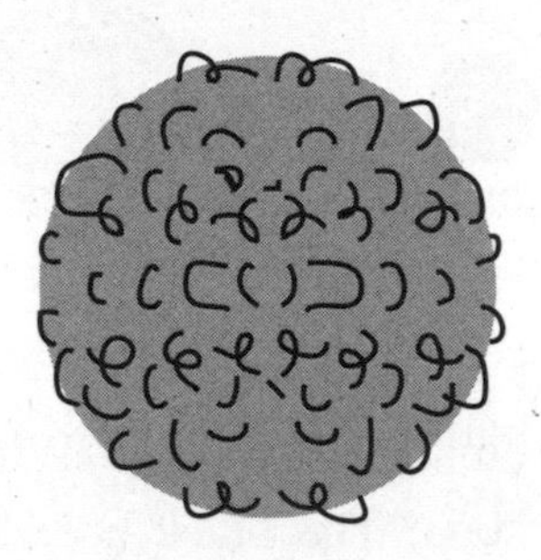

It seems that John Wheeler was wrong: black holes *are* covered with hair. The nightmare was over, and I now had a lecture to give.

When Strings Cross

Fundamental strings can pass right through each other. The following figure shows an example. Think of the closer string moving away from you and the more distant one moving toward you. At some point, they will cross, and if they were ordinary bungee cords, they would get stuck.

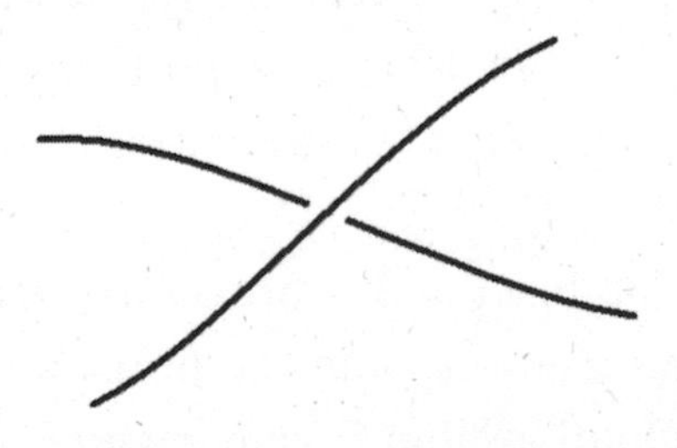

But the mathematical rules of String Theory allow them to pass right through each other and end up as in the next picture.

To do this with real bungee cords, you would have to cut one of them and then reconnect it after they passed.

But something else can happen when the strings touch. Instead of passing through each other, they can rearrange themselves and come out looking like this:

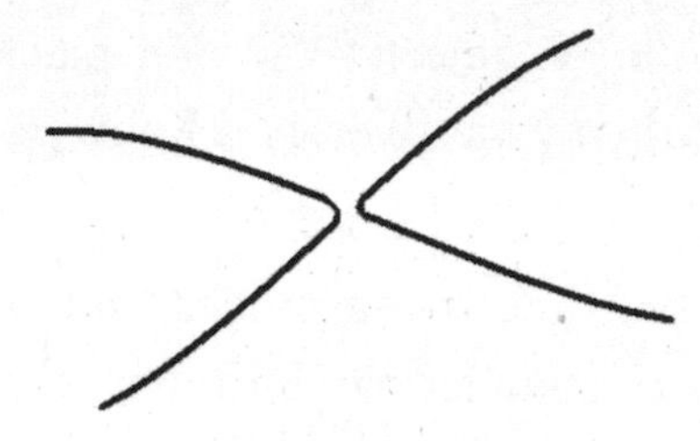

To do that with bungee cords, you would have to cut both of them and then reconnect them in a new way.

Which of the two things happens when strings cross? The answer is sometimes one and sometimes the other. Fundamental strings are quantum objects, and in Quantum Mechanics nothing is certain – all things are possible, but with definite probabilities. For example, strings might pass through each other 90 percent of the time. The remaining 10 percent of the time, they rearrange. The probability for them to rearrange is called the *string coupling constant.*

With this knowledge, let's focus on a short bit of string sticking out from the horizon of a black hole. The short segment of string is twisted and about to cross itself.

Ninety percent of the time, it passes right through itself and nothing much happens.

But 10 percent of the time, it rearranges, and when it does, something new happens. A small loop of string breaks free.

That little bit of closed string is a particle. It could be a photon, a graviton, or any other particle. Since it's on the outside of the black hole, it has a chance to escape, and when it does, the black hole loses a bit of energy. That's how String Theory explains Hawking radiation.

Back to New Jersey

The New Jersey physicists were a very tough-minded group. Edward Witten, the intellectual leader at the Institute for Advanced Study at Princeton, is not only a great physicist but also one of the world's leading mathematicians. Some people would say that small talk and idle speculation are not his greatest strength (although I find his dry wit and wide-ranging curiosity very enjoyable), but everyone would agree that intellectual rigor is. I don't mean needless mathematical rigor, but rather clear, careful, well-thought-out arguments. Talking physics with Witten can sometimes be very trying, but it is always rewarding.

At Rutgers, the intellectual discourse was also of an unusually high quality. There were six very accomplished theoretical physicists at Rutgers, each of whom was widely admired, especially by string theorists, but also in the wider world of physics. All were friends of mine, but three were particularly close. I had known Tom Banks, Steve Shenker, and Nathan "Nati" Seiberg from the time they were very young physicists, and I enjoyed their company tremendously. All six Rutgers physicists were intellectually formidable. Both institutions had reputations as places where you could not get away with half-baked claims.

Now, I knew that my own arguments were far from fully baked. Black Hole Complementarity, Alice's Airplane, and strings morphing back and forth into black holes, together with some rough estimates: my picture seemed to hold together. But the tools to turn these ideas into rigorous mathematics were not available in 1993. Nevertheless, the ideas I was advocating resonated with the tough New Jersey physicists. In particular, Witten's response was to accept, more or less straightforwardly, the proposition that a black hole horizon is composed of bits of string. He even worked out how strings evaporate in a manner similar to black hole evaporation. Shenker, Seiberg, Banks, and their colleague Michael

Douglas all had very useful suggestions about making the ideas more precise.

Also among those in New Jersey was a visiting string theorist whom I didn't know very well. Cumrun Vafa, a young professor at Harvard, had come to the United States from Iran to study physics at Princeton. By 1993 he was recognized as one of the most creative and mathematically astute theoretical physicists in the world. Primarily a string theorist, he also knew a good deal about black holes, and as it happened, he was in the audience at Rutgers when I explained how the entropy of black holes could originate from the stringy nature of the horizon. The conversation that took place between us afterward was fateful.

Extremal Black Holes

At the time of my lectures, it was understood that if an electron was dropped into a black hole, the black hole would become electrically charged. The electric charge, which quickly spread over the horizon, would cause a repulsion that pushed the horizon out a little.

But there was no reason to stop at just one electron. The horizon could be electrically charged as much as you liked. The more you charged it, the more it moved out from the singularity.

Cumrun Vafa pointed out that there is a very special kind of charged black hole that is in perfect balance between gravitational attraction and electrical repulsion. Such a black hole is called *extremal.* According to Vafa, extremal black holes would be the perfect laboratory for testing my ideas. He argued that they might be the key to a more exact calculation in which a firm equal sign (=) would replace the flabby proportional sign (~).

Let's pursue the idea of an electrically charged black hole a little further. Balls of electric charge are usually not stable. Because electrons repel each other (remember the rule: like charges repel; opposite charges attract), if a cloud of charge happens to form, it will usually be instantly torn apart by electrical repulsion. But gravity can compensate for the electrical repulsion if the ball of charge is massive enough. Since all things in the universe gravitationally attract one another, there will be a competition between the gravity and electrical repulsion — gravity pulling the charges together and the electrical force pushing them apart. A charged black hole is a tug-of-war.

If the ball of charge is very massive but has only a small amount of electric charge, gravity will win the tug-of-war, and the ball will contract. If its mass is small but it has a very large electric charge, electrical repulsion will win, and the ball will expand. There is a point of equilibrium when charge and mass are in exactly the right proportions. At this point, the electrical repulsion and gravitational attraction balance each other, and the tug-of-war is a draw. This is exactly what an extremal black hole is.

Now imagine that we have two dials, one for gravity and one for electrical force. Initially, both dials are turned on. When gravity and electrical force are in perfect balance, we have an extremal black hole. If we turn down the gravity without turning down the electrical force, the electrical force will begin to win the tug-of-war. But if we turn down both in just the right way, the balance will be preserved. Each side will get weaker, but neither side will gain an advantage.

Eventually, if we turn the dials all the way to zero, the forces of gravity and electricity disappear. What's left? A string with no forces between its parts. Throughout the entire process, the entropy didn't change. But the punch line is that the mass also didn't change. The canceling electrical and gravitational forces "do no work," which is a technical way of saying the energy remains exactly as it started.

Vafa argued that *if* we knew how to make such an extremal black hole in String Theory, we could study it with great precision as the gravity and electrical force dials were turned on and off. He said that it should then be possible, using String Theory, to compute the precise numerical factor that I had thus far been unable to compute. To mix metaphors, computing the exact numerical factor became the Holy Grail for string theorists and the way to complete the baking of my idea. But nobody knew how to assemble the appropriate kind of charged black hole out of the components that String Theory provides.

String Theory is a little like a very complicated Tinkertoy set, with lots of different parts that can fit together in consistent patterns. Later I'll tell you a little about some of these mathematical "wheels and gears," but in 1993 some important parts that were needed to build an extremal black hole had yet to be discovered.

The Indian physicist Ashoke Sen was the first to try to put together an extremal black hole and test the String Theory of black hole entropy. In 1994 he got very close, but not quite close enough to finish the story. Among theoretical physicists, Sen is held in very high esteem. He has a reputation as both a deep thinker and a technical wizard. Sen's lectures — delivered by the shy, slightly built man in a fairly heavy, lilting Bengali accent — are famous for their clarity. In perfect pedagogical technique, he writes every new concept on the blackboard. Ideas unfold in an inevitable progression that makes everything crystal clear. His scientific papers have the same perfect lucidity.

I had no idea that Sen was working on black holes. But shortly after I returned to the United States from my trip to Cambridge, someone — I think it was Amanda Peet — handed me a paper of his to read. It was long and technical, but in the last few paragraphs, Ashoke used the ideas of String Theory — the ones I had described at Rutgers — to compute the entropy of a new class of extremal black holes.

Sen's black hole was made of the parts that we knew about in 1993 — fundamental strings and the six extra compact dimensions of space. What Sen did next was a simple, but very clever, extension of my own earlier ideas. His basic innovation was to start with a string that was not only very excited but also was wound around a compact direction many times. In the simplified world of a cylinder — the fattened version of Lineland — a wound string looks like a rubber band wrapped around a piece of plastic pipe.

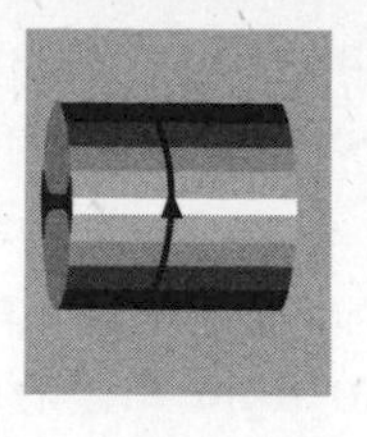
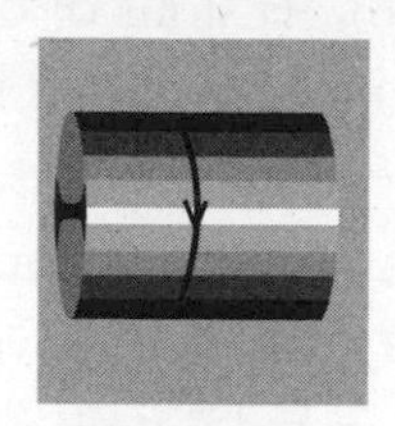

Such a string is heavier than an ordinary particle because it takes energy to stretch it around the cylinder. In typical String Theory, the mass of the wound string might be a few percent of the Planck mass.

Then Sen took a single string and looped it around the cylinder twice.

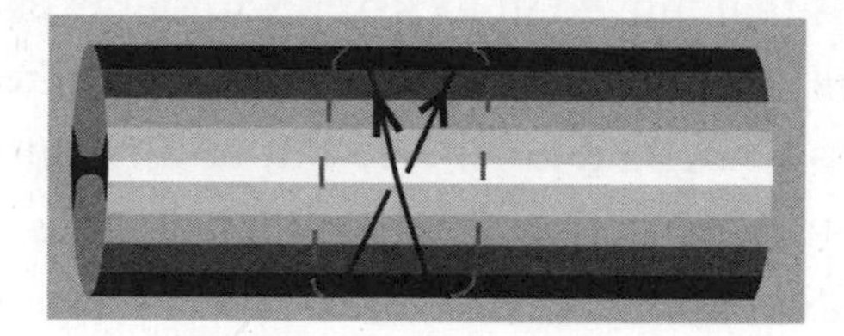

A string theorist would say that this string has *winding number 2* and is even heavier than the once-wound string. But what if the string is wound around the compact direction of space not once or twice, but billions of times?

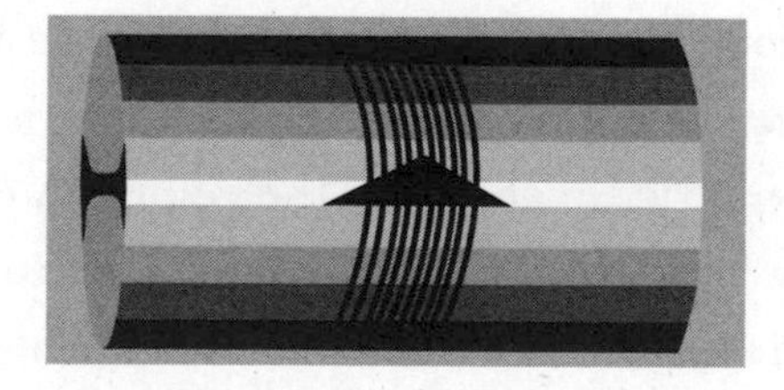

There is no limit to how many times the string can circle the compact direction of space. Eventually, it can become as heavy as a star or even a galaxy. But the room that it takes up in ordinary space —

the noncompact dimensions of ordinary three-dimensional space — is small. All that mass confined in such a small space is guaranteed to be a black hole.

Sen added one more trick, the one remaining ingredient of String Theory circa 1993: wiggles moving around the string. Information would be hidden in the details of the wiggles, just as I had argued a year earlier.

Wiggles on an elastic string don't stand still. They travel along the string like waves, some moving clockwise and others moving counterclockwise. Two wiggles moving in the same direction chase each other around the string without ever colliding. If two waves move in opposite directions, however, they will collide, yielding a complicated mess. So Sen chose to store all hidden information in clockwise waves that moved in lockstep without ever colliding.

When all the ingredients were assembled and the various dials turned up, Sen's string had no choice but to turn into a black hole. But instead of an ordinary black hole, the stretching around the circular compact direction led to a very special extremal black hole.

Extremal black holes are electrically charged. Where is the electric charge? The answer had been known for many years: wrapping a string around a compact direction gives it an electric charge. Each turn of the string gives it a single unit of charge. If the string is wound in one direction, it is positively charged; if it is wound in the other direction, it is negatively charged. Sen's giant, multiply wound string could also be viewed as a ball of electric charge held together by gravity — in other words, a charged black hole.

Area is a geometric concept, and the geometry of space and time are governed by Einstein's General Theory of Relativity. The only way to know the area of the horizon of a black hole is to work it out from Einstein's equations for gravity. Sen, being a master of the equations, had easily (easily for him) solved the equations for the special kind of black hole he had concocted and computed the area of the horizon.

Disaster! When the equations were solved and the area of the horizon was computed, the result was *zero!* In other words, instead of being a nice large shell, the horizon had shrunk to the size of a mere point of space. All the entropy stored in those wiggling, snaky strings was seemingly concentrated in a tiny point of space. This was not only trouble for black holes, but it also directly contradicted the Holographic Principle: the maximum entropy in a region of space is its area in Planck units. Something was wrong.

Sen knew exactly what the problem was. Einstein's equations are *classical,* which means that they ignore the effects of quantum fluctuations. Without quantum fluctuations, the electron in a hydrogen atom would fall into the nucleus, and the entire atom would be no bigger than a proton. But the quantum zero point motion caused by the Uncertainty Principle makes the atom 100,000 times bigger than the nucleus. Sen realized that the same thing would happen to the horizon. Although classical physics predicted that it would shrink to a point, quantum fluctuations would expand it to what I had called a *stretched horizon.*

Sen made the necessary corrections: a quick "back of the envelope" estimate demonstrated that the entropy and the area of the stretched horizon were indeed proportional to each other. This was another triumph for the String Theory of horizon entropy, but as before, the victory was incomplete. Precision was still elusive; there was uncertainty about exactly how much the horizon would be stretched by quantum fluctuations. Brilliant as it was, Sen's work still ended with a flabby ~. The best he could say was that the entropy of a black hole was *proportional* to the area of the horizon. It was close but no cigar. The "nail in the coffin" calculation had yet to be done.

There was no chance that this almost-calculation would convince Stephen Hawking — no more than my arguments had. Nevertheless, the circle was closing. To carry out Vafa's proposal and make an extremal black hole with a large classical horizon, some new Tinkertoy parts would be required. Fortunately, the necessary parts were about to be discovered in Santa Barbara.

Polchinski's D-Branes

D-branes should be called P-branes — P for Polchinski. But by the time Joe discovered his branes, the term *P-brane* was already in use for an unrelated object. So Joe called them *D-branes,* naming them for the nineteenth-century German mathematician Johann Dirichlet. Dirichlet had nothing directly to do with D-branes, but his mathematical studies of waves had some relevance.

The word *brane* is not in the dictionary except in the context of String Theory. It comes from the common term *membrane,* a two-dimensional surface that can bend and stretch. Polchinki's 1995 discovery of the properties of D-branes was one of the most important events in the recent history of physics. It would soon have profound repercussions on everything from black holes to nuclear physics.

The simplest branes are zero-dimensional objects called 0-branes. A particle or a point of space is zero-dimensional — there is nowhere to move on a point, so particles and 0-branes are synonyms. Moving up a notch, we come to 1-branes, which are one-dimensional. A fundamental string is a special case of a 1-brane. Membranes — two-dimensional sheets of matter — are 2-branes. What about 3-branes — are there such things? Think of a solid cube of rubber filling a region of space. You can call it a *space-filling 3-brane.*

It now appears that we have run out of directions. Obviously, there is no way to fit a 4-brane into three-dimensional space. But what if space has compact dimensions — six of them, for example? In that case, one of the directions in a 4-brane can extend into the compact directions. In fact, if there are a total of nine dimensions, space can hold any kind of brane up to and including 9-branes.

A D-brane is not just any kind of brane. It has a very special property — namely, that fundamental strings can end on it. Take the case of a D0-brane. The D means it is a D-brane, and the 0 means it is zero-dimensional. So a D0-brane is a particle that fundamental strings can end on.

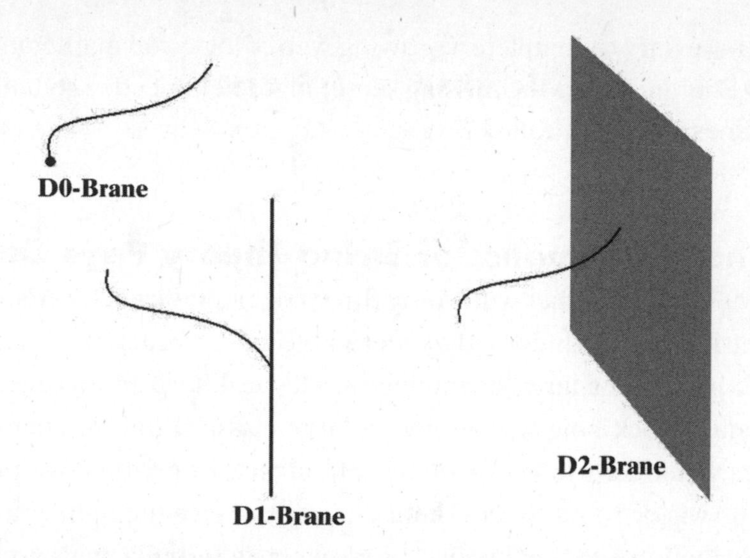

D1-branes are often called D-strings. This is because a D1-brane, being one-dimensional, is itself a kind of string, although it should not be confused with a fundamental string.[6] Typically, D-strings are much heavier than fundamental strings. D2-branes are membranes, similar to rubber sheets but, again, with the property that fundamental strings can end on them.

Were D-branes just a whim, an arbitrary addition that Polchinski added to String Theory because he could? In his first exploratory work, I think that may have been the case. Theoretical physicists often invent new concepts just to play with them and see where they lead. Indeed, back in 1994 when Joe first showed me the idea of D-branes, that was precisely the spirit of the discussion: "Look, we can add some new objects to String Theory. Isn't that fun? Let's explore their properties."

But sometime in 1995, Joe realized that D-branes filled an enormous mathematical hole in String Theory. Their existence was, in

6. It may seem odd and somewhat arbitrary that there are two kinds of string in String Theory. In fact, it is not arbitrary at all. There are powerful mathematical symmetries, known as dualities, relating fundamental strings and D-strings. These dualities are very similar to the duality that relates electric charges and the magnetic monopoles that were first hypothesized by Paul Dirac in 1931. They have had a profound influence on several subjects of pure mathematics.

fact, necessary to complete a growing web of logic and mathematics. And D-branes were the missing secret ingredient needed to build a better extremal black hole.

The Mathematics of String Theory Pays Off

In 1996, Vafa, together with Andy Strominger, pounced. By combining strings and D-branes, they were able to construct an extremal black hole with a large, unambiguous classical horizon. Because an extremal black hole was seen as a large classical object, quantum jitters would have only a negligible effect on the horizon. Now there was no wiggle room. String Theory had better give the right amount of hidden information implied by Hawking's formula, with no ambiguous factors of two or π and no proportional sign.

This was not your basic old-school black hole. The object that Strominger and Vafa built out of strings and D-branes sounded like an engineering nightmare, but it was the simplest construction that had the large classical horizon they were looking for. All the mathematical tricks of String Theory were needed, including the full set of extra dimensions, strings, D-branes, and lots more. First, they stuck in a number of D5-branes that filled up five of the six compact directions of space. On top of that, embedded in the D5-branes, they wrapped a large number of D1-branes around one of the compact directions. And then they added strings attached by both their ends to the D-branes. Once again, open bits of string would be the horizon-atoms that contained the entropy. (If you're a bit lost, don't worry. We are into the zone of things humans are not wired to easily comprehend.)

Strominger and Vafa followed the same steps that had been used earlier. First, they turned the dials to zero so that gravity and other forces would vanish. Without these forces to confuse things, it was possible to calculate exactly how much entropy was stored in the fluctuations of the open strings. The technical calculations were more complicated and subtle than anything that had come before, but in a mathematical tour de force, they succeeded.

The next step was to solve Einstein's field equations for this kind of extremal black hole. This time no uncertain stretching procedure was needed to calculate the area. To their great satisfaction (and to mine), Strominger and Vafa found that the horizon area and the entropy were not just proportional; the hidden information in the stringy wiggles attached to the branes agreed exactly with Hawking's formula. They had nailed it.

As often happens, more than one group of people came upon the new idea almost simultaneously. At the same time that Strominger and Vafa were doing their work, one of the most brilliant of the new generation of physicists was still a student at Princeton. Juan Maldacena's thesis adviser was Curt Callan (C of CGHS). Maldacena and Callan were also putting D5-branes together with D1-branes and open strings. Within a few weeks of Strominger and Vafa, Callan and Maldacena posted their own paper. Their methods were somewhat different, but their conclusions completely confirmed what Strominger and Vafa had claimed.

In fact, Callan and Maldacena were able to go a little beyond previous work and get a handle on slightly non-extremal black holes. An extremal black hole is an oddity in physics. It is an object with entropy, but with no heat or temperature. In most quantum mechanical systems, once all the energy is drained away, everything is rigidly locked in place. For example, if all the heat were removed from an ice cube, the result would be a perfect crystal with absolutely no imperfections. Any rearrangement of the water molecules would take energy and therefore add a bit of heat. Ice with all the heat drained out has no excess energy, no temperature, and no entropy.

But there are exceptions. Certain special systems have many states with exactly the same minimum value of energy. In other words, even after all the energy has been drained away, there are ways of rearranging the system to hide information, and to do so without adding energy. Physicists say that such systems have *degenerate ground states.* Systems with degenerate ground states have entropy — they can hide information — even at absolute zero. Extremal black holes are perfect examples of these unusual systems. Unlike ordinary Schwarzschild black holes, they are at absolute zero, which means they don't evaporate.

Let's go back to Sen's example. In that case, the wiggles on the string all move in the same direction, and therefore they don't bump into one another. But suppose we add some wiggles moving in the opposite direction. As you might expect, they bump into the original wiggles and create a bit of a mess. In fact, they heat up the string and raise its temperature. Unlike ordinary black holes, these almost-extremal black holes don't completely evaporate; they shed their excess energy and return to the extremal state.

Callan and Maldacena were able to use String Theory to compute the rate at which almost-extremal black holes evaporate. The way String Theory explains the evaporation process is fascinating. When two wiggles moving in opposite directions collide,

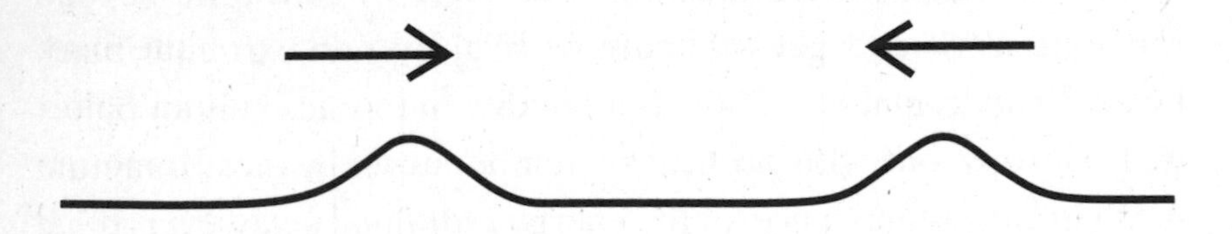

they form a single, bigger wiggle that looks something like this:

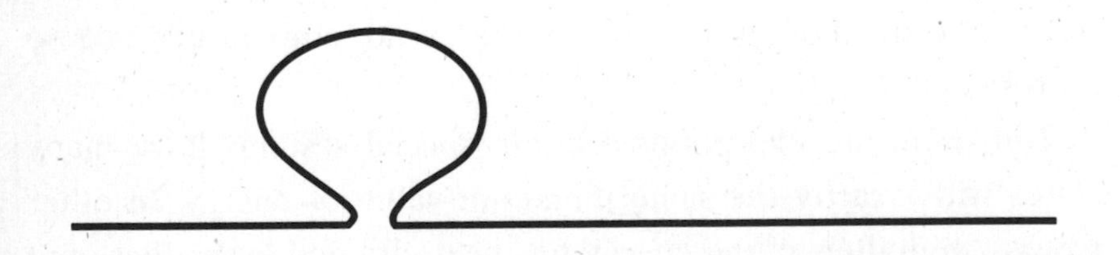

Once that bigger wiggle forms, there is nothing to prevent it from breaking off, in a fashion that is not so different from what Feynman and I had spoken about in 1972.

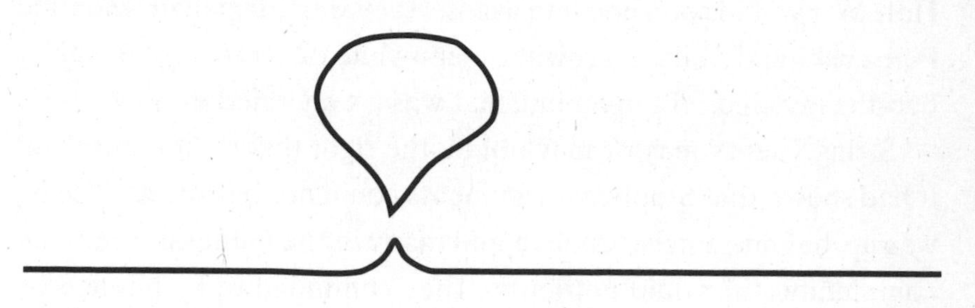

But Callan and Maldacena had done more than talk. They had made very detailed calculations of the evaporation rate. The remarkable fact was that their results exactly agreed with Hawking's twenty-year-old method, with one important difference: Maldacena and Callan had used only the conventional methods of Quantum Mechanics. As we have discussed in previous chapters, although Quantum Mechanics has a statistical element, it forbids information loss. Hence, there was no possibility that information could be lost during the evaporation process.

Again, similar ideas were being worked on by others. Quite independently, two pairs of Indian physicists — Sumit Das and Samir Mathur, and Gautam Mandal and Spenta Wadia — from Bombay's Tata Institute (also the home of Ashoke Sen) did computations with similar results.

Taken together, these works were prodigious accomplishments, and they are justly famous. That fact that black hole entropy can be accounted for by the information stored in string wiggles went strongly against the views of many relativists, including Hawking. Stephen saw black holes as *eaters of information,* not storage containers for retrievable information. The success of the Strominger-Vafa calculation showed how a single mathematical result can tip the scales. It was the beginning of the end for information loss.

The drama of the moment did not go unnoticed. Many people, including my Santa Barbara friends, abruptly jumped ship and defected to the other side. If I had any lingering doubts that the Black Hole War would soon come to a close, they were dispelled when Joe Polchinski and Gary Horowitz — erstwhile neutrals in the war — became my allies.[7] In my mind, that was a watershed event.

String Theory may or may not be the right theory of nature, but it had shown that Stephen's arguments could not be correct. The jig was up, but amazingly, Stephen and many in the General Relativity community still would not let go. They continued to be blinded by Hawking's early arguments.

7. Polchinski and Horowitz wrote a paper using the same method that I had used in 1993 to compute the entropy of the many kinds of black holes that occur in String Theory — both extremal and otherwise — and in every case, the answers agreed with the Bekenstein-Hawking area formula.

22

SOUTH AMERICA WINS THE WAR

Most people don't think of South America when they think of brilliant physicists. Even South Americans are surprised to know how many very distinguished theoretical physicists hail from Argentina, Brazil, and Chile. Daniele Amati, Alberto Sirlin, Miguel Virasoro, Hector Rubinstein, Eduardo Fradkin, and Claudio Teitelboim are just a few who have had a major impact on the subject.

Teitelboim, who has recently changed his name to Claudio Bunster (see footnote, page 145), is a remarkable character, unlike any other physicist I have ever known. His family had very close ties to the Chilean socialist president Salvador Allende and to the poet-activist and Nobel Prize winner Pablo Neruda. Claudio's brother, César Bunster, was a leading figure in the September 7, 1986, assassination attempt against former fascist dictator General Augusto Pinochet.

Claudio is a tall, dark man with a powerful, athletic body and fierce, penetrating eyes. Despite a mild stutter, he has the kind of charm and charisma that would make him a great political leader. In fact, he was the antifascist leader of a small band of scientists who helped keep science alive in Chile during the dark years. I have no doubt that his life was at risk at that time.

Claudio is a man of tremendous ability and a touch of real craziness. Though an enemy of the military regime in Chile, he loves all the trappings of military life. When he lived in Texas before returning to Chile, he frequented knife and gun shows, and even today he often wears military fatigues. The first time I went to visit him in Chile, he scared the hell out of me while playing soldier.

It was 1989, and the Pinochet dictatorship was still in full power. When my wife and I, along with our friend Willy Fischler, got off the plane in Santiago, we were brusquely herded into a long passport inspection line by heavily armed men in uniform. The clerks at passport control were military, all armed, some with large automatic weapons. Clearing passport control was not easy: the long line hardly moved, and we were exhausted.

All of a sudden, I saw a tall figure wearing sunglasses and a military uniform (or what passed for a military uniform) coming through the blockade and heading straight for us. It was Claudio, and he was giving orders to the soldiers as if he were a general.

When he got to us, he grabbed me by the arm and haughtily escorted us past the guards, waving them aside with an extraordinary air of authority. He grabbed our luggage and quickly led us out of the airport to his illegally parked, khaki-colored jeep. Then we sped out of the airport, sometimes on two wheels, and into Santiago. Every time we passed a group of soldiers, Claudio would salute. "Claudio," I whispered, "what the *bleep* are you doing? You're going to get us killed." But no one stopped us.

The last time I was in Chile,[1] well after the Pinochet regime had been replaced by a democratic government, Claudio had real contacts in the military, especially the air force. The occasion was a conference on black holes that Claudio had organized at his small institute. He had used his influence with the air force to fly a few of us, including Hawking and myself, to the Chilean Antarctic base.

1. Just as this book was in its final stages of editing, I again visited Chile, this time to help celebrate Claudio Bunster's sixtieth birthday. The photograph of Stephen and myself at the back of the book was taken at that party.

We had a lot of fun, but the most remarkable thing was the way the air force generals, including the chief of staff, entertained us. One general poured tea, another served hors d'oeuvres. Claudio was evidently a man of considerable influence in Chile.

But it was in 1989, on a tour bus during a trip into the Chilean Andes, that Claudio first told me about certain *anti de Sitter black holes.* Today they are called *BTZ black holes,* for Bañados, Teitelboim, and Zanelli. Max Bañados and Jorge Zanelli were members of Claudio's inner circle when the three of them made a discovery that would have a lasting impact on the Black Hole War.

Angels and Devils

Black hole physicists are forever fantasizing about sealing a black hole in a box, keeping it safe like a precious jewel. Safe from what? From evaporation. Sealing it in a box is just like putting a lid on a pot of water. Instead of evaporating into space, the particles would bounce off the walls of the box (or the lid of the pot) and fall right back into the black hole (or pot).

Of course, no one will ever really put a black hole in a box, but the thought experiment is interesting. A stable, unchanging black hole would be much simpler than one that evaporates. But there is a problem: no real box can surround a black hole forever. Like anything else, real boxes randomly jitter, and sooner or later an accident will occur. The box will come in contact with the black hole, and oops, it will get sucked right in.

That's where anti de Sitter Space (ADS) comes in. First of all, despite its name, anti de Sitter Space is really a space-time continuum that includes time among its dimensions. Willem de Sitter was a Dutch physicist, mathematician, and astronomer who discovered the four-dimensional solution of Einstein's equations that bears his name. Mathematically, de Sitter Space is an exponentially expanding universe that grows in much the same way our universe does.[2] De Sitter Space was long thought to be no more than a curiosity, but in recent years, it has become tremendously important to cosmologists. It is a curved space-time continuum with positive curvature, meaning that the angles of a triangle add up to something greater than 180 degrees. But all of this is beside the point. In this discussion, it is *anti de Sitter Space,* not de Sitter Space, that interests us.

Anti de Sitter Space was not discovered by de Sitter's antimatter twin. The "anti" indicates that the curvature of the space is negative, meaning that the sum of the angles of a triangle is less than 180 degrees. The most interesting thing about ADS is that it has many of the properties of the interior of a spherical box, but a box that cannot be swallowed by a black hole. This is because the spherical wall of ADS exerts a powerful force — an irresistible repulsion — on anything that approaches it, and that includes the horizon of a black hole. The repulsion is so strong that there is no possibility of contact between the wall and the black hole.

Altogether, ordinary space-time has four dimensions: three dimensions of space and one of time. Physicists sometimes call it four-dimensional, but that obscures the obvious difference between space and time. A more accurate description is to refer to space-time as (3 + 1)-dimensional.

Flatland and Lineland are also space-time continua. Flatland is a world with only two dimensions of space, but the inhabitants also experience a sense of time. They would properly call their world (2 + 1)-dimensional. Linelanders, who can move only along a single axis, but who can also keep track of time, live in (1+1)-dimensional space-time. The wonderful thing about (2 + 1) and (1 + 1) dimensions is that we can easily draw pictures of them to help our intuitions.

2. In recent years, astronomers and cosmologists have found that our universe is expanding at an accelerated rate, doubling in size about every ten billion years. This exponential expansion is believed to be due to a cosmological constant, or what the popular press calls "dark energy."

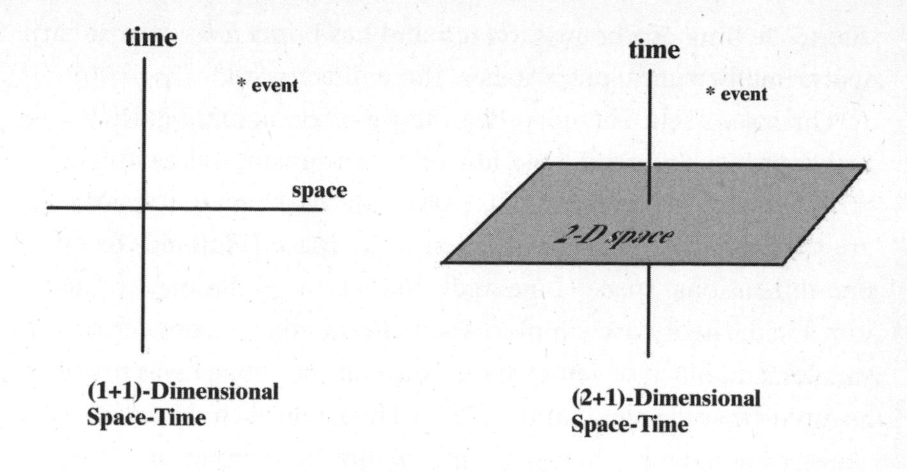

Of course, there is nothing to prevent a mathematical physicist from inventing worlds with any number of space dimensions, despite the brain's inability to visualize them. One might wonder whether it is also possible to change the number of time dimensions. In a completely abstract mathematical sense, the answer is yes, but there does not seem to be much sense of doing so from a physicist's point of view. A single time dimension seems just about the right number.

Anti de Sitter Space also comes in a variety of dimensions. It can have any number of spatial directions but only one time direction. The ADS space that Bañados, Teitelboim, and Zanelli worked with was (2 + 1)-dimensional, making it easy to explain with pictures.

Physics in Various Dimensions

Three-dimensional space (*not* space-time) is one of those things that seem to be hardwired into our cognitive systems. No one can visualize four-dimensional space without the crutch of abstract mathematics. You might think that one- or two-dimensional space is easier to picture, and in a sense it is. But if you think about it for a moment, you realize that when you visualize lines and planes, you always picture them embedded in three-dimensional space. That is almost certainly due to the way our brains evolved and has nothing to do with any special mathematical properties of three dimensions.[3]

Quantum Field Theory — the theory of elementary particles — makes as much sense in a world with fewer dimensions as it does in three-dimensional space. As far as we can tell, elementary particles are perfectly possible in two-dimensional space (Flatland) or even one-dimensional space (Lineland). In fact, the equations of Quantum Field Theory are simpler when the number of dimensions is smaller, and much of what we know about the subject was first discovered by studying Quantum Field Theory in such model worlds. Thus, it was in no way unusual for Bañados, Teitelboim, and Zanelli to be studying a universe in which the number of space dimensions was only two.

Anti de Sitter Space

The best way to explain ADS is the way that Claudio explained it in the Chilean tour bus: with pictures. Let's ignore time and begin with ordinary space inside a hollow round box. In three dimensions, a round box means the interior of a sphere; in two dimensions, it is even simpler, the interior of a circle.

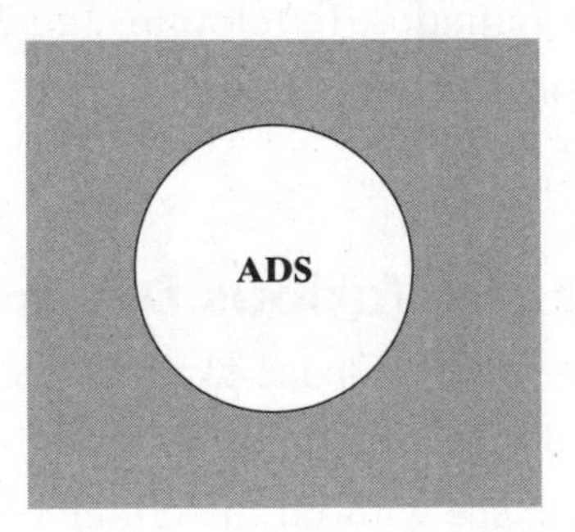

3. Could the physical world have been one- or two-dimensional (I am speaking about space, not space-time)? I don't know for sure — we don't know all the principles that might determine such issues — but from a mathematical point of view, Quantum Mechanics and Special Relativity are just as consistent in one or two dimensions as they are in three. I don't mean to say that intelligent life could exist in these alternative worlds, but only that physics of some kind seems possible.

Now let's add time. Plotting time along the vertical axis, the space-time continuum inside the box resembles the interior of a cylinder. In the following figure, the ADS is the unshaded interior of the cylinder.

Imagine slicing ADS (remember, it has a time dimension) in the same way that we sliced a black hole to make an embedding diagram. Slicing it exposes a spatial cross section that can truly be said to be space.

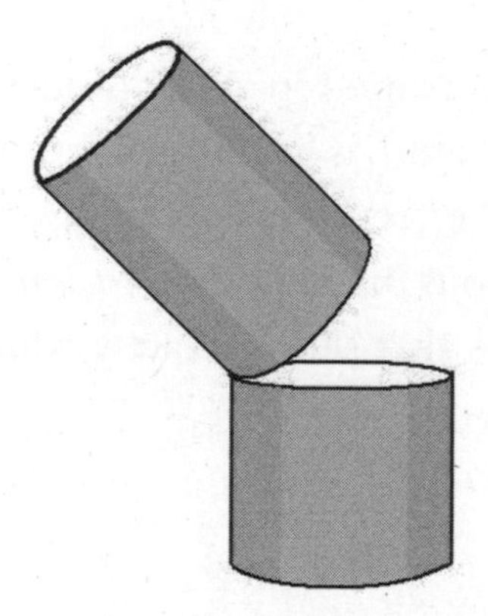

Let's examine the two-dimensional slice a little more closely. As you might expect, it is also curved, in some ways like the Earth's surface. This means that to draw it on a flat plane (a sheet of paper), you have to stretch and distort the surface. It is impossible to draw a map of the Earth on a flat sheet of paper without major distortions. Regions near the northern and southern edges of a Mercator map look much too big when compared with regions near the equator. Greenland looks as big as Africa, although the area of Africa is really about fifteen times larger.

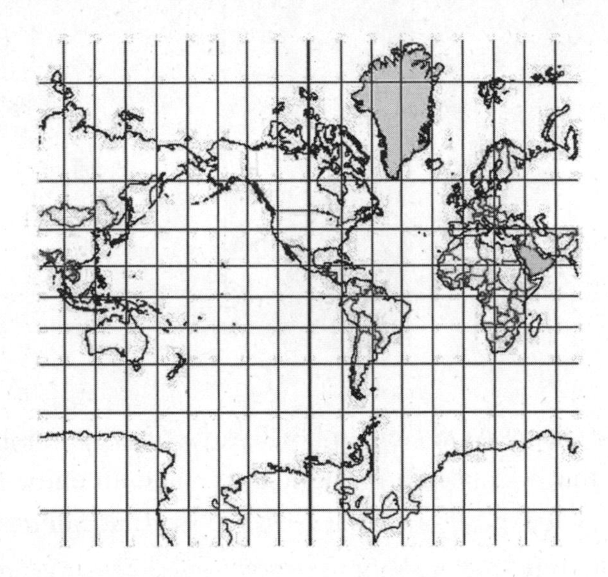

Space (and also space-time) in ADS is curved, but unlike the Earth's surface, the curvature is negative. Distorting it onto a plane has an "anti-Mercator" effect: it makes things near the edge look too small. Escher's famous drawing *Circle Limit IV* is a "map" of a negatively curved space that shows exactly what two-dimensional slices of ADS look like.

I find *Circle Limit IV* hypnotic, to say the least. (It reminds me of *The Mouse and His Child* and the characters' endless quest to see The Last Visible Dog. See chapter 20.) The angels and devils go on endlessly, fading into an infinite fractal edge. Did Escher make a bargain with the devil enabling him to draw an infinite number of angels? Or if I look hard enough, will I see the last visible angel?

Pause for a moment to rewire yourself so that you see the angels and devils as being all the same size. It's not an easy bit of mental gymnastics, but it helps to remember that Greenland is almost exactly the same size as the Arabian Peninsula, despite looking about eight times bigger on a Mercator map. Apparently, Escher was exceptionally well wired for this kind of mental exercise, but with practice you can get the hang of it, too.

Now let's add time and put it all together in a picture of anti de Sitter Space. As usual, we plot time on the vertical axis. Each horizontal slice represents ordinary space at a particular instant. Think of ADS as an infinite number of thin slices of space — a thinly sliced infinite salami — that, when stacked up, form a space-time continuum.

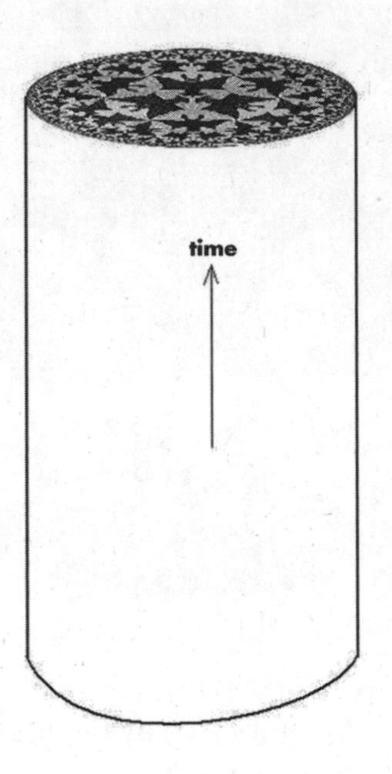

Space is weirdly warped in ADS, but no more so than time. Recall from chapter 3 that in the General Theory of Relativity, clocks located at different positions often run at different rates. For example, the slowdown of clocks near a black hole horizon allows the black hole to be used as a time machine. Clocks in ADS also behave oddly. Imagine that each Escher devil carries its own wristwatch. If the devils nearest the center looked around at their slightly more distant neighbors, they would notice something peculiar: the more distant timepieces would run about twice as fast as their own watches. Assuming that devils have a metabolism, the outer neighbors' metabolic functions also would run faster. In fact, every measure of time would seem to be speeded up, and as they looked farther away, the clocks would appear to run even faster. Each successive layer would run faster than the previous one until, out near the boundary, the clocks would run so fast that the devils near the center would see a whirling blur.

The space-time curvature in ADS creates a gravitational field that pulls objects to the center, *even if there is nothing there.* One manifestation of this ghostly gravitational field is that if a mass were displaced toward the boundary, it would be pulled back, almost as though on a spring. Left to itself, the mass would bob endlessly back and forth. A second effect is really just the opposite side of the coin. A pull toward the center is no different than a push away from the boundary. That push is the irresistible repulsion that keeps everything, including black holes, from making contact with the boundary.

Boxes are made to put things in, so let's put a few particles in the box. Wherever we place them, they will be pulled to the center. A single particle will eternally oscillate around the center, but if there are two or more, they may collide. Gravity — not the ghost gravity of ADS, but the ordinary gravitational attraction between the particles — may cause them to coalesce into a blob. Adding more particles will increase the pressure and the temperature at the center, and the blob may ignite to form a star. The addition of even more mass will eventually lead to a cataclysmic collapse: a black hole will form — a black hole trapped in a box.

Bañados, Teitelboim, and Zanelli were not the first to study black holes in ADS; that honor goes to Don Page and Stephen Hawking. But BTZ did discover the simplest example, easy to visualize because space has only two dimensions. Here is an imaginary snapshot of a BTZ black hole. The edge of the black region is the horizon.

With one exception, anti de Sitter black holes have all the features of ordinary black holes. As always, a very disagreeable singularity hides behind the horizon. Adding mass will increase the size of the black hole, pushing the horizon out closer to the boundary.

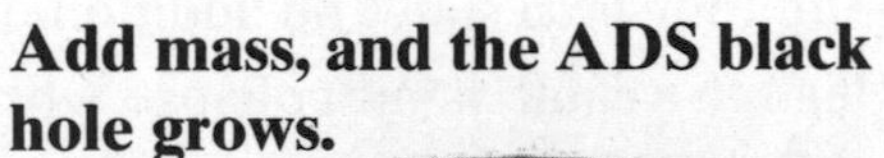

Add mass, and the ADS black hole grows.

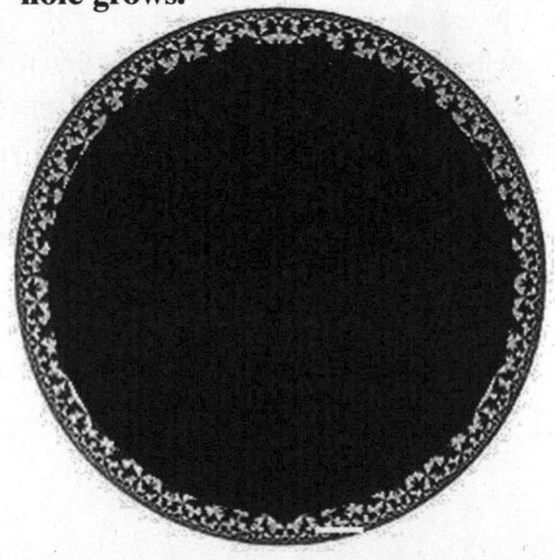

But unlike ordinary black holes, the ADS variety doesn't evaporate. The horizon is an infinitely hot surface, which continually emits photons. But the photons have nowhere to go. Instead of evaporating into empty space, they fall back into the black hole.

A Little More About ADS

Imagine zooming in on a boundary point of *Circle Limit IV* and then blowing it up so that the edge looks extremely straight.

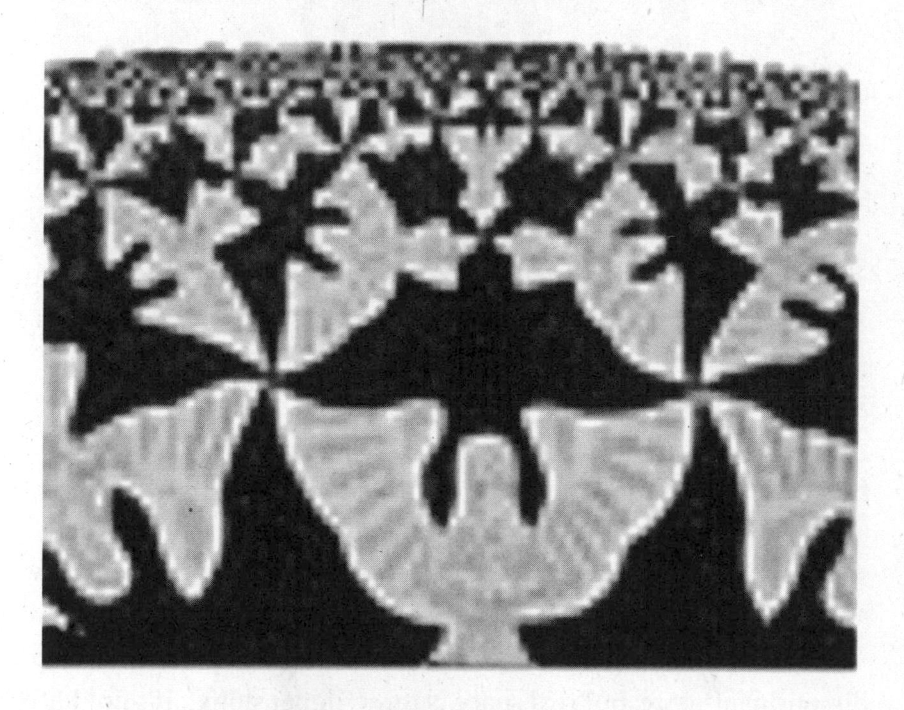

We can do this over and over, without ever running out of angels and devils, until in the limit the edge looks perfectly straight and infinite. I am no Escher, and I won't try to reproduce his elegant creatures, but if I simplify to the point where the devils are replaced by squares, the picture becomes a kind of lattice of smaller and smaller squares as we approach the boundary. Think of ADS as an infinite brick wall. As you proceed down the wall, the bricks double in size with every new layer.

Of course, there would be no real lines in anti de Sitter Space, any more than there are lines of longitude and latitude on the Earth's surface. They're just here to guide your eye and to indicate how sizes are distorted due to the curvature of space.

Escher's drawing and my crude version of it represent two-dimensional space, but real space is three-dimensional. It's not hard to imagine what space would look like if one more dimension (not time) were added. All we have to do is replace the squares with solid three-dimensional cubes. In the following picture, I show a finite piece of the 3-D "brick wall," but keep in mind that it goes on forever in the horizontal directions as well as the vertical direction.

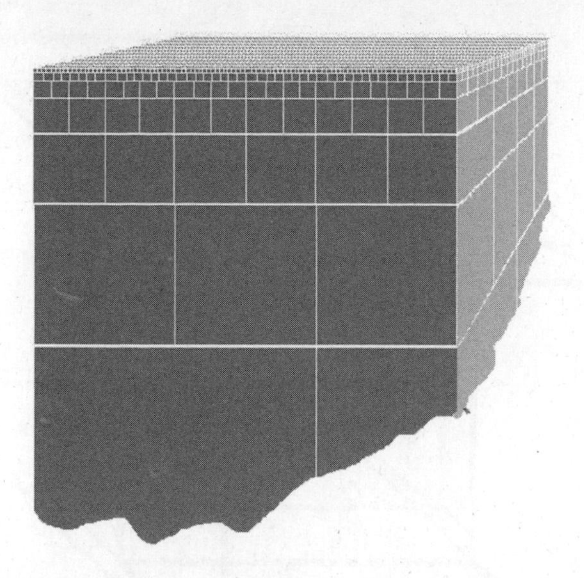

Adding time to the picture is the same as before: each square or cube is equipped with its own clock. The rate at which the clocks run depends on which layer they are in. Each time we go one layer closer to the boundary, the clocks speed up by a factor of two. Conversely, as we go down the wall, the clocks slow down.

From a mathematical perspective, there is no reason to stop at three dimensions of space. By stacking four-dimensional cubes of varying size, it is possible to construct (4 + 1)-dimensional anti de Sitter Space, or any other number of dimensions. But drawing even a single 4-D cube is complicated. Here is an attempt.

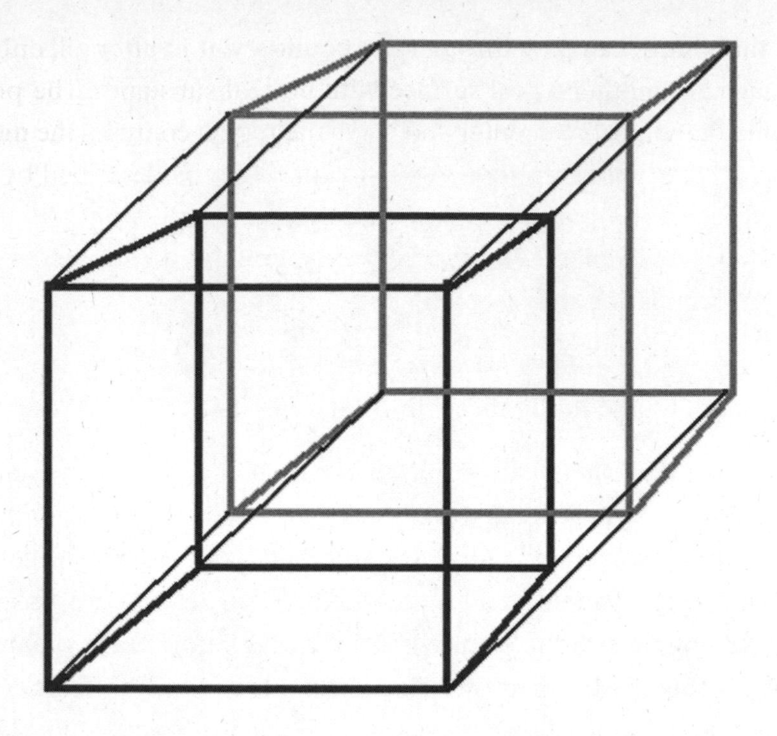

Trying to stack them together to draw a 4-D version of ADS would result in a baffling mess.

The World in a Box

Keeping black holes from evaporating is a good reason for studying physics in a box, but the idea of a world in a box is far more interesting than that. The real goal is to understand the Holographic Principle and to make it mathematically precise. Here is how I explained the Holographic Principle in chapter 18: "The three-dimensional world of ordinary experience — the universe filled with galaxies, stars, planets, houses, boulders, and people — is a hologram, an image of reality coded on a distant two-dimensional surface. This new law of physics, known as the Holographic Principle, asserts that everything inside a region of space can be described by bits of information restricted to the boundary."

Part of the imprecision in formulating the Holographic Principle is that things can pass through the boundary; it is, after all, only an imaginary mathematical surface with no real substance. The possibility that objects can enter and leave the region confuses the meaning of "everything inside a region of space can be described by bits of information restricted to the boundary." But a world in a box with perfectly impenetrable walls would not have this problem. The new formulation would be:

> *Everything inside a box with impenetrable walls can be described by bits of information stored in pixels on the walls.*

In the Chilean tour bus in 1989, I didn't understand why Claudio Teitelboim was so excited about anti de Sitter Space. Black holes in a box — so what? It took another eight years for me to get the point — eight years and another South American physicist, this time an Argentinean.

Maldacena's Amazing Discovery

Juan Maldacena is different in every way from Claudio Teitelboim. He is not as tall and is far more sober. It is impossible for me to imagine him speeding through dangerous Santiago wearing a fake military uniform. But he does not lack courage as a physicist. In 1997 he stuck his neck way out and made an extraordinarily bold claim, a claim that seems almost as crazy as my wild ride with Claudio. In effect, Maldacena argued that two mathematical worlds that seem totally dissimilar are in fact exactly the same. One world has four dimensions of space and one of time (4 + 1), while the other is (3 + 1)-dimensional, more like the usual world we experience. I am going to take a bit of license to simplify the story, and make it easier to visualize, by decreasing the number of dimensions in each case by one. Following this, I would say that a certain fictitious version of Flatland — a (2 + 1)-dimensional world — is somehow equivalent to an anti de Sitter world of (3 + 1) dimensions.

How could such a thing be possible? The most obvious thing

about space is the number of dimensions. An inability to recognize the dimensionality of space would constitute an extremely dangerous perceptual disorder. Surely it is not possible to mistake two dimensions for three, at least while sane and sober. Or so you would think.

The route to Maldacena's discovery was a convoluted, meandering path that wandered through extremal black holes, D-branes, and something called Matrix Theory,[4] and finally ended with an extraordinary confirmation of the Holographic Principle.

The starting point was Polchinski's D-branes. Recall that a D-brane is a material object that, depending on its dimension, can be a point, a line, a sheet, or a solid that fills space. The main property that distinguishes a D-brane from anything else is that fundamental strings can terminate on it. For definiteness, let's concentrate on D2-branes.[5] Think of a flat two-dimensional surface floating in three-dimensional space like a magic carpet. Open strings can attach themselves to the D-brane at both their ends. They are able to slide along the D-brane, but they can't jump free into the third dimension. The bits of string skate on a frictionless sheet of metaphorical ice, unable to lift their feet. From a distance, each piece of string looks like a particle moving in a two-dimensional world. If there is more than one string, they can collide, scatter, and even coalesce into more complicated objects.

4. Matrix Theory in this context had nothing to do with the S-matrix. It was a predecessor and close relative of Maldacena's discovery that also involved a mysterious growth of dimensions. It was the first example of a mathematical correspondence confirming the Holographic Principle. Matrix Theory was discovered by Tom Banks, Willy Fischler, Steve Shenker, and me in 1996.

5. In Maldacena's original work, he concentrated on an example involving four-dimensional space. It would be called (4 + 1)-dimensional ADS. The reason for dealing with four-dimensional space instead of the usual three dimensions is technical and not important for the rest of this chapter, but it is relevant to part of the epilogue.

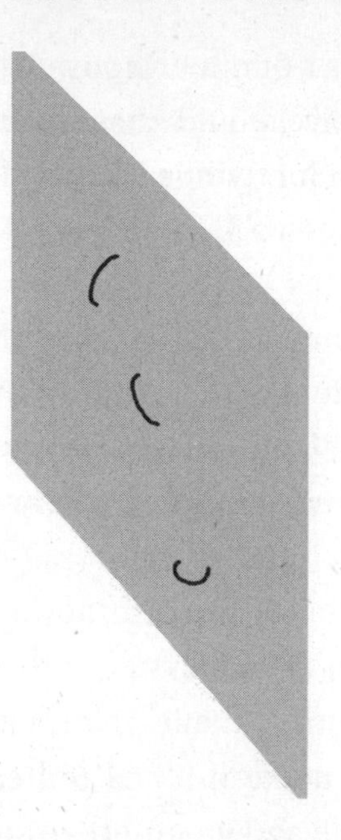

D-branes can exist individually, but they are sticky. If gently brought together, they will bond and form a composite brane of several layers, as in the following illustration.

I've shown the D-branes as slightly separated from one another, but when they bind together, the gaps disappear. A group of D-branes bound together is called a *D-brane stack.*

Open strings moving on a D-brane stack have richer properties and more variety than the strings that move on a single D-brane. The two ends of a string can attach to different members of the stack, as if one skate moved on a slightly different plane than the other. To keep track of the different branes, we can give them names. For example, in the stack shown above, we could call them red, green, and blue.

The strings that skate on the D-brane stack must always have their ends attached to a D-brane, but now there are several possibilities. For example, a string could have both ends attached to the red brane. That would make it a red-red string. In a similar way, there would be blue-blue and green-green strings. But it is also possible for the two ends of a string to be attached to different branes. Thus, there could be red-green strings, red-blue strings, and so on. In fact, there are nine distinct possibilities for the strings that move on this D-brane stack.

Interesting things happen if several strings are attached to the branes.

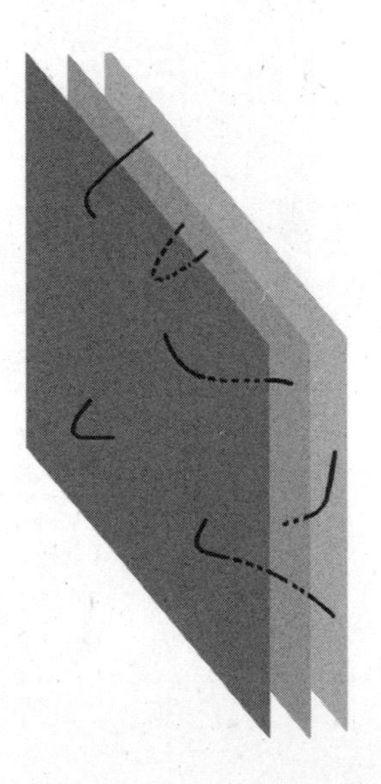

Strings on a D2-brane stack look a lot like ordinary particles, albeit in a world with only two dimensions of space. They interact with one another, scatter when they collide, and exert forces on nearby strings. One string can also break up into two strings. Here is a sequence depicting a string on a single brane splitting to become two strings. The time sequence proceeds from top to bottom.

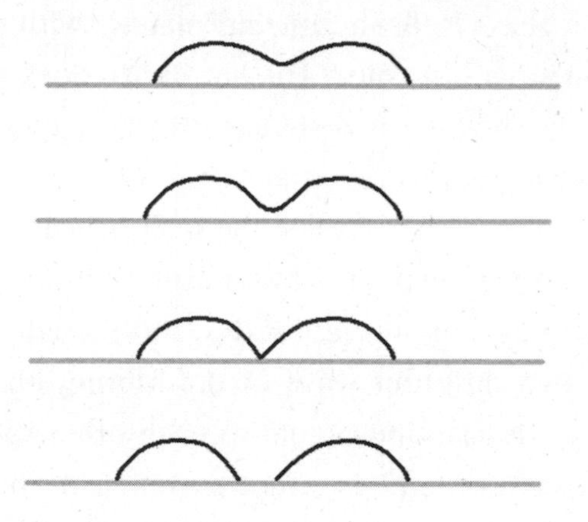

A point on the initial string comes in contact with the brane, allowing the string to split in two, but always in such a way that all ends are attached to the brane. This figure can also be read from bottom to top, so that a pair of strings coalesce to form a single string.

Here is a sequence involving strings on a stack of three D-branes. The sequence depicts a red-green string colliding with a green-blue string. The two strings coalesce to form a single red-blue string.

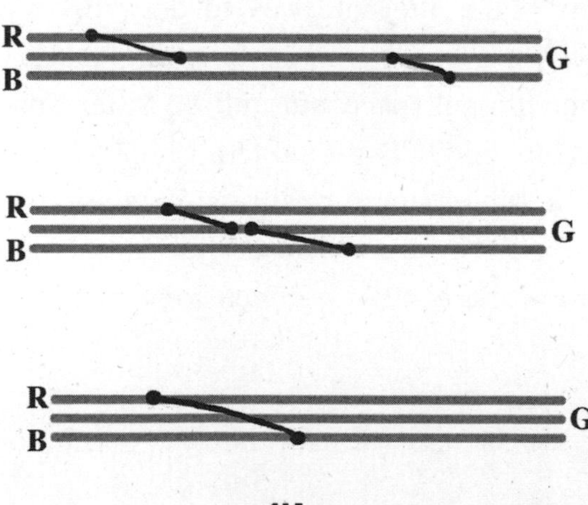

A red-red string could not coalesce with a green-green string because their ends would never touch.

Do you have the feeling that you've seen all this before? You have, assuming you read chapter 19. The rules governing strings attached to a D-brane stack are exactly the same as the rules governing gluons in Quantum Chromodynamics (QCD). In chapter 19, I explained that a gluon is like a little bar magnet with two ends, each end being labeled with a color. The similarity does not end there. The figure above, showing two strings combining to form a single string, is just like a gluon vertex diagram of QCD.

This parallel between "physics on a D-brane" and the usual world of elementary particles is a fascinating fact that, as we will see in the next chapter, has proved to be enormously useful. When physicists find two different ways of describing the same system, they call the two descriptions "dual to each other." An example is the dual description of light as either waves or particles. Physics is full of *dualities,* and there was nothing especially surprising or new in the fact that Maldacena had discovered two dual descriptions of strings on a D-brane. What was new, almost unheard-of,[6] was that the two descriptions described worlds with *different numbers of spatial dimensions.*

I've already hinted at one description: the (2 + 1)-dimensional Flatland version of QCD. It describes flat protons, mesons, and glueballs, but like real QCD, it contains no hint of gravity. The other half of the duality – the alternate way to describe exactly the same thing – describes a world of *three-dimensional* space, and not just any three-dimensional space, but anti de Sitter Space. Maldacena argued that Flatland QCD is dual to a (3 + 1)-dimensional anti de Sitter universe. Moreover, in this three-dimensional world, matter and energy exert gravitational forces just as in the real world. In other words, a world of (2 + 1) dimensions that includes QCD but

6. Almost unheard-of, but not quite. Matrix Theory was an earlier example.

not gravity is equivalent to a universe of (3 + 1) dimensions *with gravity*.

How did that happen? Why would a world with only two dimensions be exactly the same as one with three dimensions? Where did the extra dimension of space come from? The key is the distortion of anti de Sitter Space, that makes objects near the boundary look small by comparison with the same objects deep in the interior of the space. The distortion affects the imaginary devils, but also real objects as they move through the space. For example, if one projected a one-meter letter *A* onto the boundary by creating a shadow, the image would shrink and grow as the object approached and receded from the boundary.

From the point of view of the three-dimensional interior, this is an illusion with no more reality than the large size of Greenland on a Mercator map. But in the dual description – the Flatland theory – there is no notion of distance in the perpendicular third dimension. Instead, it is replaced by a notion of size. This is a very surprising mathematical connection: growing and shrinking in the Flatland half of the duality is exactly the same as moving back and forth along the third direction in the other half of the duality.

Again, this should have a familiar ring, this time from chapter 18, where we discovered that the world is a hologram of sorts. Maldacena's two dual descriptions were the Holographic Principle in action. Everything that takes place in the interior of anti de Sitter Space "is a hologram, an image of reality coded on a distant two-dimensional surface." A three-dimensional world with gravity is equivalent to a two-dimensional quantum hologram on the boundary of space.

I don't know whether Maldacena made the connection between his discovery and the Holographic Principle, but Ed Witten soon did. Just two months after Maldacena's paper, Witten placed his own paper on the Internet, giving it the title "Anti De Sitter Space and Holography."

Of all the things in Witten's paper, the one that especially caught my eye was a section on black holes. Anti de Sitter Space — the original version, not the flattened brick wall version — is like a can of soup. Horizontal slices through the can represent space; the vertical axis of the can is time. The label on the outside of the can is the boundary, and the interior is the space-time continuum itself.

Pure ADS is like an empty can, but it can be made more interesting by filling it with "soup" — that is, matter and energy. Witten explained that by injecting enough mass and energy into the can, a black hole could be created. That raised a question. According to Maldacena, there must be a second description — a dual description — that makes no reference to the inside of the can. The alternate description would be in terms of a two-dimensional Quantum Field Theory of particles similar to gluons that move on the label. The existence of a black hole in the soup must be equivalent to something on the boundary hologram, but what was that something? In the Boundary Theory, Witten argued that the black hole in the soup was equivalent to an ordinary hot fluid of elementary particles — basically just gluons.

The moment I saw Witten's paper, I knew that the Black Hole War was finished. Quantum Field Theory is a special case of Quantum Mechanics, and information in Quantum Mechanics can never be destroyed. Whatever else Maldacena and Witten had done, they had proved beyond any shadow of a doubt that information would never be lost behind a black hole horizon. The string theorists could understand this immediately; the relativists would take longer. But the war was over.

Although the Black Hole War should have come to an end in early 1998, Stephen Hawking was like one of those unfortunate soldiers who wander in the jungle for years, not knowing that the hostilities have ended. By this time, he had become a tragic figure. Fifty-six years old, no longer at the height of his intellectual powers, and almost unable to communicate, Stephen didn't get the point. I am certain that it was not because of his intellectual limitations. From the interactions I had with him well after 1998, it was obvious that his mind was still extremely sharp. But his physical abilities had so badly deteriorated that he was almost completely locked within his own head. With no way to write an equation and tremendous obstacles to collaborating with others, he must have found it impossible to do the things that physicists ordinarily do to understand new, unfamiliar work. So Stephen went on fighting for some time.

Not long after the publication of Witten's paper, another conference took place in Santa Barbara, this one a celebration of holography and of Maldacena's discovery. The after-dinner speaker was Jeff Harvey (H of CGHS), but instead of giving a speech, he got everyone singing and dancing to a victory song, "The Maldacena," to be sung and danced to the tune of "Macarena."[7]

You start with the brane
and the brane is BPS[8]

Then you go near the brane
and the space is ADS

Who knows what it means
I don't I confess

Ehhhh! Maldacena!

Super Yang Mills
With very large N

Gravity on a sphere
flux without end

Who says they're the same
holographic he contends

Ehhhh! Maldacena!

Black holes used to be
a great mystery

7. "Macarena" was a popular Latin dance tune of the mid-1990s.
8. BPS is a technical property of D-branes. BPS stands for the three authors — Bogomol'nyi, Prasad, and Sommerfield — who discovered this property.

Now we use D-brane
to compute D-entropy

And when D-brane is hot
D-free energy

Ehhhh! Maldacena!

M-theory is finished
Juan has great repute

The black hole we have mastered
QCD we can compute

Too bad the glueball spectrum
is still in some dispute

Ehhhh! Maldacena![9]

9. Lyrics © Jeff Harvey.

23

NUCLEAR PHYSICS? YOU'RE KIDDING!

Skeptics will point out that everything I have told you about the quantum properties of black holes — from entropy, temperature, and Hawking radiation to Black Hole Complementarity and the Holographic Principle — is pure theory, with not an ounce of experimental data to confirm it. Unfortunately, they may be right for a very long time.

That said, a totally unexpected connection has recently turned up — a connection between black holes, quantum gravity, the Holographic Principle, and experimental nuclear physics that may once and for all belie the claim that these theories are beyond scientific confirmation. On the face of it, nuclear physics seems a most unpromising place to test ideas such as the Holographic Principle and Black Hole Complementarity. Nuclear physics is not usually deemed to be part of the cutting edge. It's an old subject, and most physicists, including me, thought it had exhausted its capacity to teach us anything new about fundamental principles. From the viewpoint of modern physics, nuclei are like soft marshmallows — giant squish balls that are mostly full of empty space.[1] What could they possibly

1. It is interesting to compute the mass density of a nucleon in Planck units. The radius of a proton is about 10^{20} and the mass is about 10^{-19}. That makes the mass per unit volume about 10^{-79}.

teach us about physics at the Planck scale? Surprisingly, it seems, quite a lot.

String theorists have always had an interest in nuclei. The prehistory of String Theory was all about hadrons: protons, neutrons, mesons, and glueballs. Like nuclei, these particles are big, soft composites made of quarks and gluons. Yet it seems that on a scale a hundred billion billion times larger than the Planck scale, nature repeats itself. The mathematics of hadron physics turns out to be almost the same as the mathematics of String Theory. That seems extremely surprising in view of the fact that the scales are so different: nucleons may be 10^{20} times larger in size than fundamental strings, and they oscillate 10^{20} times more slowly. How can these theories be the same, or even remotely similar? Nevertheless, in a way that will become clear, they are. And if the ordinary subnuclear particles are really similar to fundamental strings, why not test the ideas of String Theory in nuclear physics laboratories? In fact, it's been going on for almost forty years.

The connection between hadrons and strings is one of the pillars of modern particle physics, but up until very recently, it was not possible to test the nuclear analog of black hole physics. That situation is now changing.

Out on Long Island, about seventy miles from Manhattan, nuclear physicists at the Brookhaven National Laboratory are slamming heavy atomic nuclei together just to see what happens. The Relativistic Heavy Ion Collider (RHIC) accelerates gold nuclei to almost the speed of light — fast enough that when they collide, they create a huge splash of energy a hundred million times hotter than the surface of the Sun. The physicists at Brookhaven are not interested in nuclear weapons or any other nuclear technology. Their motivation is pure curiosity — curiosity about the properties of a new form of matter. How does this hot nuclear material behave? Is it a gas? A liquid? Does it hold together, or does it instantly evaporate into separate particles? Do jets of extremely high-energy particles zip out of it?

As I said, nuclear physics and quantum gravity take place on vastly different scales, so how can they have anything to do with each other? The best analogy that I know involves one of the worst movies ever, an old horror flick from the era of the drive-in movie. The movie features a monstrous fly. I don't know how the film was made, but I imagine that an ordinary housefly was filmed and then magnified to fill the entire screen. The image is projected in very slow motion, which gives the fly the ominous feel of a huge, hideous bird. The result is horrifying, but more to the point, it almost perfectly illustrates the connection between gravitons and glueballs. Both are closed strings, but the graviton is much smaller and faster than the glueball – about 10^{20} times smaller and faster. It seems that hadrons are a lot like images of fundamental strings blown up and slowed down not a few hundred times like the fly, but a fantastic 10^{20} times.

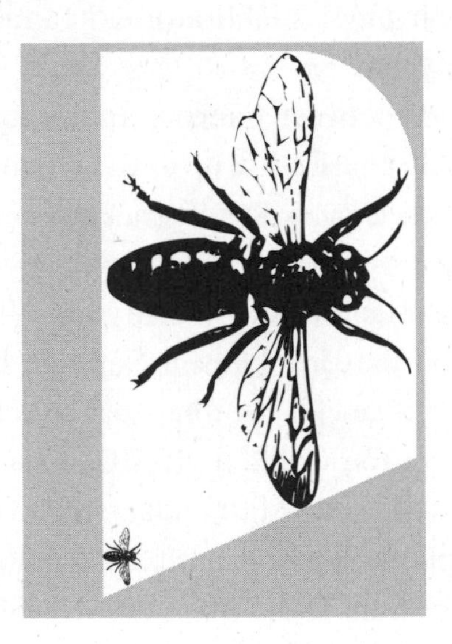

So if we can't collide Planck-sized particles at stupendous energies to make black holes, perhaps we can collide their blown-up versions – glueballs, mesons, or nucleons – and create a magnified version of a black hole. But wait – doesn't that also require prodigious amounts of energy? No, it does not, and to understand why, we need to recall from chapter 16 the counterintuitive twentieth-century connection between size and mass: *small is heavy, big is light.* The fact that nuclear physics takes place on an immensely larger scale than fundamental String Theory implies that the corresponding phenomena require far less energy concentrated in a hugely larger volume. When the numbers are plugged in and the calculations are carried out, something very similar to a slow-motion, magnified black hole should form when ordinary nuclei collide in RHIC.

To understand in what sense black holes are created by RHIC, we must return to the Holographic Principle and to Juan Maldacena's discovery. In a way that no one had foreseen, Maldacena found that two different mathematical theories were really the same — "dual to each other," in String Theory jargon. One theory was String Theory, with its gravitons and black holes, albeit in (4 + 1)-dimensional anti de Sitter Space (ADS). (In chapter 22, for purposes of visualization, I took the liberty of decreasing the dimension of space. In this chapter, I restore the missing dimension.)

Four dimensions of space is one too many for nuclear physics, but remember the Holographic Principle: everything that takes place in ADS must be completely describable by a mathematical theory with one less dimension of space. Because Maldacena started with four space dimensions, the holographic dual theory has only three dimensions, the same number as everyday space. Could this holographic description be similar to any of the theories that we use to describe conventional physics?

It turns out that the answer is yes: the holographic dual is mathematically quite similar to Quantum Chromodynamics (QCD), the theory of quarks, gluons, hadrons, and nuclei.

Quantum gravity in ADS ↔ QCD

For me, the main interest in Maldacena's work was the way it confirmed the Holographic Principle and shed light on the workings of quantum gravity. But Maldacena and Witten saw another opportunity. They realized — brilliantly, I must say — that the Holographic Principle is a two-way street. Why not read it backward? That is, use what we know about gravity — in this case, gravity in (4 + 1)-dimensional ADS — to teach us things about ordinary Quantum Field Theory. For me this was a totally unexpected twist, a bonus of the Holographic Principle that I had never thought of.

A little work was required to accomplish this. QCD is not quite the same as Maldacena's theory, but the main difference can easily be taken into account by modifying ADS in a simple way. Let's review ADS, as seen from a point very near the boundary (where the last visible devil shrinks to zero). I'm going to call that boundary the *UV-brane.*[2] UV stands for ultraviolet — the same term we use for very short-wavelength light. (Over the years, the term *ultraviolet* has come to stand for any phenomenon that takes place on small scales. In the present context, the word refers to the fact that the angels and devils near the boundary of Escher's drawing shrink to infinitesimal size.) The word *brane* is really a misnomer in *UV-brane,* but since it has stuck, I will use it. The UV-brane is a surface close to the boundary.

Imagine moving away from the UV-brane into the interior, where the square devils expand and clocks slow down without limit. Objects that are small and fast near the UV-brane become big and slow as we move deeper into ADS. But ADS is not quite the right thing for describing QCD. Although the difference is not great, the modified space deserves its own name; let's call it *Q-space.* Like ADS, Q-space has a UV-brane where things shrink and speed up, but unlike ADS, there is a second boundary called the *IR-brane.* (IR stands for infrared, a term used for very long-wavelength light.) The IR-brane is a second boundary — a kind of impenetrable barrier where the angels and devils reach a maximum size. If the UV-brane

2. Much of what I am describing in these few paragraphs is explained with great clarity in Lisa Randall's excellent book *Warped Passages.*

is the ceiling of a bottomless chasm, Q-space is an ordinary room with a ceiling and floor. Ignoring the time direction and drawing only two space directions, ADS and Q-space look like this:

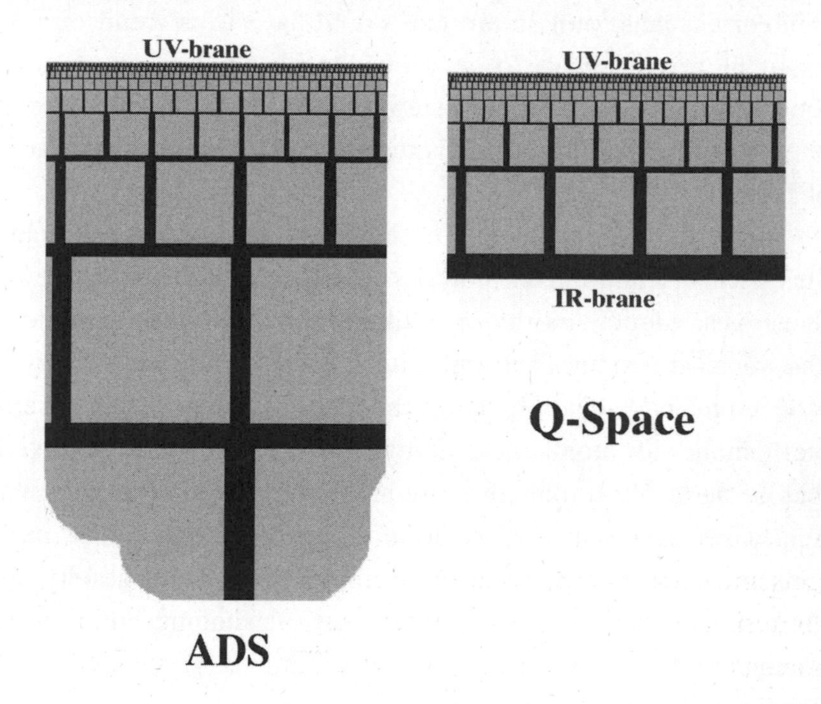

Imagine putting a stringlike particle into Q-space by first placing it near the UV-brane. Like the angels and devils surrounding it, it will appear to be very small – possibly Planck sized – and very rapidly vibrating. But if the same particle is moved toward the IR-brane, it will appear to grow, almost as if it were being projected onto a receding screen. Now watch the string as it vibrates. The vibrations define a kind of clock, and like all clocks, it runs fast when it is near the UV-brane and slow as it moves toward the IR-brane. A string near the IR end will not only look like a gigantic blown-up version of its shrunken UV self, but it will also oscillate far more slowly. This difference sounds a lot like the difference between real flies and their cinematic images – or the difference between fundamental strings and their nuclear counterparts.

If the super-small Planck-sized particles of String Theory "live" near the UV-brane and their blown-up versions — the hadrons — live near the IR-brane, just how far apart are they from each other? In a certain sense, not so far; you would have to descend through about 66 square devils to get from Planck-sized objects to hadrons. But remember that each step is twice as high as the previous one. Doubling in size 66 times is the same as expanding by a factor of 10^{20}.

There are two views of the similarity between fundamental String Theory and nuclear physics. The more conservative view is that it is accidental, more or less like the similarity between atoms and the solar system. That similarity was useful in the early days of atomic physics. Niels Bohr, in his theory of the atom, used the same mathematics for atoms that Newton had used for the solar system. But neither Bohr nor anyone else really thought that the solar system was a blown-up version of an atom. According to this more conservative view, the connection between quantum gravity and nuclear physics is also just a mathematical analogy, but a useful analogy that allows us to use the mathematics of gravity to explain certain features of nuclear physics.

The more exciting view is that nuclear strings really are the same objects as fundamental strings, except seen through a distorting lens that stretches their image and slows them down. According to this view, when a particle (or string) is located near the UV-brane, it appears small, energetic, and rapidly oscillating. It looks like a fundamental string; it behaves like a fundamental string; so it must be a fundamental string. For example, a closed string located at the UV-brane would be a graviton. But the same string, if it moves to the IR-brane, slows down and grows in size. In every way, it looks and behaves like a glueball. In this view of things, gravitons and glueballs are exactly the same objects, except for their location in the brane sandwich.

Imagine a pair of gravitons (strings near the UV-brane) about to collide with each other.

Two Particles About to Collide Near the UV-Brane

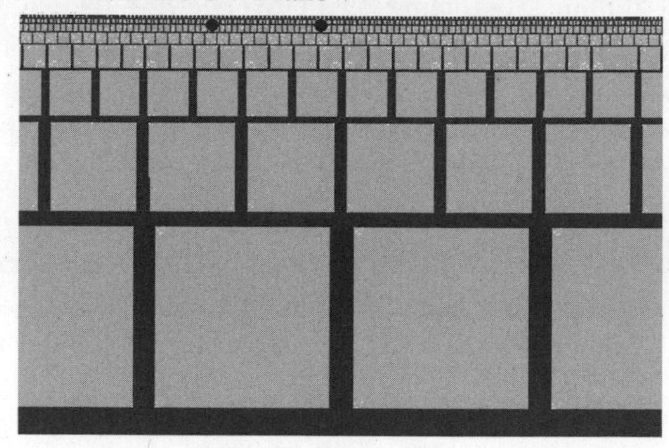

If they have enough energy, when they meet near the UV-brane, an ordinary small black hole will form: a blob of energy stuck to the UV-brane. Think of it as a drop of fluid hanging from the ceiling. The bits of information that make up its horizon are Planck sized.

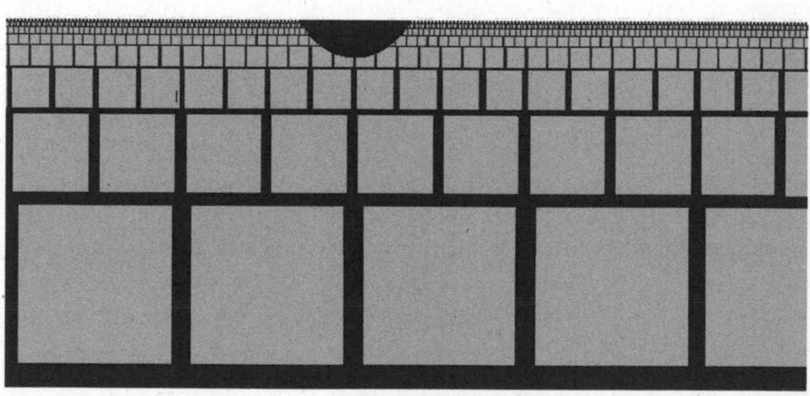

This is, of course, exactly the experiment that we will probably never be able to do.

But now replace the gravitons with two nuclei (near the IR-brane) and smash *them* together.

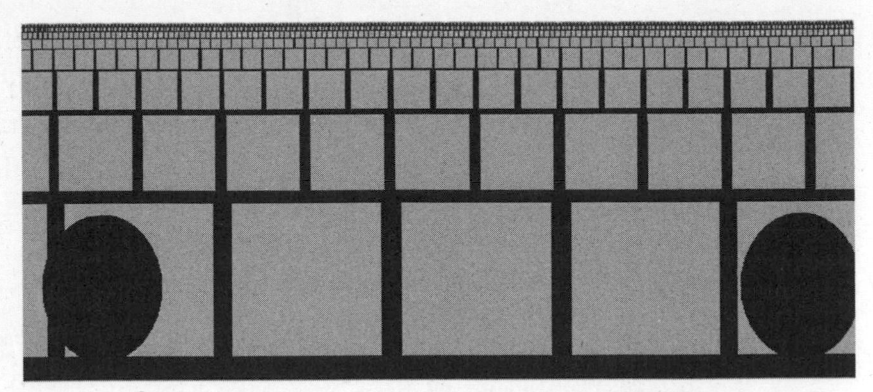

Two Nuclei About to Collide Near the IR-Brane

Here is where the power of duality makes itself felt. On the one hand, we can think of it in the four-dimensional version, in which two objects collide and form a black hole. This time the black hole will be near the IR-brane — a big puddle on the floor. How much energy is required? Far less than when the black hole forms near the UV-brane. In fact, the energy is easily within the range of RHIC.

UV

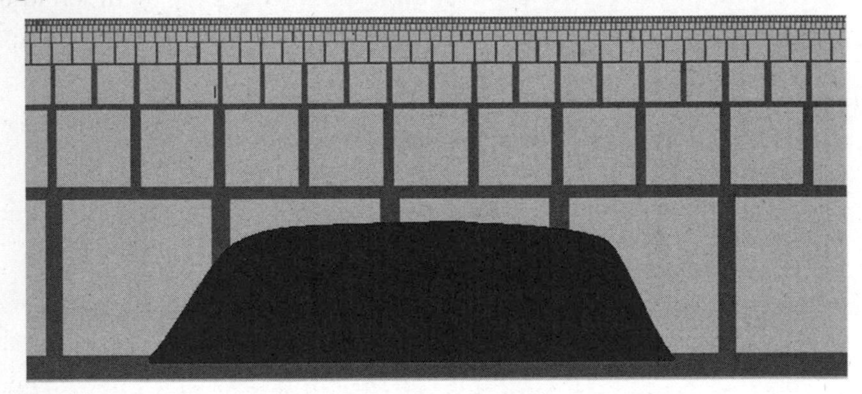

IR

We can also view it from the three-dimensional viewpoint. In that case, hadrons or nuclei collide and make a splash of quarks and gluons.

Originally, before anyone realized QCD's potential connection with black hole physics, QCD experts had expected the energy of the collision to reappear as a gas of particles that would quickly fly apart with very little resistance. But what they saw was different: the energy holds together in what looks much more like a blob of fluid — call it *hot quark soup.* Hot quark soup is not just any fluid; it has some very surprising flow properties that resemble nothing so much as the horizon of a black hole.

All fluids are viscous. Viscosity is a type of friction that acts between the layers of a fluid when they slide over each other. Viscosity is what distinguishes a very viscous fluid such as honey from a much less viscous fluid such as water. Viscosity is not just a qualitative concept. Instead, for every fluid, there is a precise numerical measure called *shear viscosity.*[3]

Theorists had initially applied standard approximation methods and concluded that hot quark soup would have a very high viscosity. Everyone was quite surprised when it turned out to have an astonishingly small viscosity[4] — everyone, that is, except for a few nuclear physicists who happened to know a bit about string theory.

According to a certain quantitative measure of viscosity, hot quark soup is the least viscous fluid known to science — much less viscous than water. Even superfluid liquid helium (the previous champion of low viscosity) is a good deal more viscous.

Is there anything in nature that might rival the low viscosity of hot quark soup? There is, but it's not an ordinary fluid. A black hole horizon behaves like a fluid when it is disturbed. For example, if a small black hole falls into a bigger black hole, it temporarily creates a bulge on the horizon, similar to the bulge that a blob of honey leaves if dropped onto the surface of a pool of honey. The blob on the horizon spreads out just as a viscous fluid does. Long ago, black hole physicists calculated the viscosity of a horizon, and when it was translated to fluid terms, it easily beat out superfluid helium. When

3. The word *shear* refers to the sliding of one layer past another.
4. Strictly speaking, it is the viscosity divided by the entropy of the fluid that is so small.

string theorists began to suspect a connection between black holes and nuclear collisions,[5] they realized that of all things, hot quark soup is most like the horizon of a black hole.

What eventually becomes of the blob of fluid? Like a black hole, it evaporates — into a variety of particles, including nucleons, mesons, photons, electrons, and neutrinos. Viscosity and evaporation are just two of several properties that horizons and hot quark soup share.

Nuclear fluid is now under intense study to find out whether other properties show similar connections to black hole physics. If that trend continues, it will mean that we have been granted an extraordinary opportunity — a remarkable window into the world of quantum gravity, blown up in size and slowed down in frequency, so that the Planck distance becomes not much smaller than a proton — to confirm the theories of Hawking and Bekenstein, as well as Black Hole Complementarity and the Holographic Principle.

It has been said that peace is nothing but the brief interlude between wars. But in science, Thomas Kuhn has rightly said, the opposite is true: most "ordinary science" takes place during the long, peaceful, humdrum periods between upheavals. The Black Hole War led to a violent restructuring of the laws of physics, but now we are seeing it work its way into the day-to-day activities of the more mundane side of physics. Like so many earlier revolutionary ideas, the Holographic Principle is evolving from radical paradigm shift to everyday working tool of — surprisingly — nuclear physics.

5. Pavel Kovtun, Dam T. Son, and Andrei O. Starinets — three theoretical physicists at the University of Washington in Seattle—were the first to recognize the implications of the Holographic Principle for the viscous properties of hot quark soup.

24

HUMILITY

We are just an advanced breed of monkeys on a minor planet of a very average star. But we can understand the Universe. That makes us something very special.

— STEPHEN HAWKING

Rewiring ourselves for relativity was hard enough, and for Quantum Mechanics it was much harder. Predictivity or determinism had to go, and the failed classical rules of logic had to be replaced by quantum logic. Uncertainty and complementarity were expressed in terms of abstract, infinite, dimensional Hilbert spaces, mathematical commutation relations, and other bizarre inventions of the mind.

Throughout all of the rewiring of the twentieth century, at least till the mid-1990s, the reality of space-time, and the objectivity of events, went almost unquestioned. It was universally assumed that quantum gravity would play no role when it came to the large-scale properties of space-time. Stephen Hawking, with his information paradox, was the one who unwittingly, and rather unwillingly, forced us out of that frame of mind.

The new views of the physical world that evolved over a little more than a decade involve a new kind of relativity and a new kind of quantum complementarity. The objective meaning of simultaneity (of two events) failed in 1905, but the concept of an event itself remained rock solid. If a nuclear reaction takes place in the Sun, all observers will agree that it happened in the Sun. No one will detect it taking place on Earth. But something new happens in the powerful gravitation of a black hole, something that undermines the objectivity of events. Events that a falling observer reckons to be deep inside an enormous black hole, another observer detects outside the horizon, scrambled among the Hawking radiation photons. An event cannot be both behind the horizon *and* in front of it. The same event is either behind the horizon *or* in front of the horizon depending on which experiment the observer does. But even the utter strangeness of complementarity is dwarfed by the bizarre Holographic Principle. It seems that the solid three-dimensional world is an illusion of a sort, the real thing taking place out at the boundaries of space.

For most of us, the breakdown of concepts such as simultaneity (in Special Relativity) and determinism (in Quantum Mechanics) are no more than obscure oddities that only a few physicists are interested in. But in reality the opposite is true: it is the agonizing slowness of human motion and the ponderous mass of the 10^{28} atoms in the human body that are the odd exceptions of nature. There are roughly 10^{80} elementary particles in the universe for every human. Most of them move at close to the speed of light and are very uncertain, if not about where they are, then about how fast they move.

The weakness of gravity that we experience on Earth is also an exception. The universe was born in a state of violent expansion; every point of space was surrounded on every side by horizons within a distance smaller than a single proton. The most notable inhabitants of the universe — the galaxies — are built around giant black holes that are continually gobbling up stars and planets. Out of every 10,000,000,000 bits of information in the universe, 9,999,999,999 are associated with the horizons of black holes. It should be evident that our naive ideas about space, time, and information are wholly inadequate to understand most of nature.

The rewiring for quantum gravity is far from complete. I don't think we have a proper framework yet to replace the older paradigm of objective space-time. The powerful mathematics of String Theory is a help. It allows us a rigorous framework to test ideas that we would otherwise be able to argue about only philosophically. But String Theory is an incomplete work in progress. We don't know its defining principles, or whether it is the deepest level of reality or just another temporary theory along the way. The Black Hole War has taught us some very important and unexpected lessons, but they are only a hint of how different reality is from our mental model, even after rewiring the model for relativity and Quantum Mechanics.

Cosmic Horizons

The Black Hole War is over (this claim may upset a handful of people who are still fighting it), but just as it ended, nature, the great spoiler, threw us another curveball. At about the time of Maldacena's discovery, physicists started to become convinced (by cosmologists) that we live in a world with a nonvanishing *cosmological constant.* An astonishingly small constant of nature,[1] smaller by far than any other physical constant, the cosmological constant is the main determinant of the future history of the universe.

The cosmological constant, also known as dark energy, has been a thorn in the side of physics for almost a century. In 1917 Einstein speculated about a kind of antigravity that would cause everything

1. The numerical value of the cosmological constant is approximately 10^{-123} in Planck units. The suspicion that a cosmological constant exists began in the mid-1980s among a few cosmologists who looked closely at astronomical data. But it didn't really get much traction in the physics community for more than a decade. The incredible smallness of its value had fooled almost all physicists into believing that it didn't exist.

in the universe to repel everything else, counteracting the usual pull of gravity. The speculation was by no means an idle one; it was firmly based on the mathematics of General Relativity. There was room in the equations for an extra term that Einstein called the cosmological term. The strength of the new force was proportional to a new constant of nature — the so-called cosmological constant — denoted by the Greek letter lambda (Λ). If Λ is positive, the cosmological term creates a repulsive force that increases with distance; if it is negative, the new force is attractive; if Λ is zero, there is no new force and we can ignore it.

At first Einstein guessed that Λ would be positive, but he soon grew to dislike the whole idea, famously calling it his worst mistake. For the rest of his life, he set Λ to zero in all his equations. Most physicists agreed with Einstein, although they didn't understand why Λ should be absent from the equations. But over the past decade, the astronomical case for a small, positive cosmological constant has become persuasive.

The cosmological constant, and all the puzzles and paradoxes that it has created, are the subject of my book *The Cosmic Landscape.* Here I will just tell you its most important consequence: the repulsive force, acting at cosmological distances, causes space to expand *exponentially.* There is nothing new about the universe expanding, but without a cosmological constant, the rate of expansion would gradually slow down. Indeed, it could even reverse itself and begin to contract, eventually imploding in a giant cosmic crunch. Instead, as a consequence of the cosmological constant, the universe appears to be doubling in size about every fifteen billion years, and all indications are that it will do so indefinitely.

In an expanding universe, or for that matter an expanding balloon, the greater the distance between two points, the faster they recede from each other. The relation between distance and velocity is called Hubble's Law, and it says that the recessional velocity between any two points is proportional to the distance separating them. Any observer, no matter where he is stationed, looks around and sees the distant galaxies moving away, their velocity proportional to their distance.

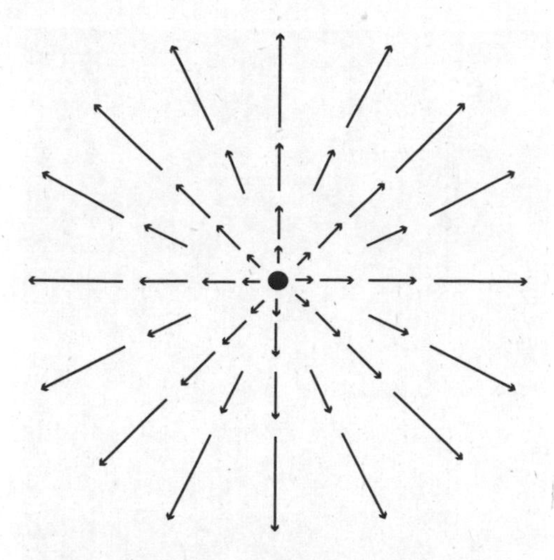

If you look out far enough in such an expanding universe, you will come to a point where the galaxies are moving away from you with the speed of light. One of the most remarkable properties of an exponentially expanding universe is that the distance to that point never changes. It appears that in our own universe, at a distance of about fifteen billion light-years, things are moving away with the speed of light, but even more important, it will always be that way for all eternity.

There is something familiar yet different about this. It brings to mind the pollywog lake in chapter 2. At some point Alice, if she goes with the flow, will pass the point of no return, and recede away from Bob with the speed of sound. Something like that is taking place on a grand scale. In every direction that we look, galaxies are passing the point at which they are moving away from us faster than light can travel. Each of us is surrounded by a *cosmic horizon* – a sphere where things are receding with the speed of light – and no signal can reach us from beyond that horizon. When a star passes the point of no return, it is gone forever. Far out, at about fifteen billion light-years, our cosmic horizon is swallowing galaxies, stars, and probably even life. It is as if we all live in our own private inside-out black hole.

Are there really worlds like our own that long ago passed through our horizon and became completely irrelevant to anything we can ever detect? Even worse, is most of the universe forever beyond our knowledge? This is extremely disturbing to some physicists. There is a philosophy that says that if something is unobservable — unobservable in principle — it is not part of science. If there is no way to falsify or confirm a hypothesis, it belongs to the realm of metaphysical speculation, together with astrology and spiritualism. By that standard, most of the universe has no scientific reality — it's just a figment of our imaginations.

But it is hard to dismiss most of the universe as nonsense. There is no evidence that the galaxies thin out or come to an end at the horizon. Astronomical observation indicates that they go on as far as the eye, or the telescope, can see. What are we to make of this situation?

There have been other circumstances in the past in which "unobservable" things have been dismissed as being unscientific. Other people's emotions are a notable example. An entire school of psychology — behaviorism — was founded on the principle that emotions and internal states of consciousness are not observable, and therefore should never be invoked in a scientific discussion. Only the observable behaviors of experimental subjects — their body movements, facial gestures, temperature, blood pressure — were fair game for behaviorist psychology. Behaviorism exerted enormous influence during the mid-twentieth century, but today most people consider it an extreme point of view. Perhaps we should simply accept worlds beyond the horizon in the same way that we accept that other people have an impenetrable interior life.

However, there may be a better answer. The properties of cosmic horizons seem to be very similar to those of black holes. The mathematics of an accelerating (exponentially expanding) universe imply that as things approach the cosmic horizon, we see them slow down. If we could send a thermometer attached to the end of a long cable to the vicinity of the cosmic horizon, we would discover that the temperature increases, eventually approaching the infinite temperature at the horizon of a black hole. Does that mean that all the people on those distant planets are being roasted? The answer is no more, or no less, than they would be if they were near a black hole. To the observers traveling with the flow, passing the cosmic horizon is a non-event, a mathematical point of no return. But our own observations, supplemented with some mathematical analysis, would indicate that they are approaching a region of incredible temperature.

What happens to their bits of information? The same arguments that Hawking used to prove that black holes radiate black body radiation tell us that cosmic horizons also radiate. In this case, the radiation is not outward but inward, as if we lived in a room with warm, radiant walls. From our perspective, it would appear that as things move toward the horizon, they are heated and radiated back as photons. Could it be that there is a Principle of Cosmic Complementarity?

To an observer inside a cosmic horizon, the horizon is a hot layer composed of horizon-atoms that absorb, scramble, and then return all bits of information. To a freely moving observer who passes through the cosmic horizon, the passing is a non-event.

At the present time, however, we understand very little about cosmic horizons. The meaning of the objects behind the horizon — whether they are real and what role they play in our description of the universe — may be the deepest question of cosmology.

Falling stones and orbiting planets are pale hints of what gravity is really all about. Black holes are where gravity takes its rightful place. Black holes are not merely dense stars; rather they are the ultimate information reservoirs, where bits are packed as tightly as a two-dimensional stack of cannonballs, but on a scale thirty-four orders of magnitude smaller. That's what quantum gravity is all about: information and entropy, densely packed.

Hawking may have given the wrong answer to his own question, but the question itself was one of the most profound in the recent history of physics. It may be that he was too classically wired — too prone to seeing space-time as a preexisting, though flexible, canvas that physics is painted on — to recognize the profound implications of reconciling quantum information conservation with gravitation. But the question itself may have opened the way for the next major conceptual revolution in physics. Not many physicists can make that claim.

As to Hawking's legacy, it is bound to be very large. Others before him knew that the mismatch between gravity and quantum theory would have to be bridged someday, but Bekenstein and Hawking were the first to enter a remote country and bring back gold. I hope that future historians of science will say that they started it all.

> *He who has never failed somewhere, that man can not be great.*
>
> — Herman Melville

Physics in a Nutshell

Confusion and disorientation reign; cause and effect break down; certainty evaporates; all the old rules fail. That's what happens when the dominant paradigm breaks down.

But then new patterns emerge. They make no sense at first, but they are patterns. What to do? Take the patterns and classify, quantify, and codify them in new mathematics, even new laws of logic, if necessary. Replace the old wiring with new and become familiar with it. Familiarity breeds contempt, or at least acceptance.

Very likely, we are still confused beginners with very wrong mental pictures, and ultimate reality remains far beyond our grasp. The old cartographer's term *terra incognita* comes to mind. The more we discover, the less we seem to know. That's physics in a nutshell.

EPILOGUE

In 2002 Stephen Hawking reached his sixtieth birthday. No one thought he would do it, least of all his doctors. The event was worth a great celebration — a really grand birthday party — and so I found myself once again in Cambridge, along with hundreds of others — physicists, journalists, rock stars, musicians, a Marilyn Monroe imitator, cancan dancers — as well as a great deal of food, wine, and liquor. It was a giant media event, side by side with a serious physics conference. Everyone who was anyone in Stephen's scientific life gave a speech, including Stephen himself. Here's a brief excerpt from mine.

> Stephen, as we all know, is by far the most stubborn and infuriating person in the universe. My own scientific relation with him I think can be called adversarial. We have disagreed profoundly about deep issues concerning black holes, information, and all that kind of thing. At times he has caused me to pull my hair out in frustration — and you can plainly see the result. I can assure you that when we began to argue more than two decades ago, I had a full head of hair.

At this point, I could see Stephen in the rear of the auditorium with his mischievous grin. I went on:

> I can also say that of all the physicists I have known he has had the strongest influence on me and on my thinking. Just about everything I have thought about since 1983 has in one

> way or another been a response to his profoundly insightful question about the fate of information that falls into a black hole. While I firmly believe his answer was wrong, the question, and his insistence on a convincing answer, has forced us to rethink the foundations of physics. The result is a wholly new paradigm that is now taking shape. I am deeply honored to be here to celebrate Stephen's monumental contributions and especially his magnificent stubbornness.

I meant every word of it.

I recall only three other speeches. Two of them were by Roger Penrose. I can't remember why Roger gave two talks, but he did. In the first, he argued that information has to be lost in black hole evaporation. The arguments were the original ones that Stephen had made twenty-six years earlier, and Roger maintained that both he and Stephen continued to believe them. I was surprised, since as far as I (and anyone who had been following the recent developments) was concerned, Matrix Theory, Maldacena's discovery, and Strominger and Vafa's entropy calculations had finally put the question to rest.

But in his second talk, Roger maintained that the Holographic Principle and Maldacena's work were based on a series of misconceptions. Simply stated, his argument was, "How could it possibly be that physics, in more dimensions, can be described by a theory in fewer dimensions?" I think he hadn't thought about it hard enough. Roger and I have been friends for forty years, and I know he is a rebel, always running against the standard wisdom. I shouldn't have been surprised that he was being contrary.

The other lecture that has stuck in my memory was Stephen's — not for what he said, but for what he didn't say. He briefly recalled the notable high points of his career — cosmology, Hawking radiation, excellent cartoons — but offered not a single word about information loss. Could it be that he was beginning to waver? I imagine so.

Then, in a press conference in 2004, Hawking announced that he had changed his mind. His most recent investigations, Stephen said, had finally solved his own paradox: it seems that, after all, information does leak out of black holes and ultimately winds up in the evaporation products. Somehow, according to Stephen, the mechanism had been overlooked for all this time, but he had finally identified it and would report his new conclusions at an upcoming conference in Dublin. The media was alerted, and the conference was breathlessly awaited.

The newspapers also reported that Stephen would pay off a bet with John Preskill (who had worried me in Santa Barbara with his ingenious thought experiment). In 1997 John had wagered Stephen that information *did* escape from black holes. The payoff was a baseball encyclopedia.

Very recently, I learned that in 1980, Don Page had made a similar bet with Stephen. As I suspected from Don's talk in Santa Barbara, he had been skeptical about Stephen's claim all along. On April 23, 2007, two days before I wrote this paragraph, Stephen formally conceded. Don was kind enough to send me a photocopy of the original contract — a bet of one British pound against one U.S. dollar — along with Stephen's signed concession. The dark blob at the end is Stephen's thumbprint.

How Predictable Is Quantum Gravity?

Don Page bets Stephen Hawking one pound Sterling that strong quantum cosmic censorship holds, namely, that a pure initial state composed entirely of regular field configurations on complete, asymptotically flat hypersurfaces will have a unique S-matrix evolution under the laws of physics to a pure final state composed entirely of regular field configurations on complete, asymptotically flat hypersurfaces.

Stephen Hawking bets Don Page $1.00 that in quantum gravity the evolution of such a pure initial state can be given in general only by a $-matrix to a mixed final state and not always by an S-matrix to a pure final state.

"I concede in light of the weakness of the $"
Stephen Hawking, 23 April 2007

Don N. Page
Stephen Hawking

What was in Stephen's lecture? I don't know; I wasn't there. But a subsequent paper, written several months later, gave the details. There weren't many: a brief history of the paradox, a wordy description of some of Maldacena's arguments, and a final tortured explanation of how everyone had been right all along.

But everyone hadn't been right.

Over the past few years, we have seen some remarkably contentious arguments disguised as scientific debates, but they are really political squabbles. They include disputes about intelligent design; whether global warming is really occurring, and if so, whether it is man-made; the value of expensive missile defense systems; and even String Theory. Fortunately, however, not all scientific debates are polemical. From time to time, real differences of opinion about substantive issues turn up and lead to new insights, or even paradigm shifts. The Black Hole War is an example of a debate that was never polemical; it involved genuine differences of opinion about clashing scientific principles. Although the issue of whether information is lost in black holes was certainly a matter of opinion at first, scientific opinion has now largely coalesced around a new paradigm. But even though the original war is over, I doubt that we have learned all of its important lessons. String Theory's most troubling loose end is how to apply it to the real universe. The Holographic Principle was spectacularly confirmed by Maldacena's theory of anti de Sitter Space, but the geometry of the real universe is not anti de Sitter Space. We live in an expanding universe that, if anything, is more like de Sitter Space, with its cosmic horizons and bubbling pocket universes. At the moment, no one knows how to apply String Theory, the Holographic Principle, or other lessons about black hole horizons to cosmic horizons, but the connections are likely to be very deep. My own guess is that these connections are at the root of many cosmological puzzles. Someday I hope to write another book explaining how all this eventually plays out, but I don't think it will be too soon.

Claudio Teitelboim (Bunster), Gerard 't Hooft, author, John Wheeler, and François Englert, Valparaiso, 1994.

Stephen and author, Valdivia, Chile, 2008.

ACKNOWLEDGMENTS

I am very grateful to many people for their help in bringing this book to completion. My agent, John Brockman, was, as always, a source of wisdom and good advice. To all the people at Little, Brown who worked so hard — Geoff Shandler, freelancer Barbara Jatkola, Karen Landry, and Junie Dahn — I would like to express my deepest thanks.

I am also deeply indebted to Stephen Hawking and Gerard 't Hooft for many years of friendship, and for the extraordinary and exhilarating experience that made this book possible.

GLOSSARY

anti de Sitter Space — A space-time continuum with uniform negative curvature that resembles a spherical box.

antipodal — Pertaining to the opposite side of the Earth.

bit — The basic unit of information.

black body radiation — Electromagnetic radiation emitted by a nonreflecting body due to its own heat.

black hole — An object so massive and dense that nothing can escape its gravity.

Black Hole Complementarity — Bohr's principle of complementarity applied to black holes.

Boundary Theory — The mathematical theory on the boundary of a region of space that describes everything inside that region.

Brownian Motion — The random motion of a grain of pollen suspended in water. The cause is the constant bombardment by water molecules that have been excited by heat.

classical physics — Physics that does not take into account Quantum Mechanics. Usually refers to deterministic physics.

closed string — A string with no ends, similar to a rubber band.

corpuscles — Newton's term for the hypothetical particles of light.

curvature — The bending of space or space-time.

dark star — A star so heavy and dense that light cannot escape from it. Now called a *black hole*.

D-brane — A surface in space-time where a fundamental string can end.

determinism — The principle of classical physics that says that the future is completely determined by the present. Undermined by Quantum Mechanics.

Dollar-matrix — Hawking's attempt to replace the S-matrix.

duality — The relation between two apparently different descriptions of the same system.

dumb hole — A drain hole where the velocity of the flow exceeds the speed of sound (in water) close to the drain.

electric field — The force field surrounding electric charges.

electromagnetic waves — Wavelike disturbances of space consisting of vibrating electric and magnetic fields. Light is an electromagnetic wave.

embedding diagram — A representation of space-time at a moment of time created by "slicing" the space-time continuum.

entropy — A measure of hidden information, often the information stored in things too small and numerous to keep track of.

escape velocity — The minimum velocity with which a projectile will escape the gravitational pull of a massive object.

Equivalence Principle — Einstein's principle that gravity is indistinguishable from acceleration — for example, in an elevator.

event — A point in space-time.

extremal black hole — An electrically charged black hole that has reached its lowest mass for a given charge.

First Law of Thermodynamics — The law of the conservation of energy.

fundamental strings — The strings that make up gravitons. The typical size of a fundamental string is thought to be not much bigger than the Planck length.

gamma rays — The shortest-wavelength and most energetic electromagnetic waves.

General Theory of Relativity — Einstein's theory of gravity based on curved space-time.

geodesic — The closest thing to a straight line in a curved space; the shortest path between points.

glueball — A hadron composed only of gluons with no quarks. Glueballs are closed strings.

gluons — The particles that combine to form the strings that bind quarks.

grok — To understand something in a deeply intuitive way, at a gut level.

ground state — The state of a quantum system with the least possible energy. Often identified as the state at absolute zero temperature.

hadrons — The particles closely related to the nucleus: nucleons, mesons, and glueballs. Hadrons are made up of quarks and gluons.

Hawking radiation — Black body radiation emitted by a black hole.

Hawking temperature — The temperature of a black hole seen from a distance.

Heisenberg Uncertainty Principle — The principle of Quantum Mechanics that limits one's ability to determine position and velocity simultaneously.

hertz — A unit of frequency that measures the number of complete oscillations per second.

hologram — A two-dimensional representation of three-dimensional information. A type of photograph from which a three-dimensional image can be reconstructed.

Holographic Principle — The principle that says that all information lies at the boundary of a region of space.

horizon — The surface within which nothing can escape the singularity of a black hole.

information — The data that distinguish one state of affairs from another. Measured in bits.

infrared radiation — Electromagnetic waves of wavelength somewhat longer than visible light.

interference — A wave phenomenon in which waves from two separate sources cancel or reinforce each other at certain places.

IR — Infrared. Often used to indicate large distances.

magnetic field — The force field surrounding magnets and electric currents.

microwaves — Electromagnetic waves of wavelength somewhat shorter than radio waves.

neutron star — The final stage of a star too big to form a white dwarf but not big enough to collapse into a black hole.

Newton's constant — The numerical constant, G, in Newton's law of gravitational forces; $G = 6.7 \times 10^{-11}$ in metric units.

No-Quantum-Xerox Principle — A theorem of Quantum Mechanics that forbids the possibility of a machine that can perfectly copy quantum information. Also called *No-Cloning Principle.*

nucleon — A proton or neutron.

open string — A string with two ends. A rubber band is a closed string, but if it is cut with scissors, it becomes an open string.

oscillator — Any system that undergoes periodic vibrations.

photons — Indivisible quanta (particles) of light.

Planck length — The unit of length when the three fundamental constants of nature — c, h, and G — are set equal to one. Often thought to be the smallest meaningful length, 10^{-33} centimeters.

Planck mass — The unit of mass in Planck units; 10^{-8} kilograms.

Planck's constant — The numerical constant, h, that governs quantum phenomena.

Planck time — The unit of time in Planck units; 10^{-42} seconds.

point of no return — An analog for the horizon of a black hole.

proper time — Time elapsed according to a moving clock; a measure of distance along a world line.

QCD — Quantum Chromodynamics.

QCD strings — The strings made of gluons that bind quarks together to form hadrons.

Quantum Chromodynamics — The Quantum Field Theory describing quarks and gluons and how they form hadrons.

Quantum Field Theory — The mathematical theory that unifies the particle and wave characteristics of matter. The basis for elementary particle physics.

quantum gravity — The theory that unifies Quantum Mechanics with Einstein's General Relativity; the quantum theory of gravity. At present an incomplete theory.

radio waves — The longest-wavelength electromagnetic waves.

RHIC — Relativistic Heavy Ion Collider. An accelerator that accelerates heavy nuclei to almost the speed of light and collides them to create a splash of very hot nuclear material.

Schwarzschild radius — The radius of the horizon of a black hole.

Second Law of Thermodynamics — Entropy always increases.

simultaneity — Referring to events that take place at the same time. Since the Special Theory of Relativity, simultaneity is no longer considered an objective property.

singularity — The infinitely dense point at the center of a black hole where tidal forces become infinite.

S-matrix — A mathematical description of the collision between particles. The S-matrix is a list of all possible inputs and the probability amplitudes for all outcomes.

space-time — All of space and time united into a single, four-dimensional manifold.

Special Theory of Relativity — Einstein's 1905 theory dealing with the paradoxes of the velocity of light. The theory says that time is the fourth dimension.

speed of light — The speed at which light moves, approximately 186,000 miles per second; denoted by the letter *c*.

String Theory — A mathematical theory in which elementary particles are seen as microscopic, one-dimensional strings of energy. A candidate for quantum gravity.

temperature — The increase in the energy of a system if one bit of entropy is added.

tidal forces — Distorting forces due to spatial variations in the strength of gravity.

tunneling — A quantum mechanical phenomenon in which a particle passes through a barrier even though it doesn't have enough energy to do so classically.

ultraviolet radiation — Electromagnetic waves of a wavelength somewhat shorter than visible light.

UV – Ultraviolet. Often used to refer to very small sizes.

viscosity – Friction between the layers of a fluid when they move past each other.

wavelength – The distance occupied by one full wave from crest to crest.

white dwarf – The last stage of a star not much more massive than the Sun.

world line – The trajectory of a particle in space-time.

X-rays – Electromagnetic waves of somewhat shorter wavelength than ultraviolet radiation but not as short as gamma rays.

zero point motion – The residual motion of a quantum system that can never be eliminated because of the Uncertainty Principle. Also called *quantum jitters*.

INDEX

ABOUT THE AUTHOR

Leonard Susskind is the Felix Bloch Professor of theoretical physics at Stanford University. The author of *The Cosmic Landscape,* he is a member of the National Academy of Sciences and the American Academy of Arts and Sciences.